Freshwater Mollusks of the World

Freshwater Mollusks of the World

A Distribution Atlas

Edited by

Charles Lydeard and Kevin S. Cummings

 JOHNS HOPKINS UNIVERSITY PRESS   BALTIMORE

Johns Hopkins University Press
2715 North Charles Street
Baltimore, Maryland 21218-4363
www.press.jhu.edu

Library of Congress Cataloging-in-Publication Data

Names: Lydeard, Charles, editor. | Cummings, Kevin, editor.
Title: Freshwater mollusks of the world : a distribution atlas / edited
    by Charles Lydeard and Kevin S. Cummings.
Description: Baltimore : Johns Hopkins University Press, 2019. |
    Includes bibliographical references and index.
Identifiers: LCCN 2018020728 | ISBN 9781421427317 (hardcover :
    alk. paper) | ISBN 1421427311 (hardcover : alk. paper) | ISBN
    9781421427324 (electronic) | ISBN 142142732X (electronic)
Subjects: LCSH: Mollusks—Geographical distribution.
Classification: LCC QL407.5 .F74 2019 | DDC 594—dc23
LC record available at https://lccn.loc.gov/2018020728

A catalog record for this book is available from the British Library.

*Special discounts are available for bulk purchases of this book. For more
information, please contact Special Sales at 410-516-6936 or specialsales@
press.jhu.edu.*

Contents

Acknowledgments

We would like to thank the contributing authors for their knowledge and valuable time in writing the chapters and making this book possible. Thanks also for their patience while we worked to complete it. Special thanks to Vince Burke for recognizing the value of such a scholarly volume and to Doug Eernisse for reviewing the entire document. Our research over the years has been supported by the National Science Foundation; the US Fish and Wildlife Service; the Bureau of Land Management; the Illinois Natural History Survey (INHS); Prairie Research Institute, at the University of Illinois; the University of Alabama; Morehead State University; and others. Special thanks also to Danielle Ruffatto (INHS) for her expert preparation of the distribution maps. The base map used was from Natural Earth, which provides free vector and raster map data at naturalearthdata .com. Special thanks also to Guido Poppe (Conchology, Inc., www.conchology.be/) for his generosity in allowing us to use many of his images. Thanks also to Liath Appleton, formerly of the University of Michigan, for the use of the photo of *Burnupia,* to Ken Hayes for the photo of *Pomacea,* to Dr. Anders Hallan for the photos of *Austropyrgus* and *Glacidorbis,* and to Yasunori Kano for the photos of *Acochlidium* and *Tantulum.*

Kevin would like to send special thanks to his wife, Kathy Harden, and friends and family for continued support. Chuck would like to thank Rhonda Lydeard and their children, Andrew, Emily, and Catherine Lydeard and Khalil Caldwell, for support and encouragement over the years as he pursued a fulfilling career in biology. Thanks to Ms. Meredith Harris, Lydeard's ADS, for keeping the Department of Biology and Chemistry running smoothly while the book was being completed.

Freshwater Mollusks of the World

1 Introduction and Overview

CHARLES LYDEARD AND KEVIN S. CUMMINGS

The study of distributions of past and present organisms is referred to as biogeography. The study of animal distributions is called zoogeography. The "father" of zoogeography is Alfred Russel Wallace, whose tome *The Geographical Distribution of Animals* (1962; originally published in 1876), synthesized what was known about animal distributions at the time. As Wallace indicated in the introduction,

> it is a fact within the experience of most persons, that the various species of animals are not uniformly dispersed over the surface of the country. If we have a tolerable acquaintance with any district, be it a parish, a county, or a larger extent of territory, we soon become aware that each well-marked portion of it has some peculiarities in its animal productions. If we want to find certain birds or certain insects, we have not only to choose the right season but to go to the right place. If we travel beyond our district in various directions we shall almost certainly meet with something new to us; some species which we were accustomed to see almost daily will disappear, others which we have never seen before will make their appearance. If we go very far, so as to be able to measure our journey by degrees of latitude and longitude and to perceive important changes of climate and vegetation, the differences in the forms of animal life will become greater; till at length we shall come to a country where almost everything will be new, all the familiar creatures of our own district being replaced by others more or less differing from them. (p. 3)

The observation that different regions were made up of distinct animals and plants was first noted by Georges-Louis Leclerc, Comte de Buffon (1707–1788), and became known as Buffon's Law. Philip Schlater (1858) described six biogeographic regions based on the distribution of passerine birds: the Palearctic Region (Europe, temperate Asia, and North Africa to the Atlas mountains); the Ethiopian Region (Africa south of the Atlas, Madagascar, and the Mascarene Islands, with southern Arabia); the Indian Region (India south of the Himalayas, to southern China and to Borneo and Java); the Australian Region (Celebes and Lombock, eastward to Australia and the Pacific Islands); the Nearctic Region (Greenland, and North America to northern Mexico); and the Neotropical Region (South America, the Antilles, and southern Mexico).

Wallace (1962) recognized the same six regions (although he renamed the Indian region as Oriental) based on the distribution patterns of mammals, birds, reptiles and amphibians, freshwater fishes, butterflies and moths, beetles, and land snails. He showed that countries exceedingly similar in climate and physical features may yet have distinct animal populations: "The equatorial parts of Africa and South America, for example, are very similar in climate and are both covered with luxuriant forests, yet their animal life is widely different; elephants, apes, leopards, guinea-fowls and touracos in the one (Ethiopian), are replaced by tapirs, prehensile-tailed monkeys jaguars, curassows and toucans in the other (Neotropical)" (p. 5).

Wallace supposed that each region came to have distinct groups of species because there were barriers to dispersal that isolated them from their source population and allowed them to adapt to the unique environment, and over time new varieties formed via natural selection. Indeed, the geographical distribution of animals provided important data for both Wallace and Charles Darwin in the development and description of their theory of evolution via natural selection, which they independently described in papers that were read at the Linnean Society of London 1858 (Darwin and Wallace, 1858).

During Wallace's time, it was believed that the continents were fixed in their current positions and that they had always been in those positions. Therefore, the only way species could get from one place to another was by means of dispersal. Some animals were better dispersers than others, and when there were formidable barriers to dispersal like mountain ranges or oceans, different species were typically found on opposite sides of a barrier. Wallace and Darwin went to great lengths to explore the means by which organisms could disperse, especially across major oceanic barriers. They hypothesized that although it may be a rare occurrence, over enough time poorly dispersing species could colonize new areas accidentally via floating, swimming, rafting, or air drift during hurricanes, and so on. Other biologists were not convinced; they hypothesized that, at some time in the past, temporary land bridges had existed that organisms could cross and that these structures later subsided under the sea again. Because there were no known mechanisms to explain such bridges, most biologists did not support this viewpoint at the time.

In 1912, Alfred Wegener (1912a, b) presented and later expounded (Wegener, 1915) on a hypothesis that would become the theory of continental drift, which proposed that lighter continental blocks float and move horizontally and that at one time all were joined in a single landmass called Pangaea. Wegener presented data from historical geology, geomorphology, paleontology, paleoclimatology, and other areas of science to explain his new theory. Because Wegener was a German meteorologist and a relative outsider to the mainstream scientific community, his concept met much resistance, and it was not until the 1960s that additional evidence from geology, paleomagnetism, and the development of plate tectonics as a mechanism to explain how continents could move horizontally that the theory was widely accepted. It is now known that during the Triassic, about 250 million years ago, the continents were joined in a single landmass called Pangaea and over time eventually separated into a northern landmass Laurasia and a southern landmass Gondwana. Over millions

of years, the continents drifted to their current positions, and they continue to drift today.

With the acceptance of continental drift by means of plate tectonics, biologists began to realize that geologic history could shape biotic history. Rather than dispersing from one major landmass to another, organisms "rafted" on continents that later separated from one another. Indeed, as Pangaea, Laurasia, and Gondwana began to break up, the fauna that was found on these landmasses would have had their distributions severed and through subsequent isolation would have radiated into unique species. This revelation led to the development of a whole new discipline called vicariance biogeography (Nelson and Rosen, 1981; Nelson and Platnick, 1981). Vicariance biologists supposed that geologic history shaped biotic history and that so-called vicariant events of separating landmasses resulted in divergent species or groups of species (vicariants). Vicariance biogeography coupled the study of evolutionary relationships of organisms (systematics) with biogeography and greatly advanced the study of biogeography. However, over time, many advocates of vicariance began to assume that dispersal had never played a role in shaping the distribution of organisms and supposed that all organisms' distributions were a product of vicariance. The point of contention was timing. For example, platyrrhine monkeys are restricted to the Neotropics, while their closely related, and purported sister group, the catarrhine monkeys, are found in the Ethiopian region. Because Africa and South America were at one time connected and subsequently separated, vicariance biogeographers supposed the observed distribution pattern was due to vicariance; however, based on fossil evidence, it is thought that New World and Old World monkeys are not old enough to have existed at the time when the landmasses split sometime approximately 100 million years ago, so they could not be a product of vicariance. Vicariance biogeographers would ordinarily counter that the appropriate aged fossil evidence simply has not been found yet, but such fossils would actually predate the supposed origin time for the entire order of Primates. Another approach to the solution, one advocated by Alan de Queiroz (2014), is to construct an evolutionary tree of monkeys using DNA sequence data (molecular data) and calibrate the tree based on known fossil evidence to determine likely dates of divergence. Such an approach indicates that New World and Old World monkeys diverged no earlier than 51 million years ago, which postdates the separation of South America and Africa and, together with an evolutionary tree of Primates, indicates that South America was colonized from Africa via dispersal. Today, many studies indicate that both dispersal and vicariance have played roles in shaping the distribution patterns of organisms. This is typically done by examining the systematics of the group of interest and their distribution patterns.

The phylum Mollusca, which includes familiar creatures such as bivalves, snails, slugs, squid, and octopuses and less familiar ones such as chitons, monoplacophorans, and tusk shells, is thought to be the second-largest group of animals behind the arthropods, with recent estimates of total diversity being more than two hundred thousand living species (Bouchet et al., 2016). Mollusks live in almost every conceivable habitat, including deep-sea vents, coral reefs, estuaries, rocky and sandy shorelines, deserts and rainforests, and freshwater lakes and rivers. Freshwater mollusks are restricted to two classes, the Bivalvia and the Gastropoda (snails and slugs). Freshwater mollusks, which tend to be restricted to particular lake or river systems, are an excellent group for studying biogeography because they are ancient (the fossil record of bivalves and gastropods dates back more than 500 million years ago) and their distribution patterns are a reflection of past continental and climate changes.

Although theoretically freshwater mollusks are excellent candidates for biogeographic studies, they have not figured prominently in studies of distribution relative to vertebrates, including freshwater fishes (Berra, 2001; Lévêque et al., 2008). This may be in part attributable to the paucity of biologists who study mollusks relative to those who study other groups of animals. However, with the concomitant development of rigorous, objective methodological approaches for constructing evolutionary trees based on morphological characters and the revolution in molecular biological techniques that enable investigators to generate molecular data like DNA sequences to examine relationships, there has been a proliferation of systematic studies of mollusks, including many

freshwater groups, over the past 20 years. These studies have shed considerable light on the evolutionary relationships of the species and higher-level family groups and have helped delimit taxonomic and distributional boundaries.

Gastropods are one of the two classes of mollusks into which freshwater species have evolved. Of the 476 families of Recent gastropods currently recognized (Bouchet et al., 2017), 34 are composed of species that are wholly or partially restricted to freshwater (Strong et al., 2008; see Table 1.1, p. 7). Three largely marine families (Littorinidae, Buccinidae, and Marginellidae) include isolated genera that have invaded freshwater; they are not treated in this volume. Strong et al. (2008) made a meritorious attempt to describe global patterns of freshwater gastropod species diversity as it relates to the zoogeographic regions of the world. The Palearctic led the way with 1408-1711 species, followed by the Nearctic with 585 species and the Oriental, Neotropical, and Australasian with similar ranges of 509-606, 440-533, and 490-514, respectively. Lower levels of diversity are found in the Afrotropical region (366 species) and the Pacific Oceanic Islands (154-169 species). Admittedly, determining species richness in a region is difficult to accomplish. In North America, for example, for some groups like the Physidae and Pleuroceridae, there was a tendency to describe virtually every morphotype as a distinct species, resulting in a proliferation of nominal species—more than 450 for the Physidae (Taylor, 2003) and more than 1000 for the Pleuroceridae (Graf, 2001). Far fewer of these nominal species are considered valid today, and that number is likely to change when modern monographic treatments analyzing morphology and molecules are examined. For example, Wethington and Lydeard (2007) examined the systematics of the Physidae using molecular data and concluded that of the ca. 29 species and subspecies examined, there was support for the recognition of only 12 phylogenetic species. In contrast, the Hydrobiidae of North America were initially treated as few widely distributed species, but when anatomical and molecular studies were conducted, it was discovered that the group was much more diverse than initially realized. Other regions like the Neotropics are not especially diverse, but it remains to be seen whether this indeed results from not being a biotically rich region for freshwater mollusks or results from a lack of detailed investigation of the fauna using the same taxonomic lens applied to other regions. In spite of the shortcomings of estimating species diversity of freshwater gastropods, Strong et al. (2008) estimated that there are approximately 4000 valid described species of freshwater gastropods and approximately 8000 total species.

Some freshwater gastropods play a role as disease vectors, acting as intermediate hosts for parasitic trematodes (flukes) that infect humans and their livestock. Schistosomiasis is an acute and chronic disease caused primarily by five species of the blood fluke, *Schistosoma* (*S. mansoni, S. japonicum, S. mekongi, S. guineensis,* and *S. haematobium*) (World Health Organization, www.who.int/mediacentre/factsheets/fs115 /en/). Schistosomiasis is prevalent in tropical and subtropical regions of Africa and Southeast Asia, and *S. mansoni* was introduced into the Caribbean and South America, where susceptible snails serve as intermediate hosts, perhaps during the slave trade (Peters and Pasvol, 2002; Morgan et al., 2005). The World Health Organization estimates that 258 million people required preventive treatment in 2014 (www.who .int/mediacentre/factsheets/fs115/en/). Foodborne trematodiases are caused by flukes, of which the most common affecting humans are the liver flukes of the genera *Clonorchis, Fasciola, Opisthorchis* and the lung flukes of the genus *Paragonimus.* Foodborne trematodiases are estimated to affect more than 56 million people throughout the world (World Health Organization, www.who.int/mediacentre/factsheets /fs368/en/). The freshwater families that play the primary role as disease vectors for the aforementioned infections are the Planorbidae and Pomatiopsidae for schistosomiasis and the Bithyniidae, Lymnaeidae, Pachychilidae, Semisulcospiridae, and Thiaridae for liver and lung flukes (Malek and Cheng, 1974; Davis, 1980, 1992; Peters & Pasvol, 2002; Strong et al., 2008). Understanding disease transmission and epidemiology has stimulated a number of studies on the systematics of various snail host groups and/or their coevolution (e.g., Brown, 1994; Davis, 1980, 1992; Bandoni et al., 1995).

Bivalves are the second class of Mollusca to invade freshwater. There are at least 21 families that have at least one representative living in freshwater, but only 16 actually live and reproduce in inland waterways (Graf, 2013). Seven largely marine families (Arcidae, Cardiidae, Corbulidae, Donacidae, Mytilidae, Pharidae, and Pholadidae) include one to five species that have invaded freshwater (not treated in this volume). Of the approximately 1200 species of freshwater bivalves, 97% belong to eight primary freshwater families (Unionidae, Margaritiferidae, Hyriidae, Etheriidae, Mycetopodidae, and Iridinidae (all Unionoida) and Sphaeriidae and Cyrenidae (both Order Venerida). Freshwater mussels (Unionida) make up the vast majority (~70%) of the diversity in freshwater bivalves, with about 890 species (Graf and Cummings, 2007; chaps. 38–43, this volume; Graf, 2013).

Graf (2013, table 1) provided a table showing the patterns of global species richness of all the families containing bivalves reported from freshwater. The area containing the highest diversity of primary freshwater bivalves is the Nearctic (347 species), closely followed by the Indotropics (337). These two regions support more than half (57%) of the diversity. The Neotropics support a diverse fauna of about 250 species, followed by the Palearctic (133), Afrotropics (121), and Oceana/Australasia (62 species). As with the gastropods, it remains to be seen whether these numbers and percentages will hold up after additional surveys and taxonomic studies have been completed in understudied regions.

Invasive bivalves have caused considerable ecological damage and have resulted in huge economic costs. Modification to the natural hydrology via the creation of canals, channelization of large rivers, and increased commerce has mediated the spread of brackish-water mussels (Dreissenidae and Mytilidae) and Asian clams (Cyrenidae) around the world. Zebra mussels, *Dreissena* spp., invaded Europe in the 1800s and North America in the early 1980s. The Golden mussel *Limnoperna fortunei* (Dunker, 1857) invaded South America around 1991 and now inhabits five countries (Argentina, Uruguay, Paraguay, Bolivia, and Brazil) from the Rio Sao Francisco in Pernambuco, Brazil, in the north, to the Pantanal in central South America and the northernmost tributaries of the Rio Paraná, to the Rio de la Plata estuary in the south (Boltovskoy & Correa, 2015; Oliveira et al., 2015; Barbosa et al., 2016).

The ecological impact of the exotic brackish-water

mussels *Limnoperna fortunei* and *Dreissena* spp. in native bivalves has been substantial and well documented (Ricciardi 1998; Ricciardi et al., 1998; Burlakova et al., 2000; Darrigran et al., 2012; Krebs et al., 2015).

In trying to quantify the annual and cumulative economic impact of zebra mussels in North America, from the first full year after their introduction in 1989 to 2004, Connelly et al. (2007) surveyed water treatment and electric power generation facilities in infested areas. The dollar amounts they came up with were surprisingly low, compared to early predictions of $1 billion to $5 billion per year (Roberts, 1990; Aldridge et al., 2004). The mean annual expenditure per facility in 1989–1995 was $52,000, decreasing to $29,000 in 1996–2000. Average annual costs per facility since 2000 have fluctuated between $25,000 and $35,000. The total costs for 1989–2004 were $267 million (with the 95% confidence interval comprising $167 million to $467 million). The ecological and economic impact of Asian clams (*Corbicula* spp.) is less well known.

Surface freshwater habitats contain only about 0.01% of the world's water and cover only about 0.8% of the earth's surface (Gleick, 1996), yet they are thought to harbor ca. 100,000 out of 1.75 million described species (Hawksworth and Kalin-Arroyo, 1995). Although freshwater gastropods represent only approximately 5% of overall gastropod diversity, and freshwater bivalves constitute roughly 10% of overall bivalve diversity (Graf, 2013), these species are among the most imperiled species of all animal groups (Lydeard et al., 2004; Cowie et al., 2017). For example, of the 732 species listed as extinct by the 2016 IUCN Red List of Threatened Species (www.redlist.org), 310 species are mollusks, and of those, 72 are freshwater gastropods and 29 are freshwater bivalve species; many more are threatened with extinction. Threats to freshwater biodiversity outlined by Dudgeon et al. (2006) include overexploitation, water pollution, flow modification, destruction or degradation of habitat, and invasion by exotic species. Although overexploitation mainly affects fishes, unionoid bivalves were overexploited during the early 1900s for the button industry and later (1960–1990) for the cultured pearl industry; this no longer is much of a threat (Bräutigam, 1990). Perhaps the most profound impact has been flow modification from the construction of dams. For example, the middle reaches of the Tennessee River in northern Alabama, in a region called Muscle Shoals, and the Coosa watershed, which drains parts of Tennessee, Georgia, and eastern Alabama, lost 32 of 69 and 33 of 51 species of unionoid mussels, respectively. The Coosa drainage also lost 26 species and 4 entire genera (*Clappia, Gyrotoma, Amphigyra,* and *Neoplanorbis*) of freshwater gastropods (Bogan et al., 1995; Lydeard et al., 2004). Destruction or degradation of habitat, as well as flow modification, has also impacted hydrobiid gastropods in the arid southwestern United States and the Great Artesian Basin of Australia (Lydeard et al., 2004). The introduction of the exotic zebra mussel (*Dreissena polymorpha*) into the United States has also negatively impacted native freshwater mussels.

Conserving imperiled freshwater mollusks has been done at the species level for those especially threatened with extinction, by propagation techniques in a fisheries facility and individual species recovery plans, and these must continue, but what really is needed is ecosystem-level protection of hotspots. For example, Strong et al. (2008) identified 25 gastropod species hotspots of diversity categorized by primary habitat as follows: *springs and groundwater* (the southwestern United States; Cuatro Cienegas basin, Mexico; Florida, United States; mountainous regions in southern France and Spain; the southern Alps and Balkans region; the Great Artesian basin, Australia; western Tasmania, Australia; New Caledonia), *ancient oligotrophic lakes* (Titicaca, Ohrid and the Ohrid basin, Victoria, Tanganyika, Malawi, Baikal, Biwa, Inle and the Inle watershed, the Sulawesi lakes), *large rivers and their first and second order tributaries* (the Mobile Bay basin; the lower Uruguay River and Rio de la Plata, Argentina-Uruguay-Brazil; the western lowland forest of Guinea and Ivory Coast; the Lower Zaire basin; Zrmanja in Croatia; northwestern Ghats; the lower Mekong river in Thailand, Laos, and Cambodia), and *monsoonal wetlands* (northern Australia). Protection of these hotspots will ensure the long-term survival and viability of hundreds of endemic species.

Starobogatov (1970) and later Bănărescu (1990) provided distribution maps for the families of freshwater mollusks, but these predated many modern systematic treatments of the groups, and Bănărescu (1990) admitted that due to the poor understanding of systematic relationships of many of the groups at the time, biogeographic analyses could be proposed only with hesitation. The purpose of this volume

is to provide an up-to-date synthesis of studies on evolutionary relationships of the freshwater molluscan families, which will be useful foundational knowledge from which to build and encourage future investigations of the groups.

Table 1.1. Currently recognized freshwater mollusk families, genera and the estimated number of species in each. The type genus (if extant) of each Family or Subfamily is listed first. The higher classification of the freshwater gastropods is from Bouchet et al. (2017) and that of the bivalves from Graf & Cummings (2007), the Mussel Project (http://mussel-project.uwsp .edu/index.html) and MolluscaBase (http://molluscabase.org/). The list for each family was compiled directly from the authors of the chapters in this volume and should be cited as the source.

Phylum Mollusca (716 freshwater genera, ~5600 species)

  Class Gastropoda (34 families, 535 genera, and ~4370 species)

    Subclass Neritimorpha

      Order Cycloneritida

        Superfamily Helicinoidea Férussac, 1822

          Family Neritiliidae Schepman, 1908 (6 species)

            Neritilia Martens, 1875

            Platynerita Kano & Kase, 2003

        Superfamily Neritoidea Rafinesque, 1815

          Family Neritidae Rafinesque, 1815 (~200 species)

            Neritina Lamarck, 1816

            Clithon Montfort, 1810

            Dostia Gray, 1842

            Fluvinerita Pilsbry, 1932

            Nereina de Cristofori & Jan, 1832

            Neritodryas Martens, 1869

            Septaria Férussac, 1807

            Theodoxus Montfort, 1810

    Subclass Caenogastropoda

      Order Architaenioglossa

        Superfamily Ampullarioidea Gray, 1824

          Family Ampullariidae Gray, 1824 (186 species)

            Afropomus Pilsbry & Bequaert, 1927

            Asolene d'Orbigny, 1838

            Felipponea Dall, 1919

            Forbesopomus Bequaert & Clench, 1937

            Lanistes Montfort, 1810

            Marisa Gray, 1824

            Pila Röding, 1798

            Pomacea Perry, 1810

            Saulea Gray, 1868

        Superfamily Viviparoidea Gray, 1847

          Family Viviparidae Gray, 1847 (~150 species)

          Subfamily Viviparinae Gray, 1847

            Viviparus Montfort, 1810

            Rivularia Heude, 1890

            Tulotoma Haldeman, 1840

           Subfamily Bellamyinae Rohrbach, 1937

            Bellamya Jousseaume, 1886

            Angulyagra Rao, 1931

            Anularya Zhang & Chen, 2015

            Anulotaia Brandt, 1968

            Cipangopaludina Hannibal, 1912

            Eyriesia Fischer, 1885

            Filopaludina Habe, 1964

            Idiopoma Pilsbry, 1901

            Larina A. Adams, 1855

            Margarya Nevill, 1877

            Mekongia Crosse & Fischer, 1876

(continued)

Table 1.1. *continued*

Neothauma Smith, 1880
Notopala Cotton, 1935
Sinotaia Haas, 1939
Taia Annandale, 1918
Torotaia Haas, 1939
Trochotaia Brandt, 1974

Subfamily Lioplacinae Gill, 1863
Lioplax Troschel, 1857
Campeloma Rafinesque, 1819

Cohort Sorbeoconcha
Superorder Cerithiimorpha
Superfamily Cerithioidea Flemming, 1822

Family Hemisinidae Fischer & Crosse, 1891 (40 species)
Hemisinus Swainson, 1840
Aylacostoma Spix, 1827
Cubaedomus Thiele, 1928
Pachymelania E.A. Smith, 1893

Family Melanopsidae H. Adams & A. Adams, 1854 (~30 species)
Melanopsis Férussac, 1807
Esperiana Bourguignat, 1877
Holandriana Bourguignat, 1884
Microcolpia Bourguignat, 1884
Zemelanopsis Finlay, 1926

Family Pachychilidae Fischer & Crosse, 1892 (260 species)
Pachychilus I. Lea & H.C. Lea, 1851
Brotia H. Adams, 1866
Doryssa Swainson, 1840
Faunus Montfort, 1810
Jagora Köhler & Glaubrecht, 2003
Madagasikara Köhler & Glaubrecht, 2010
Paracrostoma Cossmann, 1900
Potadoma Swainson, 1840
Pseudopotamis Martens, 1894
Sulcospira Troschel, 1858
Tylomelania Sarasin & Sarasin, 1897

Family Paludomidae Stoliczka, 1868 (~100 species)
Paludomus Swainson, 1840
Anceya Bourguignat, 1885
Bathanalia Moore, 1898
Bridouxia Bourguignat, 1885
Chytra Moore, 1898
Cleopatra Troschel, 1857
Lavigeria Bourguignat, 1888
Limnotrochus Smith, 1880
Martelia Dautzenberg, 1908
Mysorelloides Leloup, 1953
Paramelania Smith, 1881
Philopotamis Layard, 1855
Potadomoides Leloup, 1953
Pseudocleopatra Thiele, 1928
Reymondia Bourguignat, 1885
Spekia Bourguignat, 1879
Stanleya Bourguignat, 1885
Stomatodon Benson, 1856

Table 1.1.

Stormsia Leloup, 1953
Syrnolopsis Smith, 1880
Tanalia Gray, 1847
Tanganyicia Crosse, 1881
Tiphobia Smith, 1880
Vinundu Michel, 2004
Family Pleuroceridae Fischer, 1885 (167 species)
Pleurocera Rafinesque, 1818
Athearnia Morrison, 1971
Elimia H. Adams & A. Adams, 1854
Gyrotoma Shuttleworth, 1845
Io Lea, 1831
Leptoxis Rafinesque, 1819
Lithasia Haldeman, 1840
Lithasiopsis Pilsbry, 1910
Family Semisulcospiridae Morrison, 1952 (~75 species)
Semisulcospira Boettger, 1886
Hua Chen, 1943
Juga H. Adams & A. Adams, 1854
Koreoleptoxis Burch & Jung, 1988
Namrutua Abbott, 1948
Senckenbergia Yen, 1939
Family Thiaridae Gill, 1871 (1823) (~135 species)
Thiara Röding, 1798
Balanocochlis Fischer, 1885
Fijidoma Morrison, 1952
Melania Lamarck, 1799
Melanoides Olivier, 1804
Melasma A. Adams & G.F. Angas, 1864
Mieniplotia Low & Tan, 2014
Neoradina Brandt, 1974
Plotiopsis Brot, 1874
Ripalania Iredale, 1943
Sermyla H. Adams & A. Adams, 1854
Stenomelania Fischer, 1885
Tarebia H. Adams & A. Adams, 1854
Superorder Hypsogastropoda
Order Littorinimorpha
Superfamily Truncatelloidea Gray, 1840
Family Amnicolidae Tryon, 1862 (~120 species)
Amnicola Gould & Haldeman, 1840
Akiyoshia Kuroda & Habe, 1954
Antroselates Hubricht, 1963
Baicalia Martens, 1876
Chencuia Davis, 1997
Colligyrus Hershler, 1999
Dasyscias F.G. Thompson & Hershler, 1991
Erhaia Davis & Kuo, 1985
Godlewskia Crosse & Fischer, 1879
Kolhymamnicola Starobogatov & Budnikova, 1976
Korotnewia Kozhov, 1936
Liobaicalia Martens, 1876
Lyogyrus Gill, 1863

(continued)

Table 1.1. *continued*

        Maackia Clessin, 1880

        Marstoniopsis van Regteren Altena, 1936

        Moria Kuroda & Habe, 1958

        Parabaikalia Lindholm, 1909

        Parabythinella Radoman, 1973

        Pseudobaikalia Lindholm, 1909

        Pyrgobaicalia Starobogatov, 1972

        Spirogyrus Thompson & Hershler, 1991

        Taylorconcha Hershler, et al., 1994

        Teratobaikalia Lindholm, 1909

        Terrestribythinella Sitnikova, et al., 1992

    Family Assimineidae H. Adams & A. Adams, 1856 (~20 species)

        Assiminea Fleming, 1828

        Angustassiminea Habe, 1943

        Austroassiminea Solem, Girardi, et al., 1982

        Aviassiminea Fukuda & Ponder, 2003

        Balambania Crosse, 1891

        Cavernacmella Habe, 1942

        Ditropisena Iredale, 1933

        Eussoia Preston, 1912

        Omphalotropis Pfeiffer, 1851

        Pseudogibbula Dautzenberg, 1890

        Rugapedia Fukuda & Ponder, 2004

        Solenomphala Martens, 1883

        Suterilla Thiele, 1927

        Taiwanassiminea Kuroda & Habe, 1950

        Tutuilana Hubendick, 1952

  Incerta sedis Assimineidae H. Adams & A. Adams, 1856

        Litthabitella Boeters, 1970

    Family Bithyniidae Gray, 1857 (~130 species)

        Bithynia Leach, 1818

        Codiella Locard, 1894

        Congodoma Mandahl-Barth, 1968

        Digoniostoma Annandale, 1920

        Digyrcidium Locard, 1882

        Emmericiopsis Thiele, 1928

        Funduella Mandahl-Barth, 1968

        Gabbia Tryon, 1865

        Gabbiella Mandahl-Barth, 1968

        Hydrobioides Nevill, 1885

        Incertihydrobia Verdcourt, 1958

        Jubaia Mandahl-Barth, 1968

        Liminitesta Mandahl-Barth, 1974

        Milletelona Beriozkina & Starobogatov, 1994

        Mysorella Godwin-Austen, 1919

        Neosataria Kulkarni & Khot, 2015

        Opisthorchophorus Beriozkina & Starobogatov, 1994

        Parabithynia Pilsbry, 1928

        Paraelona Beriozkina & Starobogatov, 1994

        Parafossarulus Annandale, 1924

        Petroglyphus Moellendorff, 1894

        Pseudobithynia Glöer & Pešić, 2006

        Pseudovivipara Annandale, 1918

        Sierraia Connolly, 1929

Table 1.1.

<table>
<tr><td colspan="2">Soapilia Binder, 1961</td></tr>
<tr><td colspan="2">Wattebledia Crosse, 1886</td></tr>
</table>

Family Cochliopidae Tryon, 1866 (260 species)

    Cochliopa Stimpson, 1865

    Antrobia Hubricht, 1971

    Aphaostracon F.G. Thompson, 1968

    Aroapyrgus H.B. Baker, 1931

    Balconorbis Hershler & Longley, 1986

    Coahuilix D.W. Taylor, 1966

    Cochliopina Morrison, 1946

    Durangonella Morrison, 1945

    Emmericiella Pilsbry, 1909

    Eremopyrgus Hershler, 1999

    Heleobia Stimpson, 1865

    Heleobops F.G. Thompson, 1968

    Ipnobius Hershler, 2001

    Juturnia Hershler, Liu & Stockwell, 2002

    Lithococcus Pilsbry, 1911

    Littoridina Souleyet, 1852

    Littoridinops Pilsbry, 1952

    Lobogenes Pilsbry & Bequaert, 1927

    Mesobia F.G. Thompson & Hershler, 1991

    Mexipyrgus D.W. Taylor, 1966

    Mexithauma D.W. Taylor, 1966

    Minkleyella Hershler, Liu & Landye 2011

    Nanivitrea Thiele, 1928

    Onobops F.G. Thompson, 1968

    Paludiscala D.W. Taylor, 1966

    Phreatoceras Hershler & Longley, 1987

    Phreatodrobia Hershler & Longley, 1986

    Pseudotryonia Hershler, 2001

    Pyrgophorus Ancey, 1888

    Spurwinkia Davis & Mazurkiewicz, 1982

    Stygopyrgus Hershler & Longley, 1986

    Subcochliopa Morrison, 1946

    Tepalcatia Thompson & Hershler, 2002

    Texadina Abbott & Ladd, 1951

    Texapyrgus Thompson & Hershler, 1991

    Tryonia Stimpson, 1865

    Zetekina Morrison, 1947

Family Helicostoidae Pruvot-Fol, 1937 (1 species)

    Helicostoa Lamy, 1926

Family Hydrobiidae Stimpson, 1865 (156 genera, >900 species)

    Hydrobia Hartmann, 1821

    Adriohydrobia Radoman, 1974

    Agrafia Szarowska & Falniowski, 2011

    Alzoniella Giusti & Bodon, 1984

    Anadoludamnicola Şahin, Koca &Yıldırım, 2012

    Anagastina Radoman, 1978

    Andrusovia Brusina in Westerlund, 1903

    Antibaria Radoman, 1973

    Arganiella Giusti & Pezzoli, 1980

    Avenionia Nicolas, 1882

(continued)

Table 1.1. *continued*

Balkanica Georgiev, 2011

Balkanospeum Georgiev, 2012

Belgrandia Bourguignat, 1870

Belgrandiella Wagner, 1928

Birgella Baker, 1926

Boetersiella Arconada & Ramos, 2001

Bracenica Radoman, 1973

Bucharamnicola Izzatullaev, Sitnikova & Starobogatov, 1985

Bullaregia Khalloufi, Béjaoui & Delicado, 2017

Caspia Clessin & Dybowski, 1888

Caspiohydrobia Starobogatov, 1970

Cavernisa Radoman, 1978

Chilopyrgula Brusina, 1896

Chirgisia Glöer, Boeters & Pešić, 2014

Chondrobasis Arconada & Ramos, 2001

Cilgia Schütt, 1968

Cincinnatia Pilsbry, 1891

Corbellaria Callot-Girardi & Boeters, 2012

Corrosella Boeters, 1970

Costellina Kuščer, 1933

Dalmatella Velkovrh, 1970

Dalmatinella Radoman, 1973

Daphniola Radoman, 1973

Daudebardiella Boettger, 1905

Devetakia Georgiev & Glöer, 2011

Devetakiola Georgiev, 2017

Dianella Gude, 1913

Diegus Delicado, Machordom & Ramos, 2015

Dolapia Radoman, 1973

Ecrobia Stimpson, 1865

Euxinipyrgula Sitnikova & Starobogatov, 1999

Falniowskia Bernasconi, 1991

Falsibelgrandiella Radoman, 1973

Falsipyrgula Radoman, 1973

Fissuria Boeters, 1981

Floridobia Thompson & Hershler, 2002

Ginaia Brusina, 1896

Globuliana Paladhile, 1866

Gloeria Georgiev, Dedov & Cheshmedjiev, 2012

Gocea Hadžišče, 1956

Graecoanatolica Radoman, 1973

Graecoarganiella Falniowski & Szarowska, 2011

Graecorientalia Radoman, 1973

Graziana Radoman, 1975

Grossuana Radoman, 1973

Guadiella Boeters, 2003

Hadziella Kuščer, 1932

Hauffenia Pollonera, 1898

Heideella Backhuys & Boeters, 1974

Heraultiella Bodon, Manganelli & Giusti, 2002

Horatia Bourguignat, 1887

Iberhoratia Arconada & Ramos, 2007

Iglica A.J. Wagner, 1928

Insignia Angelov, 1972

Table 1.1.

Intermaria Delicado, Pešić & Glöer, 2016
Isimerope Radea & Parmakelis, 2013
Islamia Radoman, 1973
Istriana Velkovrh, 1971
Iverakia Glöer & Pešić, 2014
Josefus Arconada & Ramos, 2006
Karucia Glöer & Pešić, 2013
Kaskakia Glöer & Pešić, 2012
Kerkia Radoman, 1978
Kirelia Radoman, 1973
Kolevia Georgiev & Glöer, 2015
Lanzaia Brusina, 1906
Lyhnidia Hadžišče, 1956
Macedopyrgula Radoman, 1973
Malaprespia Radoman, 1973
Marstonia Baker, 1926
Martensamnicola Izzatullaev, Sitnikova & Starobogatov, 1985
Mercuria Boeters, 1971
Meyrargueria Girardi, 2009
Micropyrgula Polinski, 1929
Microstygia Georgiev & Glöer, 2015
Milesiana Arconada & Ramos, 2006
Montenegrospeum Pešić & Glöer, 2013
Motsametia Vinarski, Palatov & Glöer, 2014
Myrtoessa Radea, 2016
Narentiana Radoman, 1973
Navarriella Boeters, 2000
Neofossarulus Polinski, 1929
Nicolaia Glöer et al., 2016
Notogillia Pilsbry, 1953
Ochridopyrgula Radoman, 1955
Ohridohauffenia Hadžišče, 1956
Ohridohoratia Hadžišče, 1956
Ohrigocea Hadžišče, 1956
Pauluccinella Giusti & Pezzoli, 1990
Peringia Paladilhe, 1874
Persipyrgula Delicado, Pešić & Glöer, 2016
Pezzolia Bodon & Giusti, 1986
Plagigeyeria Tomlin, 1930
Plesiella Boeters, 2003
Polinskiola Radoman, 1973
Pontobelgrandiella Radoman, 1978
Pontohoratia Vinarski, Palatov & Glöer, 2014
Prespolitorea Radoman, 1973
Prespopyrgula Radoman, 1973
Probythinella Thiele, 1928
Pseudamnicola Paulucci, 1878
Pseudavenionia Bodon & Giusti, 1982
Pseudohoratia Radoman, 1967
Pseudoislamia Radoman, 1979
Pseudopaludinella Mabille, 1877
Pseudorientalia Radoman, 1973
Pyrgohydrobia Radoman, 1955

(*continued*)

Table 1.1. *continued*

    Pyrgorientalia Radoman, 1973

    Pyrgula Cristofori & Jan, 1832

    Pyrgulopsis Call & Pilsbry, 1886

    Radomaniola Szarowska, 2006

    Rhapinema Thompson, 1970

    Sadleriana Clessin, 1890

    Salenthydrobia Wilke, 2003

    Sarajana Radoman, 1975

    Sardohoratia Manganelli et al., 1998

    Sarkhia Glöer & Pešić, 2012

    Saxurinator Schütt, 1960

    Shadinia Akramowski, 1976

    Sheitanok Schütt & Şeşen, 1991

    Sivasi Şahin, Koca & Yıldırım, 2012

    Sogdamnicola Izzatullaev, Sitnikova & Starobogatov, 1985

    Spathogyna Arconada & Ramos, 2002

    Spilochlamys Thompson, 1968

    Stankovicia Polinski, 1929

    Stiobia Thompson & McCaleb, 1978

    Stoyanovia Georgiev, 2017

    Strandzhia Georgiev & Glöer, 2013

    Strugia Radoman, 1973

    Sumia Glöer & Mrkvicka, 2015

    Tadzhikamnicola Izzatullaev, 2004

    Tanousia Bourguignat in Servain, 1881

    Tarraconia Ramos & Arconada, 2000

    Tefennia Schütt & Yıldırım, 2003

    Terranigra Radoman, 1978

    Torosia Glöer & Georgiev, 2012

    Trachyochridia Polinski, 1929

    Trichonia Radoman, 1973

    Turkmenamnicola Izzatullaev, Sitnikova & Starobogatov, 1985

    Turkorientalia Radoman, 1973

    Turricaspia Dybowski & Grochmalicki, 1917

    Valvatamnicola Izzatullaev, Sitnikova & Starobogatov, 1985

    Vinodolia Radoman, 1973

    Xestopyrgula Polinski, 1929

    Zaumia Radoman, 1973

    Zavalia Radoman, 1973

Family Lithoglyphidae Tryon, 1866 (~200 species)

    Lithoglyphus Pfeiffer, 1828

    Antrorbis Hershler & F.G. Thompson, 1990

    Benedictia Lindholm, 1927

    Clappia Walker, 1909

    Dabriana Radoman, 1974

    Fluminicola Stimpson, 1865

    Gillia Stimpson, 1865

    Kobeltocochlea Lindholm, 1909

    Lepyrium Dall, 1896

    Lithoglyphopsis Thiele, 1928

    Pseudobenedictia Sitnikova, 1987

    Pterides Pilsbry, 1909

    Somatogyrus Gill, 1863

Table 1.1.

Tanousia Servain, 1881

Yaroslawiella Sitnikova, 2001

Family Moitessieriidae Bourguignat, 1863 (~120 species)

Moitessieria Bourguignat, 1863

Atebbania Ghamizi, et al., 1999

Baldufa Alba, et al., 2010

Bosnidilhia Boeters, Glöer & Pešić, 2013

Bythiospeum Bourguignat, 1882

Clameia Boeters & Gittenberger, 1990

Corseria Boeters & Falkner, 2009

Henrigirardia Boeters & Falkner, 2003

Palacanthilhiopsis Bernasconi, 1988

Paladilhia Bourguignat, 1865

Palaospeum Boeters, 1999

Sardopaladilhia Manganelli, et al., 1998

Sorholia Boeters & Falkner, 2009

Spiralix Boeters, 1972

Incerta sedis Moitessieriidae Bourguignat, 1863

Iglica A.J. Wagner, 1928

Zeteana Glöer & Pešić, 2014

Family Pomatiopsidae Stimpson, 1865 (~170) species)

Pomatiopsis Tryon, 1862

Blanfordia A. Adams, 1863

Cecina A. Adams, 1861

Coxiella E.A. Smith, 1894

Delavaya Heude, 1889

Fenouilia Heude, 1889

Fukuia Abbott & Hunter, 1949

Gammatricula Davis & Liu, 1990

Halewisia Davis, 1979

Hemibia Heude, 1890

Hubendickia Brandt, 1968

Hydrorissoia Bavay, 1895

Idiopyrgus Pilsbry, 1911

Jinhongia Davis, 1990

Jullienia Crosse & Fischer, 1876

Karelainia Davis, 1979

Kunmingia Davis & Kuo, 1984

Lacunopsis Deshayes, 1876

Manningiella Brandt, 1970

Neoprososthenia Davis & Kuo, 1984

Neotricula Davis, 1986

Oncomelania Gredler, 1881

Pachydrobia Crosse & Fischer, 1876

Pachydrobiella Thiele, 1928

Parapyrgula Annandale & Prashad, 1919

Rehderiella Brandt, 1974

Robertsiella Davis & Greer, 1980

Saduniella Brandt, 1970

Taihua Annandale, 1924

Tomichia Benson, 1851

Tricula Benson, 1843

Wuconchona Kang, 1983

(continued)

Table 1.1. *continued*

Family Stenothyridae Tryon, 1866 (~60 species)
　　Stenothyra Benson, 1856
　　Farsithyra Glöer & Pešić, 2009
　　Gangetia Ancey, 1890
Family Tateidae Thiele, 1925 (490 species)
　　Tatea Tenison-Woods, 1879
　　Ascorhis Ponder & Clark, 1988
　　Austropyrgus Cotton, 1942
　　Beddomeia Petterd, 1889
　　Caledoconcha Haase & Bouchet, 1998
　　Catapyrgus Climo, 1974
　　Caldicochlea Ponder, 1997
　　Crosseana Zielske & Haase, 2015
　　Fluvidona Iredale, 1937
　　Fluviopupa Pilsbry, 1911
　　Fonscochlea Ponder, et al., 1989
　　Hadopyrgus Climo, 1974
　　Halopyrgus Haase, 2008
　　Hemistomia Crosse, 1872
　　Heterocyclus Crosse, 1872
　　Jardinella Iredale & Whitley, 1938
　　Kanakyella Haase & Bouchet, 1998
　　Keindahan Haase & Bouchet, 2006
　　Kuschelita Climo, 1974
　　Leiorhagium Haase & Bouchet, 1998
　　Leptopyrgus Haase, 2008
　　Meridiopyrgus Haase, 2008
　　Nanocochlea Ponder & Clark, 1993
　　Novacaledonia Zielske & Haase, 2015
　　Obtusopyrgus Haase, 2008
　　Opacuincola Ponder, 1966
　　Paxillostium Gardner, 1970
　　Phrantela Iredale, 1943
　　Pidaconomus Haase & Bouchet, 1998
　　Platypyrgus Haase, 2008
　　Posticobia Iredale, 1943
　　Potamolithus Pilsbry, 1896
　　Potamopyrgus Stimpson, 1865
　　Pseudotricula Ponder, 1992
　　Rakipyrgus Haase, 2008
　　Rakiurapyrgus Haase, 2008
　　Selmistomia Bernasconi, 1995
　　Sororipyrgus Haase, 2008
　　Sulawesidrobia Ponder & Haase, 2005
　　Tongapyrgus Haase, 2008
　　Trochidrobia Ponder, et al., 1989
　　Victodrobia Iredale, 1943
　　Westrapyrgus Ponder, et al., 1999
Subclass Heterobranchia
　　Grade "Lower Heterobranchia"
　　Superfamily Valvatoidea Gray, 1840
　　　　Family Valvatidae Gray, 1840 (71 species)
　　　　　　Valvata Müller, 1774
　　　　　　Borysthenia Lindholm, 1914

Table 1.1.

Infraclass Euthyneura
  Cohort Tectipleura
    Superorder Pylopulmonata
      Superfamily Glacidorboidea Ponder, 1986
        Family Glacidorbidae Ponder, 1986 (22 species)
            Glacidorbis Iredale, 1943
            Benthodorbis Ponder & Avern, 2000
            Gondwanorbis Ponder, 1986
            Patagonorbis Rumi & Gutiérrez-Gregoric, 2015
            Striadorbis Ponder & Avern, 2000
            Tasmodorbis Ponder & Avern, 2000
    Superorder Acochlidimorpha
      Superfamily Acochlidioidea Küthe, 1935
        Family Acochlidiidae Küthe, 1935 (7 species)
            Acochlidium Strubell, 1892
            Palliohedyle Rankin, 1979
            Strubellia Odhner, 1937
        Family Tantulidae Rankin, 1979 (1 species)
            Tantulum Rankin, 1979
    Superorder Hygrophila
      Superfamily Chilinoidea Dall, 1870
        Family Chilinidae Dall, 1870 (~50 species)
            Chilina Gray, 1828
        Family Latiidae Hutton, 1882 (4 species)
            Latia Gray, 1850
      Superfamily Lymnaeoidea Rafinesque, 1815
        Family Lymnaeidae Rafinesque, 1815 (~100 species)
          Subfamily Lymnaeinae Rafinesque, 1815
            Lymnaea Lamarck, 1799
            Acella Haldeman, 1841
            Aenigmomphiscola Kruglov & Starobogatov, 1981
            Bakerilymnaea Weyrauch, 1964
            Bulimnaea Haldeman, 1841
            Erinna H. Adams & A. Adams, 1855
            Galba Schrank, 1803
            Hinkleyia F.C. Baker, 1928
            Ladislavella Dybowski, 1913
            Omphiscola Rafinesque, 1819
            Pseudoisidora Thiele, 1931
            Pseudosuccinea F.C. Baker, 1908
            Sibirigalba Kruglov & Starobogatov, 1985
            Sphaerogalba Kruglov & Starobogatov, 1985
            Stagnicola Jeffreys, 1830
            Walterigalba Kruglov & Starobogatov, 1985
          Subfamily Lancinae Hannibal, 1914
            Lanx Clessin, 1880
            Fisherola Hannibal, 1912
            Idaholanx Clark, et al., 2017
          Subfamily Amphipepleinae Pini, 1877
            Myxas Sowerby, 1822
            Austropeplea Cotton, 1942
            Bullastra Bergh, 1901
            Lantzia Jousseaume, 1872

(continued)

Table 1.1. *continued*

Limnobulla Kruglov & Starobogatov, 1985

Orientogalba Kruglov & Starobogatov, 1985

Pectinidens Pilsbry, 1911

Radix Montfort, 1810

Family Acroloxidae Thiele, 1931 (~50 species)

Acroloxus Beck, 1838

Baicalancylus Starobogatov, 1967

Frolikhiancylus Sitnikova & Starobogatov, 1993

Gerstfeldtiancylus Starobogatov, 1989

Pseudancylastrum Lindholm, 1909

Family Bulinidae Fischer & Crosse, 1880 (~40 species)

Subfamily Bulininae Fischer & Crosse, 1880

Bulinus Müller, 1781

Indoplanorbis Annandale & Prashad, 1921

Subfamily Plesiophysinae Bequaert & Clench, 1939

Plesiophysa Fischer, 1883

Family Burnupiidae Albrecht, 2017 (~20 species)

Burnupia Walker, 1912

Family Physidae Fitzinger, 1833 (43 species)

Physa Draparnaud, 1801

Aplexa Fleming, 1820

Physella Haldeman, 1842

Stenophysa Martens, 1898

Family Planorbidae Rafinesque, 1815 (~150 species)

Subfamily Planorbinae Rafinesque, 1815

Planorbis Müller, 1774

Africanogyrus Özdikmen & Darılmaz, 2007

Anisus Studer, 1820

Bathyomphalus Charpentier, 1837

Ceratophallus Brown & Mandahl-Barth 1973

Choanomphalus Gerstfeldt, 1859

Gyraulus Charpentier, 1837

Subfamily Ancylinae Rafinesque, 1815

Ancylus Müller, 1774

Ferrissia Walker, 1903

Gundlachia Pfeiffer, 1850

Hebetancylus Pilsbry, 1914

Laevapex Walker, 1903

Rhodacmea Walker, 1917

Sineancylus Gutiérrez Gregoric, 2014

Subfamily Camptoceratinae Dall, 1870

Camptoceras Benson, 1843

Subfamily Coretinae Gray, 1847

Planorbarius Duméril, 1805

Subfamily Drepanotrematinae Zilch, 1959

Drepanotrema Fischer & Crosse 1880

Subfamily Helisomatinae F.C. Baker, 1928

Helisoma Swainson, 1840

Acrorbis Odhner, 1937

Biomphalaria Preston, 1910

Menetus H. & A. Adams, 1855

Planorbella Haldeman, 1842

Planorbula Haldeman, 1840

Table 1.1.

Subfamily Miratestinae P. & F. Sarasin, 1897
Miratesta P. & F. Sarasin, 1897
Amerianna Strand, 1928
Ancylastrum Bourguignat, 1853
Bayardella Burch, 1977
Glyptophysa Crosse, 1872
Isidorella Tate, 1896
Kessneria Walker & Ponder, 2001
Leichhardtia Walker, 1988
Patelloplanorbis Hubendick, 1957
Protancylus P. & F. Sarasin, 1898
Subfamily Neoplanorbinae Hannibal, 1912
Neoplanorbis Pilsbry, 1906
Amphigyra Pilsbry, 1906
Subfamily Segmentininae F.C. Baker, 1945
Segmentina Fleming, 1818
Helicorbis Benson, 1850
Hippeutis Charpentier, 1837
Lentorbis Mandahl-Barth, 1954
Segmentorbis Mandahl-Barth, 1954
Unassigned valid planorbid genera
Pecosorbis Taylor, 1985
Promenetus F.C. Baker, 1935
Theratodocion Brown & Verdcourt, 1998
Vorticifex Meek (in Dall), 1870
Class Bivalvia (9 Families and 180 genera, ~1230 species)
Subclass Heterodonta
Infraclass Euheterodonta
Superorder Imparidentia
Order Venerida
Superfamily Cyrenoidea Gray, 1840
Family Cyrenidae Gray, 1840 (~125 species)
Batissa Gray, 1853
Corbicula von Mühlfeld, 1811
Cynocyclas Blainville, 1818
Geloina Gray, 1842
Polymesoda Gray, 1853
Villorita Gray, 1833
Superfamily Sphaerioidea Deshayes, 1854 (1820)
Family Sphaeriidae Deshayes, 1854 (1820) (~227 species)
Subfamily Sphaeriinae Deshayes, 1854 (1820)
Sphaerium Scopoli, 1777
Afropisidium Kuiper, 1962
Cyclocalyx Dall, 1903
Odhneripisidium Kuiper, 1962
Pisidium Pfeiffer, 1821
Subfamily Euperinae Heard, 1965
Eupera Bourguignat, 1854
Byssanodonta d'Orbigny, 1846
Order Myida
Superfamily Dreissenoidea Gray, 1840
Family Dreissenidae Gray, 1840 (~11 species)

(continued)

Table 1.1. *continued*

Dreissena van Beneden, 1835

Congeria Partsch, 1835

Mytilopsis Conrad, 1857

Subclass Paleoheterodonta

Order Unionida (173 genera, 918 species)

Superfamily Unionoidea Rafinesque, 1820 (137 genera, 725 species)

Family Unionidae Rafinesque, 1820 (713 species)

Subfamily Unioninae Rafinesque, 1820

Tribe Unionini Rafinesque, 1820

Unio Philipsson in Retzius, 1788

Aculamprotula Wu, Liang, Wang & Shan, 1999

Acuticosta Simpson, 1900

Cuneopsis Simpson, 1900

Diaurora Cockerell, 1903

Inversiunio Habe, 1991

Lepidodesma Simpson, 1896

Nodularia Conrad, 1853

Rhombuniopsis Haas, 1920

Schistodesmus Simpson, 1900

Tribe Lanceolariini Froufe, Lopes-Lima, & Bogan, 2017

Lanceolaria Conrad, 1853

Arconaia Conrad, 1865

Tribe Anodontini Rafinesque, 1820

Anodonta Lamarck, 1799

Anemina Haas, 1969

Cristaria Schumacher, 1817

Pletholophus Simpson, 1900

Pseudanodonta Bourguignat, 1876

Simpsonella Cockerell, 1903

Sinanodonta Modell, 1945

Tribe Alasmidontini Rafinesque, 1820

Alasmidonta Say, 1818

Anodontoides F.C. Baker, 1898

Arcidens Simpson, 1900

Lasmigona Rafinesque, 1831

Pegias Simpson, 1900

Pseudodontoideus Frierson, 1927

Pyganodon Crosse & Fischer, 1894

Simpsonaias Frierson, 1914

Strophitus Rafinesque, 1820

Utterbackia F.C. Baker, 1927

Utterbackiana Frierson, 1927

Subfamily Ambleminae Rafinesque, 1820

Tribe Amblemini Rafinesque, 1820

Amblema Rafinesque, 1820

Tribe Lampsilini Ihering, 1901

Lampsilis Rafinesque, 1820

Actinonaias Crosse & Fischer, 1894

Arotonaias Martens, 1900

Cyprogenia Agassiz, 1852

Cyrtonaias Crosse & Fischer, 1894

Delphinonaias Crosse & Fischer, 1894

Disconaias Crosse & Fischer, 1894

Dromus Simpson, 1900

Table 1.1.

   Ellipsaria Rafinesque, 1820
   Epioblasma Rafinesque, 1831
   Friersonia Ortmann, 1912
   Glebula Conrad, 1853
   Hamiota Roe & Hartfield, 2005
   Lemiox Rafinesque, 1831
   Leptodea Rafinesque, 1820
   Ligumia Swainson, 1840
   Medionidus Simpson, 1900
   Obliquaria Rafinesque, 1820
   Obovaria Rafinesque, 1819
   Ortmanniana Frierson, 1927
   Pachynaias Crosse & Fischer, 1894
   Potamilus Rafinesque, 1818
   Ptychobranchus Simpson, 1900
   Toxolasma Rafinesque, 1831
   Truncilla Rafinesque, 1819
   Venustaconcha Frierson, 1927
   Villosa Frierson, 1927
  Tribe Pleurobemini Hannibal, 1912
   Pleurobema Rafinesque, 1819
   Elliptio Rafinesque, 1819
   Elliptoideus Frierson, 1927
   Eurynia Rafinesque, 1820
   Fusconaia Simpson, 1900
   Hemistena Rafinesque, 1820
   Parvispina Perkins, Johnson, & Gangloff, 2017
   Plethobasus Simpson, 1900
   Pleuronaia Frierson, 1927
  Tribe Quadrulini Ihering, 1901
   Quadrula Rafinesque, 1820
   Cyclonaias Pilsbry in Ortmann & Walker, 1922
   Megalonaias Utterback, 1915
   Theliderma Swainson, 1840
   Tritogonia Agassiz, 1852
 Incerta cedis Ambleminae
   Barynaias Crosse & Fischer, 1894
   Martensnaias Frierson, 1927
   Micronaias Simpson, 1900
   Nephritica Frierson, 1927
   Nephronaias Crosse & Fischer, 1894
   Plectomerus Conrad, 1853
   Popenaias Frierson, 1927
   Psoronaias Crosse & Fischer, 1894
   Psorula Haas, 1930
   Reginaia Campbell & Lydeard, 2012
   Reticulatus Frierson, 1927
   Sphenonaias Crosse & Fischer, 1894
   Uniomerus Conrad, 1853
  Subfamily Gonideinae Ortmann, 1916
   Gonidea Conrad, 1857
   Bineurus Simpson, 1900
   Chamberlainia Simpson, 1900

(continued)

Table 1.1. *continued*

Discomya Simpson, 1900
Gibbosula Simpson, 1900
Inversidens Haas, 1911
Lamprotula Simpson, 1900
Leguminaia Conrad, 1865
Microcondylaea Vest, 1866
Monodontina Conrad, 1853
Potomida Swainson, 1840
Pronodularia Starobogatov, 1970
Pseudodon Gould, 1844
Pseudodontopsis Kobelt, 1913
Pilsbryoconcha Simpson, 1900
Ptychorhynchus Simpson, 1900
Sinohyriopsis Starobogatov, 1970
Solenaia Conrad, 1869
Subfamily Rectidentinae Modell, 1942
Rectidens Simpson, 1900
Caudiculatus Simpson, 1900
Contradens Haas, 1911
Elongaria Haas, 1913
Ensidens Frierson, 1911
Haasodonta McMichael, 1956
Hyriopsis Conrad, 1853
Physunio Simpson, 1900
Pressidens Haas, 1910
Prohyriopsis Haas, 1914
Trapezoideus Simpson, 1900
Subfamily Modellnaiinae Brandt, 1974
Modellnaia Brandt, 1974
Subfamily Parreysiinae Henderson, 1935
Tribe Parreysiini Henderson, 1935
Parreysia Conrad, 1853
Tribe Coelaturini Modell, 1942
Coelatura Conrad, 1853
Brazzaea Bourguignat, 1885
Grandidieria Bourguignat, 1885
Nitia Pallary, 1924
Nyassunio Haas, 1936
Prisodontopsis Tomlin, 1928
Pseudospatha Simpson, 1900
Tribe Lamellidentini Modell, 1942
Lamellidens Simpson, 1900
Trapezidens Bolotov, Vikhrev, & Konopleva, 2017
Tribe Oxynaiini Starobogatov, 1970
Oxynaia Haas, 1913
Radiatula Simpson, 1900
Scabies Haas, 1911
Incerta sedis Parreysinae
Arcidopsis Simpson, 1900
Incerta sedis Unionidae Rafinesque, 1820
Ctenodesma Simpson, 1900
Germainaia Graf & Cummings, 2009
Harmandia Rochebrune, 1881

Table 1.1.

        Protunio Haas, 1913

        Pseudobaphia Simpson, 1900

        Schepmania Haas, 1910

        Unionetta Haas, 1955

     Family Margaritiferidae Henderson, 1929 (12 species)

        Margaritifera Schumacher, 1816

Superfamily Hyrioidea (15 genera, 96 species)

     Family Hyriidae Swainson, 1840 (96 species)

        Subfamily Hyriinae Swainson, 1840

           Tribe Hyriini Swainson, 1840

           Prisodon Schumacher, 1817

           Triplodon Spix in Spix & Wagner, 1827

           Tribe Castalini Swainson, 1840

           Castalia Lamarck, 1819

           Callonaia Simpson, 1900

           Castaliella Simpson, 1900

           Tribe Hyridellini McMichael, 1956

           Hyridella Swainson, 1840

           Cucumerunio Iredale, 1934

           Virgus Simpson, 1900

           Tribe Rhipidodontini Starobogatov, 1970

           Diplodon Spix in Spix & Wagner, 1827

        Subfamily Velesunioninae Iredale, 1834

           Velesunio Iredale, 1934

           Alathyria Iredale, 1934

           Lortiella Iredale, 1934

           Microdontia Tapparone Canefri, 1883

           Westralunio Iredale, 1934

   Incerta sedis Hyriidae

        Echyridella McMichael & Hiscock, 1958

Superfamily Etherioidea (21 genera, 97 species)

     Family Etheriidae Deshayes, 1832 (4 species)

        Etheria Lamarck, 1807

        Acostaea d'Orbigny, 1851

        Bartlettia H. Adams, 1867

        Pseudomulleria R. Anthony, 1907

     Family Mycetopodidae Gray, 1840 (53 species)

        Subfamily Mycetopodinae Gray, 1840

           Mycetopoda d'Orbigny, 1835

           Mycetopodella Marshall, 1927

        Subfamily Anodontitinae Modell, 1942

           Anodontites Bruguière, 1792

           Lamproscapha Swainson, 1840

        Subfamily Leiinae Morettes, 1849

           Leila Gray, 1840

        Subfamily Monocondylaeinae Modell, 1942

           Monocondylaea d'Orbigny, 1835

           Diplodontites Marshall, 1922

           Fossula Lea, 1870

           Haasica Strand, 1932

           Iheringella Pilsbry, 1893

           Tamsiella Haas, 1931

(continued)

24

Table 1.1. *continued*

Family Iridinidae Swainson, 1840 (40 species)
Subfamily Iridininae Swainson, 1840
Chelidonopsis Ancey, 1887
Mutela Scopoli, 1777
Pleiodon Conrad, 1834
Subfamily Aspathariinae Modell, 1942
Aspatharia Bourguignat, 1885
Chambardia Bourguignat in Servain, 1890
Moncetia Bourguignat, 1885

LITERATURE CITED

Aldridge, D.C., P. Elliott, and G.D. Moggridge. 2004. The recent and rapid spread of the zebra mussel (*Dreissena polymorpha*) in Great Britain. Biological Conservation 119: 253-261.

Bănărescu, P. 1990. Zoogeography of Fresh Waters, Vol. 1: General Distribution and Dispersal of Freshwater Animals. Aula-Verlag, Wiesbaden.

Bandoni, S.M., M. Mulvey, and E.S. Loker. 1995. Phylogenetic analysis of eleven species of *Biomphalaria* Preston, 1910 (Gastropoda: Planorbidae) based on comparisons of allozymes. Biological Journal of the Linnean Society 54: 1-27.

Barbosa, N.P.U., F.A. Silva, M. Divina de Oliveira, M.A. dos Santos Neto, M.D. de Carvalho, and A.V. Cardoso. 2016. *Limnoperna fortunei* (Dunker, 1857) (Mollusca, Bivalvia, Mytilidae): First record in the São Francisco River basin, Brazil. Check List 12: 1846, 1-6.

Berra, T.M. 2001. Freshwater Fish Distribution. Academic Press, Cambridge, UK.

Bogan, A.E., J.M. Pierson, and P. Hartfield. 1995. Decline in the freshwater gastropod fauna in the Mobile Bay basin. Pp. 249-252 in Our Living Resources: A Report to the Nation on the Distribution, Abundance, and Health of the U.S. Plants, Animals, and Ecosystems (E.T. LaRoe, G.S. Farris, C.E. Pucket, P.D. Doran, and M.J. Mac, eds.). US Department of the Interior, National Biological Service, Washington, DC.

Boltovskoy, D., and N. Correa. 2015. Ecosystem impacts of the invasive bivalve *Limnoperna fortunei* (golden mussel) in South America. Hydrobiologia 746: 81-95.

Bouchet, P., S. Bary, V. Héros, and G. Marani. 2016. How many species of molluscs are there in the world's oceans, and who is going to describe them? Pp. 9-24 in Tropical Deep-Sea Benthos 29. (V. Héros, E. Strong, and P. Bouchet, eds.). Mémoires du Muséum national d'Histoire naturelle, 208. Muséum national d'Histoire naturelle, Paris.

Bouchet, P., J-P. Rocroi, B. Hausdorf, A. Kaim, Y. Kano, A. Nützel, P. Parkhaev, M. Schrödl, and E.E. Strong. 2017. Revised classification, nomenclator and typification of gastropod and monoplacophoran families. Malacologia 61: 1-526.

Bräutigan, A. 1990. Trade in unionids from the United States for the cultured pearl industry. Tentacle 2: 6-7.

Brown, D. 1994. Freshwater Snails of Africa and Their Medical Importance. 2nd ed. Taylor & Francis, London.

Burlakova, L.E., A.Y. Karatayev, and D.K. Padilla. 2000. The impact of *Dreissena polymorpha* (Pallas) invasion on unionid bivalves. International Review of Hydrobiology 85: 529-541.

Connelly, N.A., C.R. O'Neill Jr., B.A. Knuth, and T.L. Brown. 2007. Economic impacts of zebra mussels on drinking water treatment and electric power generation facilities. Environmental Management 40: 105-112.

Cowie, R.H., C. Régnier, B. Fontaine, and P. Bouchet. 2017. Measuring the sixth extinction: What do mollusks tell us? Nautilus 131: 3-41.

Darrigran, G., C. Damborenea, E.C. Drago, I. Ezcurra de Drago, A. Paira, and F. Archuby. 2012. Invasion process of *Limnoperna fortunei* (Bivalvia: Mytilidae): The case of Uruguay River and emissaries of the Esteros del Iberá Wetland, Argentina. Zoologia 29: 531-539.

Darwin, C.R., and A.R. Wallace. 1858. On the tendency of species to form varieties; and on the perpetuation of varieties and species by natural means of selection. Journal of the Proceedings of the Linnean Society of London. Zoology 3: 45-50.

Davis, G.M. 1980. Snail hosts of Asian *Schistosoma* infecting man: Evolution and coevolution. Malacological Review Supplement 2: 195-238.

Davis, G.M. 1992. Evolution of Prosobranch snails transmitting Asian *Schistosoma;* coevolution with *Schistosoma:* A review. Progress in Clinical Parasitology 3: 145-204.

De Queiroz, A. 2014. The Monkey's Voyage: How Improbable Journeys Shaped the History of Life. Basic Books, New York.

Dudgeon, D., A.H. Arthington, M.O. Gessner, Z. Kawabata, D.J. Knowler, C. Lévêque, R.J. Naiman, et al. 2006. Freshwater biodiversity: Importance, threats, status and conservation challenges. Biological Reviews of the Cambridge Philosophical Society 81: 163-182.

Gleick, P.H. 1996. Water resources. Pp. 817-823 in Encyclopedia of Climate and Weather (S.H. Schneider, ed.). Oxford University Press, New York.

Graf, D.L. 2001. The cleansing of the Augean Stables, or a lexicon of the nominal species of the Pleuroceridae (Gastropoda: Prosbranchia) of recent North America, North of Mexico. Walkerana 12: 1-124.

Graf, D.L. 2013. Patterns of freshwater bivalve global diversity and the state of phylogenetic studies on the Unionoida, Sphaeriidae, and Cyrenidae. American Malacological Bulletin 31: 135-153. DOI: 10.4003/006.031.0106.

Graf, D.L., and K.S. Cummings. 2007. Review of the systematics and global diversity of freshwater mussel species (Bivalvia: Unionoida). Journal of Molluscan Studies 73: 291-314.

Hawksworth, D.J., and M.T. Kalin-Arroyo. 1995. Magnitude and distribution of biodiversity. Pp. 107-138 in Global biodiversity Assessment (V.H. Heywood, ed.). Cambridge University Press, Cambridge, UK.

Krebs, R.A., E.M. Barkett, and M.T. Bcglcy. 2015. The impact of dreissenid mussels on growth of the fragile papershell (*Leptodea fragilis*), the most abundant unionid species in Lake Erie. Canadian Journal of Zoology 93: 143-148.

Lévêque, C., T. Oberdorff, D. Paugy, M.L.J. Stiassny, and P.A. Tedesco. 2008. Global diversity of fish (Pisces) in freshwater. Hydrobiologia 595: 545-567.

Lydeard, C., R.H. Cowie, A.E. Bogan, P. Bouchet, K.S. Cummings, T.J. Frest, D.G. Herbert, et al. 2004. The global decline of nonmarine mollusks. BioScience 54: 321-330.

Malek, E.A., and T.C. Cheng. 1974. Medical and Economic Malacology. Academic Press, New York.

Morgan, J.A.T., R.J. Dejong, G.O. Adeoye, E.D.O. Ansa, C.S. Barbosa, P. Bremond, I.M. Cesari, et al. 2005. Origin and diversification of the human parasite *Schistosoma mansoni*. Molecular Ecology 14: 3889-3902.

Nelson, G., and N. Platnick. 1981. Systematics and Biogeography: Cladistics and Vicariance. Columbia University Press, New York.

Nelson, G., and D.E. Rosen. 1981. Vicariance Biogeograpy: A Critique. Columbia University Press, New York.

Oliveira, M.D., M.C.S. Campos, E.M. Paolucci, M.C.D. Mansur, and S.K. Hamilton. 2015. Colonization and spread of *Limnoperna fortunei* in South America. Pp. 333-356 in *Limnoperna fortunei:* The Ecology, Distribution and Control of a Swiftly Spreading Invasive Fouling Mussel (D. Boltovskoy, ed.). Springer Series in Invasion Ecology. Springer International.

Peters, W., and G. Pasvol. 2002. Tropical Medicine and Parasitology. 5th ed. Mosby, London.

Ricciardi, A. 1998. Global range expansion of the Asian mussel *Limnoperna fortunei* (Mytilidae): Another fouling threat to freshwater systems. Biofouling 13: 97-106.

Ricciardi, A., R.J. Neves, and J.B. Rasmussen. 1998. Impending extinctions of North American freshwater mussels (Unionoida) following the zebra mussel (*Dreissena polymorpha*) invasion. Journal of Animal Ecology 67: 613-619.

Roberts, L. 1990. Zebra mussel invasion threatens U.S. waters. Science 249: 1370-1372.

Schlater, P.L. 1858. On the general geographical distribution of the members of the class Aves. Zoological Journal of the Linnean Society 2: 130-145.

Starobogatov, Y.I. 1970. Fauna Mollyuskov i Zoogeograficheskoe Raionirovanie Kontinental'nykh Vodoemov Zemnogo Shara [in Russian; Fauna of mollusks and zoogeographic division of continental waterbodies of the globe]. Nauka, Leningrad.

Strong, E.E., O. Gargominy, W.F. Ponder, and P. Bouchet. 2008. Global diversity of gastropods (Gastropoda: Mollusca) in freshwater. Hydrobiologia 595: 149-166.

Taylor, D.W. 2003. Introduction to Physidae (Gastropoda: Hygrophila), biogeography, classification, morphology. Revista de Biologia Tropical 51, Supplement 1: 1-287.

Wallace, A.R. 1962. The Geographical Distribution of Animals with a Study of the Relations of Living and Extinct Faunas as Elucidating the Past Changes of the Earth's Surface. Reprint, Hafner, New York. Original publication 1876.

Wegener, A. 1912a. Die Entstehung der Kontinente [in German]. Petermanns Geographische Mitteilungen 58: 185-195, 253-256, 305-308.

Wegener, A. 1912b. Die Entstehung der Kontinente. Geologische Rundschau 3: 276-292.

Wegener, A. 1915. Die Entstehung der Kontinente und Ozeane. Vieweg, Brunswick, Germany. Translated into English in 1966.

Wethington, A.R., and C. Lydeard. 2007. A molecular phylogeny of Physidae (Gastropoda: Basommatophora) based on mitochondrial DNA sequences. Journal of Molluscan Studies 73: 241-257.

2 Neritiliidae Schepman, 1908

YASUNORI KANO

The living members of the Neritiliidae include 23 described species and at least a dozen unnamed ones in seven genera from low latitude areas of the world. Schepman (1908) established this family for the freshwater and brackish-water genus *Neritilia* Martens, 1875, by emphasizing its unique radular morphology and concentrically growing operculum. However, the genus was later classified as a subfamily of Neritidae (Baker, 1923; p. 31, this volume) by apparently attaching greater importance to the overall similarity to neritids in the shape of the teleoconch. This taxonomic treatment had been long maintained (Andrews, 1937; Russell, 1941; Starmühlner, 1983), until Kano and Kase (2000a, b) reassigned it to the independent Neritiliidae together with the submarine-cave genus *Pisulina* Nevill & Nevill, 1869, based on their characteristic protoconch morphology and shell microstructure. Succeeding anatomical and histological examination has revealed numerous synapomorphies between *Neritilia* and *Pisulina* that also confirmed their distant relationship to the Neritidae (Kano and Kase, 2002). A molecular phylogeny inferred from partial sequences of the 28S rRNA gene corroborated the independence of the Neritiliidae from the Neritidae within the gastropod clade Neritimorpha (Kano et al., 2002). In this phylogeny, the former family was recovered as a possible sister group of the exclusively terrestrial superfamily Helicinoidea, while the latter formed a robust clade with another aquatic family, Phenacolepadidae. This topology led Bouchet et al. (2017) to assign the former within Helicinoidea (and the Neritidae and Phenacolepadidae in Neritoidea) in their classification of the Gastropoda, but this superfamilial assignment remains to be tested based on more extensive sampling of genes and taxa. Regardless, the globose to hemispherical shell in most species of Neritiliidae and Neritidae represents the plesiomorphic condition of the Neritimorpha, as it is obvious in the Mesozoic fossil record of the clade (Kano et al., 2002; see also Kaim and Sztajner, 2005; Bandel, 2007; Uribe et al., 2016).

The Neritiliidae occupy an unusual range of habitats. Species of *Neritilia* occur in freshwater streams to brackish estuaries (Kano and Kase, 2003), ground-

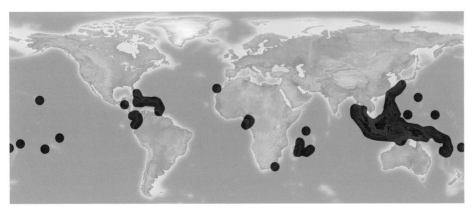

Distribution of nonmarine Neritiliidae.

waters (Sasaki and Ishikawa, 2002), and anchialine waters—bodies of haline water with subterranean connection to the sea and showing noticeable marine as well as terrestrial influences (Kano et al., 2001; Kano and Kase, 2004). Another limnic genus, *Platynerita* Kano & Kase 2003, currently contains the type species *P. rufa* from the tropical streams of the western Pacific as its sole extant, named taxon (Kano and Kase 2003). On the other hand, *Pisulina* and several other genera exclusively inhabit submerged caves in the shallow tropical waters of the Indo-West Pacific (Kano and Kase, 2000b, 2008), and curiously enough, no marine species is known outside caves. Cryptic habits have also been suggested for the Miocene genus *Pisulinella* Kano & Kase, 2000, and Oligocene *Bourdieria* Lozouet, 2004. Although the limnic *Neritilia* and *Platynerita* appeared earlier in the fossil record, both since the Eocene (Lozouet, 2004; Symonds and Tracey, 2014), a submarine cave origin of the family has been proposed from an anatomical point of view (Kano et al., 2002). In addition to the higher morphological (hence generic) diversity in the submarine caves than in limnic waters, all species of the family have open pit eyes without a vitreous body, a condition otherwise unknown among the Gastropoda except the true limpets in Patellogastropoda (Kano and Kase, 2002; see also Sumner-Rooney et al., 2016). If their eyes are degenerative in origin, the loss of the cornea and vitreous body seemingly took place in a common ancestral species that lived in dark, cryptic habitats. In this scheme, the invasion of the surface, limnic habitat is convincingly explained by the presence of underground corridors or anchialine habitats, where a number of *Neritilia* species are known to occur (Kay, 1979; Kano et al., 2001, 2003; Kano and Kase, 2004; Pérez-Dionis et al., 2010; Espinosa et al., 2017).

Species identification of freshwater and brackish-water neritiliids is relatively easy if specimens are properly dissected and observed for the female reproductive tract (Kano and Kase, 2003). Other important morphological characters for species taxonomy and identification include the teleoconch shape, measurements of the operculum, surface ornamentation of the protoconch, and number of denticles in radular teeth, the latter two of which require scanning electron microscopy for observation (Kano and

Neritilia rubida (Pease, 1865). Japan: Kaburmata River, Ishigaki Island, Okinawa. 4.1 mm. Poppe 788861.

Kase, 2003; Fukumori and Kano, 2014). Only six species occur in the freshwater reach of coastal streams as their main habitat. These include *Platynerita rufa* from the western Pacific islands, *Neritilia rubida* and *N. vulgaris* from the Indo-West Pacific (Kano and Kase, 2003), *N. panamensis* from the eastern Pacific (Morrison, 1946; Bandel, 2001), *N. succinea* from the Caribbean (Russell, 1941), and *N. manoeli* from the Gulf of Guinea, eastern Atlantic (Brown, 1980; Bandel and Kowalke, 1999). The six species tend to be abundant and geographically widespread and share a reddish- to yellowish-brown shell with a diameter of 3-6 mm, which is relatively large for the family. This contrasts with the characteristics of their congeners in such brackish environments as anchialine caves and pools as well as in subterranean waters. The latter taxa usually have a smaller (<3 mm), colorless shell and a more narrowly restricted geographic range, for example *N. cavernicola* from the Philippines, *N. hawaiiensis* from the Big Island of Hawaii, *N. mimotoi* from the western part of mainland Japan, *N. abeli* and *N. serrana* from Cuba, and *N. margaritae* from the Canary Islands (Kay, 1979; Kano et al., 2001, 2003; Kano and Kase, 2001, 2004; Pérez-Dionis et al., 2010; Espinosa et al., 2017). Many species remain undescribed from these cryptic habitats in the world.

The widespread geographic distribution of the limnic Neritiliidae is undoubtedly accomplished through their amphidromous life cycle. Amphidromy, as it is better documented in the Neritidae among mollusks (p. 31, this volume), involves both marine and freshwater phases, with planktonic larvae migrating to the sea (McDowall, 2010). All species of freshwater neritiliids (and also the brackish ones) hatch as planktotrophic larvae, as can be deduced from their multispiral protoconchs and larval opercula (Fukumori and Kano, 2014). Although they have never been identified in marine plankton collections, feeding on phytoplankton in the sea is obligatory for the completion of their life cycle. The veligers of *N. rubida* hatched upstream all died after several days in a freshwater aquarium but lived longer in brackish and fully marine waters (Kano and Kase, 2003). The longevity of the larvae in the ocean might possibly be shorter than that of amphidromous neritids (Fukumori and Kano, 2014), and yet the same species can be found on remote islands off eastern Africa and in French Polynesia. Field observations that the metamorphosed juveniles of *N. rubida* and *P. rufa* occur only in the lower courses of streams while most adults inhabit fast-flowing waters 100 m to 1 km upstream from tidal influence further corroborate the amphidromous life cycle and migration (Kano and Kase, 2003). Some individuals were found to crawl up for a surprisingly long distance, more than 8 km, to mountain rapids at an altitude of 200 m in Vanuatu, Melanesia. Such a great journey is possible because of their longevity, some reaching 10 years of age, which is unusually long for small, freshwater gastropod species (see Heller, 1990).

With the presence of the obligatory marine phase, neritiliids do not occur in lakes and pools without connection to the sea, or large rivers where their preferred fast-flowing environment is dozens of kilometers away from the mouth. They also cannot be found on mud bottom in estuaries, as hard substrata are a prerequisite for them to graze on microalgae. With regard to conservation, the downstream and upstream migration of amphidromous neritiliids can easily be hindered with such human activities as water withdrawals, channel modifications, and sewage inputs at any reach of a stream. However, their abundant occurrences on many islands, coupled with

the very high dispersal ability, seem to make them less vulnerable to species or population extinction than many direct-developing species of freshwater gastropods. More severely threatened are anchialine species, particularly the Hawaiian endemic *N. hawaiiensis* Kay, 1979. Although larval dispersal and genetic exchange between caves and islands have been documented for *N. cavernicola* in the Philippines, anchialine habitats are restricted and disjunct in distribution and susceptible to anthropogenic alteration and pollution (Kano and Kase, 2004). The remote location of Hawaii from other islands with anchialine waters seems to highlight the evolutionary distinctness and vulnerability of *N. hawaiiensis*.

LITERATURE CITED

Andrews, E.A. 1937. Certain reproductive organs in the Neritidae. Journal of Morphology 61: 525-561.

Baker, H.B. 1923. Notes on the radula of the Neritidae. Proceedings of the Academy of Natural Sciences of Philadelphia 75: 117-178, plates 9-16.

Bandel, K. 2001. The history of *Theodoxus* and *Neritina* connected with description and systematic evaluation of related Neritimorpha (Gastropoda). Mitteilungen aus dem Geologisch-Paläontologischen Institut Universität Hamburg 85: 65-164.

Bandel, K. 2007. Description and classification of Late Triassic Neritimorpha (Gastropoda, Mollusca) from the St Cassian Formation, Italian Alps. Bulletin of Geosciences 82: 215-274.

Bandel, K., and T. Kowalke. 1999. Gastropod fauna of the Cameroonian coasts. Helgoland Marine Research 53: 129-140.

Bouchet, P., J.-P. Rocroi, B. Hausdorf, A. Kaim, Y. Kano, A. Nützel, P. Parkhaev, M. Schrödl, and E.E. Strong. 2017. Revised classification, nomenclator and typification of gastropod families. Malacologia 61: 1-526.

Brown, D.S. 1980. New and little known gastropod species of fresh and brackish waters in Africa, Madagascar and Mauritius. Journal of Molluscan Studies 46: 208-223.

Espinosa, J., J. Ortea, and Y.L. Diez-García. 2017. El género *Neritilia* von Martens, 1879 (Mollusca: Gastropoda: Neritilidae) en Cuba, con la descripción de dos nuevas especies. Avicennia 20: 49-52.

Fukumori, H., and Y. Kano. 2014. Evolutionary ecology of settlement size in planktotrophic neritimorph gastropods. Marine Biology 161: 213-227.

Heller, J. 1990. Longevity in molluscs. Malacologia 31: 259-295.

Kaim, A., and P. Sztajner. 2005. The opercula of neritopsid gastropods and their phylogenetic importance. Journal of Molluscan Studies 71: 211-219.

Kano, Y., and T. Kase. 2000a. *Pisulinella miocenica,* a new genus and species of Miocene Neritiliidae (Gastropoda: Neritopsina) from Eniwetok Atoll, Marshall Islands. Paleontological Research 4: 69-74.

Kano, Y., and T. Kase. 2000b. Taxonomic revision of *Pisulina* (Gastropoda: Neritopsina) from submarine caves in the tropical Indo-Pacific. Paleontological Research 4: 107-129.

Kano, Y., and T. Kase. 2001. Reallocation of *Teinostoma* (*Calceolata*) *pusillum* to the genus *Neritilia* (Neritopsina: Neritiliidae). Venus 60: 1-6.

Kano, Y., and T. Kase. 2002. Anatomy and systematics of the submarine-cave gastropod *Pisulina* (Neritopsina: Neritiliidae). Journal of Molluscan Studies 68: 365-384.

Kano, Y., and T. Kase. 2003. Systematics of the *Neritilia rubida* complex (Gastropoda: Neritiliidae): Three amphidromous species with overlapping distributions in the Indo-Pacific. Journal of Molluscan Studies 69: 273-284.

Kano, Y., and T. Kase. 2004. Genetic exchange between anchialine-cave populations by means of larval dispersal: The case of a new gastropod species *Neritilia cavernicola.* Zoologica Scripta 33: 423-437.

Kano, Y., and T. Kase. 2008. Diversity and distributions of the submarine-cave Neritiliidae in the Indo-Pacific (Gastropoda: Neritimorpha). Organisms, Diversity & Evolution 8: 22-43.

Kano, Y., T. Sasaki, and H. Ishikawa. 2001. *Neritilia mimotoi,* a new neritiliid species from an anchialine lake and estuaries in southwestern Japan. Venus 60: 129-140.

Kano, Y., S. Chiba, and T. Kase. 2002. Major adaptive radiation in neritopsine gastropods estimated from 28S sequences and fossil records. Proceedings of the Royal Society of London, Biological Sciences, 269: 2457-2465.

Kano, Y., T. Kase, and H. Kubo. 2003. The unique interstitial habitat of a new neritiliid gastropod, *Neritilia littoralis.* Journal of the Marine Biological Association of the United Kingdom 83: 835-840.

Kay, E.A. 1979. Hawaiian Marine Shells: Reef and Shore Fauna of Hawaii. Section 4, Mollusca. Bishop Museum Press, Honolulu.

Lozouet, P. 2004. The European Tertiary Neritiliidae (Mollusca, Gastropoda, Neritopsina): Indicators of tropical submarine cave environments and freshwater faunas. Zoological Journal of the Linnean Society 140: 447-467.

McDowall, R.M. 2010. Why bc amphidromous: Expatrial dispersal and the place of source and sink population dynamics? Reviews in Fish Biology and Fisheries 20: 87-100.

Morrison, J.P.E. 1946. The nonmarine mollusks of San José Island, with notes on those of Pedro González Island, Pearl Island, Panamá. Smithsonian Miscellaneous Collections 106 (6): 1-49, plates 1-3.

Pérez-Dionis, G., J. Espinosa, and J. Ortea. 2010. Una nueva especie del género *Neritilia* Martens, 1879 (Mollusca: Gastropoda: Neritiliidae) de las islas Canarias. Vieraea 38: 117-121.

Russell, H.D. 1941. The recent molluscs of the Family Neritidae of the western Atlantic. Bulletin of the Museum of Comparative Zoology at Harvard College, in Cambridge 88: 347-404, plates 1-7.

Sasaki, T., and H. Ishikawa. 2002. The first occurrence of a neritopsine gastropod from a phreatic community. Journal of Molluscan Studies 68: 286-288.

Schepman, M.M. 1908. The Prosobranchia of the Siboga Expedition. Part 1. Rhipidoglossa and Docoglossa: Résultats des Explorations Zoologiques, Botaniques, Océanographiques et Géologiques. Entreprises aux Indes Orientales en 1899-1900, à bord du Siboga, sous le commandement de G. F. Tydeman 39: 1-107, plates 1-9.

Starmühlner, F. 1983. Results of the hydrobiological mission 1974 of the Zoological Institute of the University of Vienna. Part 8. Contributions to the knowledge of the freshwater-gastropods of the Indian Ocean islands. Annalen des Naturhistorischen Museums in Wien, Serie B, 84: 127-249.

Sumner-Rooney, L., J.D. Sigwart, J. McAfee, L. Smith, and S.T. Williams. 2016. Repeated eye reduction events reveal multiple pathways to degeneration in a family of marine snails. Evolution 70: 2268-2295.

Symonds, M.F., and S. Tracey. 2014. *Neritilia* (Gastropoda, Neritopsina, Neritiliidae): Pushing back the timeline. Cainozoic Research 14: 3-7.

Uribe, J.E., D. Colgan, L.R. Castro, Y. Kano, and R. Zardoya. 2016. Phylogenetic relationships among superfamilies of Neritimorpha (Mollusca: Gastropoda). Molecular Phylogenetics and Evolution 104: 21-31.

3 Neritidae Rafinesque, 1815

YASUNORI KANO AND HIROAKI FUKUMORI

The Neritidae are one of the most abundant groups of freshwater snails in the coastal streams of tropical and subtropical regions worldwide, as well as in the inland waters of the European continent. Neritids (nerites) seem to have originated in the shallow sea (Little, 1972, 1990; Holthuis, 1995; Uribe et al., 2016), and roughly a hundred living species of *Nerita* Linnaeus, 1758, and *Smaragdia* Issel, 1869, inhabit rocky shores and sea-grass beds (Kano et al., 2002; Frey and Vermeij, 2008). However, they are more diverse in freshwater and brackish-water environments in terms of the number of species and also variety of shell shapes. There are roughly 200 limnic species worldwide in such genera as *Neritina* Lamarck, 1816, *Clithon* Montfort, 1810, *Theodoxus* Montfort, 1810, *Neritodryas* Martens, 1869, *Dostia* Gray, 1842, and *Septaria* Férussac, 1807 (Holthuis, 1995; Fukumori and Kano, 2014). While the marine taxa all bear a plesiomorphic "neritiform" shell, which is hemispherical with a low spire and a wide semicircular aperture on the flat plane, freshwater and brackish-water species are often spined, winged, or limpet-shaped (Starmühlner,

1983; Bandel, 2001; Haynes, 2001, 2005). Fully-grown shells of these neritids may attain a diameter of 5 cm (Maciolek, 1978), but many species do not reach 2 cm at maximum.

The taxonomy of the limnic Neritidae is far from settled, notably due to numerous, ambiguous species descriptions and missing type specimens. More than a thousand species names have been proposed for neritids, more than half of which were given for freshwater and brackish-water species, mostly in the nineteenth century by Lamarck, Lesson, Récluz, Reeve, Sowerby, von Martens, and Gassies, among many other European authors (see, e.g., Martens, 1863–1879; Kabat and Finet, 1992). As is often the case in the taxonomic papers of that age, descriptions and diagnoses were based virtually exclusively on shells, and comparison with previously named species could not be made in sufficient detail. Most neritid species in tropical to subtropical coastal streams have wide-ranging geographic distributions thanks to their amphidromous life cycle (i.e., hatched larvae return to the sea) and transoceanic dispersal (see

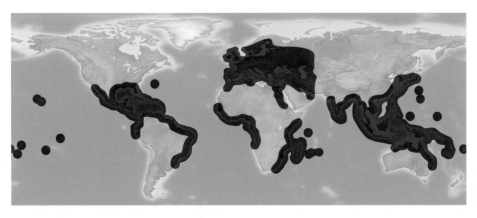

Distribution of nonmarine Neritidae.

below). However, a preconceived idea that islands harbor isolated, endemic faunas apparently led to an overestimation of global diversity and the proposal of many names until the mid-twentieth century. Examples include *Clithon corona, C. faba, C. variabilis, Septaria porcellana,* and *S. tessellata,* each of which has more than ten younger names for specimens from different localities (Fukumori, 2014). Later on and in recent years, taxonomists have tended to lump and recognize only about 50 species in the entire Indo-West Pacific region (e.g., Starmühlner, 1983; Haynes, 1985; Komatsu, 1986). This is, however, a clear underestimation. Neritid shells sometimes exhibit quite subtle differences between sympatric species in their shapes, colors, and markings (Huang, 1997; Kano, 2009), but often accompanied by considerable intraspecific variation (Gruneberg, 1982; Way et al., 1993). Thus species discrimination is difficult without understanding the exact ranges of within-species variation in these characters by employing independent criteria, most effectively by the comparison of the reproductive system (Andrews, 1937; Holthuis, 1995) and DNA sequencing (Kano et al., 2011). Labeling a biological species with the oldest existing synonym or proposing a new name is yet another task that requires considerable efforts to locate and compare (often missing) type specimens.

Generic and subfamilial classification also remains largely artificial. In her unpublished but seminal doctoral dissertation, Holthuis (1995) reconstructed a phylogeny of the Neritidae based on morphological characters, especially the reproductive organs of both sexes. This and also an unpublished molecular phylogeny of the family (Fukumori, 2014) have shown that the current use of genera and subfamilies reflects more of derived, adaptive morphology of the shell and radula, rather than phylogenetic relationships or shared ancestry. For example, the species of *Septaria* have acquired a patelliform shell, an internal operculum, and a very large foot mainly to cope with fast-flowing torrents and cascades (Vermeij, 1969). These highly modified limpets have been regarded to form an independent genus, a subfamily, or even a family (see Haynes, 2001). However, they actually represent very recent radiation within a clade of *Neritina* and *Dostia* (Holthuis, 1995; Fukumori, 2014). Another example of unwarranted generic allocation

Clithon nucleolus (Morelet, 1857). New Caledonia. 19.1 mm. Poppe 961565.

is the eastern Pacific, Caribbean, and western African species that have been classified in *Neritina,* such as *N. latissima, N. punctulata,* and *N. virginea* (Russell, 1941; Schneider and Lyons, 1993; Pyron and Covich, 2003). These American and western African species seem to collectively form a single lineage, which is distantly related to the type species *N. pulligera* in the western Pacific. They can instead be placed in a totally neglected genus, *Nereina* de Cristofori & Jan, 1832, among approximately 40 objectively valid genus-group names for the Neritidae (Holthuis, 1995). Here we need to make a subjective decision: do we want to use only a few generic names or many more for freshwater neritids with admittedly diverse morphologies via adaptive diversification but not phylogenetic antiquity? Regardless, such names as *Clithon, Theodoxus,* and *Neritodryas* would be valid in any scheme of classification. Fossil species that can be rigorously assigned to these genera have existed since the Eocene or early Oligocene time at latest (Bandel, 2001; Symonds, 2002; Symonds and Pacaud, 2010) and the names are also old enough from the nomenclatural point of view.

The geographic distribution ranges of amphidromous neritid species are relatively wide to enormous. Many species are known to distribute from Southeast Asia to Japan to Melanesia with no discernible morphological or genetic differentiation (Kano et al., 2011). Correspondingly, recent population studies have revealed genetic intermixing between distant populations (e.g., Myers et al., 2000; Bebler and Foltz, 2004; Cook et al., 2009; Crandall et al., 2010). Examples include two species in *Neritina* and *Dostia* showing high gene flow over several thousands of kilometers among South Pacific archipelagos (Crandall et al., 2010). Although the larvae of amphidromous neritids have never been noted in plankton collections from the open ocean, future studies employing DNA-based identification would illuminate their behavior and horizontal and vertical distribution in the sea.

Interestingly, the species richness of amphidromous neritids appears to peak in the Coral Triangle of Southeast Asia and to decrease with increasing distance from this diversity center. The eastern Pacific Barrier and Horn of Africa might have acted as almost impermeable biogeographic boundaries, and there are only a few or several species in each of the eastern Pacific, Caribbean, and western African regions (see above). This pattern closely approximates that of the marine genus *Nerita* (Frey, 2010) and many other coastal organisms (Roberts et al., 2002), suggesting similar processes underlying their distributions, regardless of the different habitats occupied by adults (see Crandall et al., 2010).

With their exceptional longevity, which may exceed 20 years (Shigemiya and Kato, 2001; Gorbach et al., 2012), amphidromous neritids are known to crawl upstream for very long distances, sometimes tens of kilometers (Haynes, 1985), to elevations of as high as 400 m (Maciolek, 1978). Migrating snails are often arranged in long lines and sometimes form huge aggregations (Schneider and Frost, 1986; Schneider and Lyons, 1993; Blanco and Scatena, 2005). Small juveniles of a few *Neritina* migrate by clinging to the shells of congeneric, subadult snails (Kano, 2009). Larger nerites are more capable of creeping upstream through rapids than smaller ones (Schneider and Lyons, 1993), and therefore this "hitchhiking" behavior seems to be beneficial in increasing the success rate of migration. Regarding species diversity and richness, habitat partitioning sustains dozens of species within a single river system in the western Pacific. Distance from the mouth, salinity, water velocity, and riverbed condition are the primary factors that determine species composition in a certain area of rivers and streams (Haynes, 1985; Bandel and Riedel, 1998; Okuda and Nishihira, 2002). However, several species commonly occupy exactly the same type of microhabitat in the same stream; hence their surprisingly high diversity cannot be solely explained by the resource partitioning (Kano et al., 2011). Neritids require hard substrata for their grazing and locomotion, and some species of *Dostia* and *Septaria* can exclusively be found on branches and drift logs in estuaries and mangrove swamps (Starmühlner, 1983; Haynes, 2001). These markedly euryhaline neritids have the potential to survive in fully marine conditions for some extended period of time and to disperse as benthic adults and eggs on drift logs; one species has even been found on sunken pieces of wood at depths of 105–135 m, about 3 km offshore, in Vanuatu (Kano et al., 2013).

Direct early development has evolved multiple times in the Neritidae (Holthuis, 1995; Kano, 2006). This type of ontogeny has resulted in entirely different life histories, often in more inland waters. European nerites of the genus *Theodoxus* all hatch from egg capsules as benthic, crawl-away juveniles without a marine phase (Bondesen, 1940; Holthuis, 1995; Kirkegaard, 2006). They therefore inhabit not only rivers and estuaries but also closed water systems without connection to the sea, often resulting in higher levels of population differentiation and species endemism (e.g., Brown, 1980; Bunje, 2005; Bunje and Lindberg, 2007; Fehér et al., 2009). Their longitudinal distributions range from Spain and Morocco in the west to Iran and the upstream of the Blue Nile in Ethiopia in the east (Brown, 1980; Glöer and Pešić, 2012). This lineage also has acquired exceptional resistance to cold water, which allowed the northward invasion of the type species *T. fluviatilis* to Finland and European Russia (Bunje and Lindberg, 2007; Fehér et al., 2009). Some other species inhabit deep waters in the Caspian Sea (20-100 m; Zettler, 2007) or underground waters in Turkey (Odabaşi and Arslan, 2015). Among tropical freshwater neritids, direct development has been observed in *Fluvinerita* Pilsbry, 1932, represented by the monotypic species *F. tenebricosa* in mountain streams on the Caribbean island of Jamaica (Andrews, 1935; Holthuis, 1995; Bandel, 2001).

Certain species of the Indo-West Pacific genus *Nerito-dryas* also hatch as benthic juveniles (Holthuis, 1995; Kano et al., 2011). This group has become somewhat terrestrial and even arboreal, being found on vegetation in the vicinity of streams and estuaries (Little, 1990). *Fluvinerita* and *Neritodryas* are considered to be reciprocal sister clades, despite the great geographic distance between the distribution ranges of these nonamphidromous snails (Holthuis, 1995).

Species of *Theodoxus* are known to serve as the intermediate host of trematode parasites (Zhokhova et al., 2006). However, the rate of trematode infection and public-health or veterinary significance of this and other taxa of Neritidae seems to be lower than those of such freshwater gastropods as the Planorbidae, Lymnaeidae, Pomatiopsidae, and Thiaridae (e.g., Reeves et al., 2008; Madsena and Hung, 2014). Other known parasites of freshwater neritids include turbellarian flatworms and ciliates (Raabe, 1968; Seixas et al., 2015).

The downstream and upstream migration of amphidromous neritids can easily be hindered by such human activities as water withdrawals, channel modifications, deforestation, and sewage inputs at any reach of a river or stream (Blanco and Scatena, 2006; Gorbach et al., 2012). Dams and certain types of stream crossings (e.g., culverts with pipes) can impede migration and eliminate upstream population (Resh, 2005). With their large biomass, elimination of these snails may result in food-web alternation in a stream system. Neritids are collected and eaten by humans in the tropics (Reeves et al., 2008). Large species are rare in some places, apparently due to overharvest by locals (Maciolek, 1978; Kano et al., 2011). However, the amphidromous life cycle allows recolonization of streams after local extirpation, given that adjacent populations serve as a source of larvae (Cook et al., 2009). Some species of the European *Theodoxus* are more severely threatened and are listed in the IUCN Red List as Critically Endangered or Endangered due to their low dispersal capability and endemism, coupled with habitat alternation including water pollution and withdrawal and introduction of a nonnative congener, *T. fluviatilis* (e.g., Fehér et al., 2012, 2017).

LITERATURE CITED

Andrews, E.A. 1935. The egg capsules of certain Neritidae. Journal of Morphology 57: 31-59.

Andrews, E.A. 1937. Certain reproductive organs in the Neritidae. Journal of Morphology 61: 525-561.

Bandel, K. 2001. The history of *Theodoxus* and *Neritina* connected with description and systematic evaluation of related Neritimorpha (Gastropoda). Mitteilungen aus dem Geologisch-Paläontologischen Institut Universität Hamburg 85: 65-164.

Bandel, K., and F. Riedel. 1998. Ecological zonation of gastropods in the Matutinao River (Cebu, Philippines), with focus on their life cycles. Annales de Limnologie 34: 171-191.

Bebler, M.H., and D.W. Foltz. 2004. Genetic diversity in Hawaiian stream macroinvertebrates. Micronesica 37: 119-128.

Blanco, J.F., and F.N. Scatena. 2005. Floods, habitat hydraulics and upstream migration of *Neritina virginea* (Gastropoda: Neritidae) in northeastern Puerto. Caribbean Journal of Science 41: 55-74.

Blanco, J.F., and F.N. Scatena. 2006. Hierarchical contribution of river-ocean connectivity, water chemistry, hydraulics, and substrate to the distribution of diadromous snails in Puerto Rican streams. Journal of the North American Benthological Society 25: 82-98.

Bondesen, P. 1940. Preliminary investigations into the development of *Neritina fluviatilis* L. in brackish and fresh water. Videnskabelige Meddelelser Naturhistorisk Forening i København 104: 283-318.

Brown, D.S. 1980. Freshwater Snails of Africa and Their Medical Importance. Taylor & Francis, London.

Bunje, P.M.E. 2005. Pan-European phylogeography of the aquatic snail *Theodoxus fluviatilis* (Gastropoda: Neritidae). Molecular Ecology 14: 4323-4340.

Bunje, P.M.E., and D.R. Lindberg. 2007. Lineage divergence of a freshwater snail clade associated with post-Tethys marine basin development. Molecular Phylogenetics and Evolution 42: 373-387.

Cook, B.D., S. Bernays, C.M. Pringle, and J.M. Hughes. 2009. Marine dispersal determines the genetic population structure of migratory stream fauna of Puerto Rico: Evidence for island-scale population recovery processes. Journal of the North American Benthological Society 28: 709-718.

Crandall, E.D., J.R. Taffel, and P.H. Barber. 2010. High gene flow due to pelagic larval dispersal among South Pacific archipelagos in two amphidromous gastropods (Neritimorpha: Neritidae). Heredity 104: 563-572.

Fehér, Z., M.L. Zettler, M. Bozsó, and K. Szabó. 2009. An

attempt to reveal the systematic relationship between *Theodoxus prevostianus* (C. Pfeiffer, 1828) and *Theodoxus danubialis* (C. Pfeiffer, 1828) (Mollusca, Gastropoda, Neritidae). Mollusca 27: 95-107.

Fehér, Z., C. Albrecht, A. Major, S. Sereda, and V. Krízsik. 2012. Extremely low genetic diversity in the endangered striped nerite, *Theodoxus transversalis* (Mollusca, Gastropoda, Neritidae)—a result of ancestral or recent effects? North-Western Journal of Zoology 8: 300-307.

Fehér, Z., G. Majoros, S. Ötvös, B. Bajomi, and P. Sólymos. 2017. Successful reintroduction of the endangered black nerite, *Theodoxus prevostianus* (Pfeiffer, 1828) (Gastropoda: Neritidae) in Hungary. Journal of Molluscan Studies 83: 240-242.

Frey, M.A. 2010. The relative importance of geography and ecology in species diversification: Evidence from a tropical marine intertidal snail (*Nerita*). Journal of Biogeography 37: 1515-1528.

Frey, M.A., and G.J. Vermeij. 2008. Molecular phylogenies and historical biogeography of a circumtropical group of gastropods (Genus: *Nerita*): Implications for regional diversity patterns in the marine tropics. Molecular Phylogenetics and Evolution 48: 1067-1086.

Fukumori, H. 2014. Evolutionary history, species diversity and biogeography of amphidromous neritid gastropods in the Indo-West Pacific. PhD diss., University of Tokyo.

Fukumori, H., and Y. Kano. 2014. Evolutionary ecology of settlement size in planktotrophic neritimorph gastropods. Marine Biology 161: 213-227.

Glöer, P., and V. Pešić. 2012. The freshwater snails (Gastropoda) of Iran, with descriptions of two new genera and eight new species. ZooKeys 219: 11-61.

Gorbach, K.R., M.E. Benbow, M.D. McIntosh, and A.J. Burky. 2012. Dispersal and upstream migration of an amphidromous neritid snail: Implications for restoring migratory pathways in tropical streams. Freshwater Biology 57: 1643-1657.

Gruneberg, H. 1982. Pseudo-polymorphism in *Clithon oualaniensis*. Proceedings of the Royal Society of London, Biological Sciences, 216: 147-157.

Haynes, A. 1985. The ecology and local distribution of non-marine aquatic gastropods in Viti Levu, Fiji. Veliger 28: 204-210.

Haynes, A. 2001. A revision of the genus *Septaria* Férussac, 1803 (Gastropoda: Neritimorpha). Annalen des Naturhistorischen Museums in Wien, Series B, 103: 177-229.

Haynes, A. 2005. An evaluation of members of the genera *Clithon* Montfort, 1810 and *Neritina* Lamarck, 1816 (Gastropoda: Neritidae). Molluscan Research 25: 75-84.

Holthuis, B. 1995. Evolution between marine and freshwater habitats: A case study of the gastropod suborder Neritopsina. PhD diss., University of Washington.

Huang, Q. 1997. Morphological, allozymic, and karyotypic distinctions between *Neritina* (*Dostia*) *violacea* and *N.* (*D.*) *cornucopia* (Gastropoda: Neritoidea). Journal of Zoology 241: 343-369.

Kabat, A.R., and Y. Finet. 1992. Catalogue of the Neritidae (Mollusca: Gastropoda) described by constant A. Récluz including the location of the type specimens. Revue Suisse de Zoologie 99: 223-253.

Kano, Y. 2006. Usefulness of the opercular nucleus for inferring early development in neritimorph gastropods. Journal of Morphology 267: 1120-1136.

Kano, Y. 2009. Hitchhiking behaviour in the obligatory upstream migration of amphidromous snails. Biology Letters 5: 465-468.

Kano, Y., S. Chiba, and T. Kase. 2002. Major adaptive radiation in neritopsine gastropods estimated from 28S sequences and fossil records. Proceedings of the Royal Society of London, Biological Sciences, 269: 2457-2465.

Kano, Y., E.E. Strong, B. Fontaine, O. Gargominy, M. Glaubrecht, and P. Bouchet. 2011. Focus on freshwater snails. Pp. 257-264 in The Natural History of Santo (P. Bouchet, H. Le Guyader, and O. Pascal, eds.). Patrimoines Naturels, vol. 69. Muséum National d'Histoire Naturelle, Paris.

Kano, Y., H. Fukumori, B. Brenzinger, and A. Warén. 2013. Driftwood as a vector for the oceanic dispersal of estuarine gastropods (Neritidae) and an evolutionary pathway to the sunken-wood community. Journal of Molluscan Studies 79: 378-382.

Kirkegaard, J. 2006. Life history, growth and production of *Theodoxus fluviatilis* in Lake Esrom, Denmark. Limnologica 36: 26-41.

Komatsu, S. 1986. Taxonomic revision of the neritid gastropods. Special Publication of the Mukaishima Marine Biological Station 1986: 1-69.

Little, C. 1972. The evolution of kidney function in the Neritacea (Gastropoda, Prosobranchia). Journal of Experimental Biology 56: 249-261.

Little, C. 1990. The Terrestrial Invasion: An Ecophysiological Approach to the Origin of Land Animals. Cambridge University Press, Cambridge, UK.

Maciolek, J.A. 1978. Shell character and habitat of non-marine Hawaiian neritid snails. Micronesica 14: 209-214.

Madsena, H., and N.M. Hung. 2014. An overview of freshwater snails in Asia with main focus on Vietnam. Acta Tropica 140: 105-117.

Martens, E. von. 1863-1879. Die Gattung *Neritina*. In Systematisches Conchylien-Cabinet von Martini und Chemnitz, vol. 2 (10) (H.C. Küster, W. Kobelt, and H.C. Weinkauff, eds.). A. Bauer und Raspe, Nürnberg, Germany. 1–303 + plates 1–23.

Myers, M.J., C.P. Meyer, and V.H. Resh. 2000. Neritid and thiarid gastropods from French Polynesian streams: How reproduction (sexual, parthenogenetic) and dispersal (active, passive) affect population structure. Freshwater Biology 44: 535–545.

Odabaşi, D.A., and N. Arslan. 2015. Description of a new subterranean nerite: *Theodoxus gloeri* n. sp. with some data on the freshwater gastropod fauna of Balıkdamı Wetland (Sakarya River, Turkey). Ecologica Montenegrina 2: 327–333.

Okuda, N., and M. Nishihira. 2002. Ecological distribution and assemblage structure of neritid gastropods in an Okinawan mangrove swamp, southern Japan. Benthos Research 57: 31–44.

Pyron, M., and A.P. Covich. 2003. Migration patterns, densities, growth of *Neritina punctulata* snails in Rio Espiritu Santo and Rio Mameyes, Northeastern Puerto Rico. Caribbean Journal of Science 39: 338–347.

Raabe, Z. 1968. Two new species of *Thigmotricha* (Ciliata, Holotricha) from *Theodoxus fluviatilis*. Acta Protozoologica 6: 169–173.

Reeves, W.K., R.T. Dillon Jr., and G.A. Dasch. 2008. Freshwater snails (Mollusca: Gastropoda) from the Commonwealth of Dominica with a discussion of their roles in the transmission of parasites. American Malacological Bulletin, 24: 59–63.

Resh, V.H. 2005. Stream crossings and the conservation of diadromous invertebrates in South Pacific island streams. Aquatic Conservation: Marine and Freshwater Ecosystems 15: 313–317.

Roberts, C.M., C.J. McClean, J.E.N. Veron, J.P. Hawkins, G.R. Allen, D.E. McAllister, C.G. Mittermeier, et al. 2002. Marine biodiversity hotspots and conservation priorities for tropical reefs. Science 295: 1280–1284.

Russell, H.D. 1941. The recent molluscs of the Family Neritidae of the western Atlantic. Bulletin of the Museum of Comparative Zoology at Harvard College, in Cambridge 88: 347–404, plates 1–7.

Schneider, D.W., and T.M. Frost. 1986. Massive upstream migrations by a tropical freshwater neritid snail. Hydrobiologia 137: 153–157.

Schneider, D.W., and J. Lyons. 1993. Dynamics of upstream migration in two species of tropical freshwater snails.

Journal of the North American Benthological Society 12: 3–16.

Seixas, S.A., J.F.R. Amato, and S.B. Amato. 2015. A new species of Temnocephala (Platyhelminthes, Temnocephalida) ectosymbiont on *Neritina zebra* (Mollusca, Neritidae) from the Brazilian Amazonia. Neotropical Helminthology 9: 41–53.

Shigemiya, Y., and M. Kato. 2001. Age distribution, growth, and lifetime copulation frequency of a freshwater snail, *Clithon retropictus* (Neritidae). Population Ecology 43: 133–140.

Starmühlner, F. 1983. Results of the hydrobiological mission 1974 of the Zoological Institute of the University of Vienna. Part 8. Contributions to the knowledge of the freshwater-gastropods of the Indian Ocean islands. Annalen des Naturhistorischen Museums in Wien, Serie B, 84: 127–249.

Symonds, M.F. 2002. The Neritidae of the Barton Group (Middle Eocene) of the Hampshire Basin. Tertiary Research 21: 1–10.

Symonds, M.F., and J.-M. Pacaud. 2010. New species of Neritidae (Neritimorpha) from the Ypresian and Bartonian of the Paris and Basse-Loire Basins, France. Zootaxa 2606: 55–68.

Uribe, J.E., D. Colgan, L.R. Castro, Y. Kano, and R. Zardoya. 2016. Phylogenetic relationships among superfamilies of Neritimorpha (Mollusca: Gastropoda). Molecular Phylogenetics and Evolution 104: 21–31.

Vermeij, G.J. 1969. Observations on the shells of some fresh-water neritid gastropods from Hawaii and Guam. Micronesica 5: 155–162.

Way, C.M., A.J. Burky, and M.T. Lee. 1993. The relationship between shell morphology and microhabitat flow in the endemic Hawaiian stream limpet (Hihiwai), *Neritina granosa* (Prosobranchia: Neritidae). Pacific Science 47: 263–275.

Zettler, M.L. 2007. A redescription of *Theodoxus schultzii* (Grimm, 1877), an endemic neritid gastropod of the Caspian Sea. Journal of Conchology 39: 245–251.

Zhokhova, A.E., N.M. Molodozhnikovab, and M.N. Pugachevaa. 2006. Dispersal of invading trematodes *Nicolla skrjabini* (Iwanitzky, 1928) and *Plagioporus skrjabini* Kowal, 1951 (Trematoda: Opecoelidae) in the Volga. Russian Journal of Ecology 37: 363–365.

4 Ampullariidae Gray, 1824

ROBERT H. COWIE AND KENNETH A. HAYES

The family Ampullariidae Gray, 1824 (junior synonym Pilidae Preston, 1915: Cowie, 1997; ICZN, 1999) includes nine Recent genera: *Asolene* d'Orbigny, 1838, *Felipponea* Dall, 1919, *Marisa* Gray, 1824, and *Pomacea* Perry, 1810, which are New World taxa; *Afropomus* Pilsbry & Bequaert, 1927, *Lanistes* Montfort, 1810, and *Saulea* Gray, 1868, which are African; *Pila* Röding, 1798 (*Ampullaria* Lamarck, 1799 and *Ampullarius* de Montfort, 1810 are junior synonyms of *Pila:* Cowie, 1997; ICZN, 1999), which is African and Asian; and *Forbesopomus* Bequaert & Clench, 1937, which is Asian (Cowie, 2015; Hayes et al., 2015). The family includes 186 Recent species considered valid as of the publications of Cowie (2015) and Hayes et al. (2015), with the majority in the three genera *Pomacea* (96 species), *Lanistes* (43 species), and *Pila* (29 species).

Berthold (1991) tentatively included the African fossil genus *Pseudoceratodes* Wenz, 1928 in the family. Other genera introduced for fossil species have been synonymized with one of the above Recent genera, e.g., *Kwangsispira* Hsü, 1935, synonymized with *Pila* by Berthold (1991), and *Prolanistes* Schütt, 1973,

synonymized with *Lanistes* by Berthold (1991). Many fossil species have been incorrectly described in *Ampullaria* (and *Ampullarius*) and placed in Ampullariidae, beginning with those of Lamarck (1804), most of whose *Ampullaria* species in that publication are probably placed correctly in the marine family Naticidae Guilding, 1834 (Cowie et al., 2001). Other genera have been introduced for fossil species but are not now considered to be Ampullariidae, e.g., *Mesolanistes* Yen, 1945, now placed in Physidae Fitzinger, 1833 (see Perrilliat et al., 2008). These fossil taxa are not considered further in this chapter, which focuses on the Recent species.

Ampullariids, commonly known as "apple snails" for the large round shells possessed by many species, notably in the general *Pomacea* and *Pila,* are basal members of the Caenogastropoda Cox, 1960. Bouchet et al. (2017) placed the Ampullariidae in the superfamily Ampullarioidea within the grade Architaenioglossa Haller, 1892, which also includes the Cyclophoroidea Gray, 1847 and Viviparoidea Gray, 1847. However, the Architaenioglossa has been resolved as paraphyletic

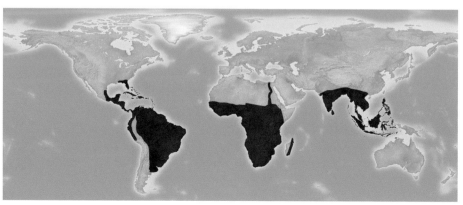

Distribution of Ampullariidae.

in higher-order analyses using both molecular and morphological data. The Campaniloidea Douvillé, 1904, also may be closely related. Relationships among these four superfamilies and the resolution of the base of the Caenogastropoda remain unresolved (see Hayes et al., 2015, and references therein).

Ampullariid species-group taxonomy has relied almost exclusively on shell morphology, with around 500 nominal species described (Cowie and Thiengo, 2003; Cowie, 2015; Hayes et al., 2015). However, the taxonomy has been extremely confused because of the gross morphological similarity within major groups accompanied by considerable intraspecific variation. The taxonomy and systematics of most species have not been adequately studied since their original descriptions, and the most recent but far from comprehensive conchologically based monographic treatment, focused on *Pomacea* and *Pila* (referring to the two genera together as *Ampullaria*), was published more than 90 years ago (Alderson, 1925). Consequently, species boundaries, other than for a few distinctive taxa, are very difficult to assess, without additional anatomical and molecular information.

Berthold (1991) published very detailed descriptions not only of the shells but also of the soft anatomy of a selection of 36 species spanning the family. This work remains the most detailed analysis of ampullariid morphology to date, although the anatomy of *Pomacea canaliculata* and *P. maculata* has been redescribed in detail by Hayes et al. (2012), with significant modification to the interpretations of Berthold (1991). Additional anatomical data have been provided for a number of species of *Pomacea* by Thiengo et al. (2011). Detailed anatomical analyses, guided by molecular phylogenetic analyses identifying monophyletic lineages, are beginning to bring order to ampullariid taxonomy, as exemplified by the revision of the invasive species *Pomacea canaliculata* (Lamarck, 1822) and *P. maculata* Perry, 1810, by Hayes et al. (2012). Relationships among the genera and among groups of lineages within genera are now better understood (e.g., Hayes et al., 2009) and provide a basis for systematic revision, e.g., the synonymization of *Pomella* Gray, 1847 with *Pomacea* and of a number of nominal species with *Pomacea*

Pomacea maculata Perry, 1810. Argentina: Pananá, Entre Riós, Province. 80 mm. NHMUK 1854.12.4.313.

canaliculata and *P. maculata* by Hayes et al. (2012), and future description of cryptic species distinguished by Hayes (2009).

In around 1980, species of *Pomacea* were introduced from their native South America to Southeast Asia, primarily as a human food resource (Mochida, 1991). These species have since become major agricultural pests, notably of rice (Cowie, 2002; Joshi and Sebastian, 2006), have threatened natural ecosystems (Carlsson et al., 2004), possibly have preyed on or out-competed native ampullariids, i.e., *Pila* species (Ng et al., 2014), and have become known as important vectors of human disease, notably rat lungworm disease or angiostrongyliasis (Lv et al., 2011). They are now widespread in southern and eastern Asia, from Thailand to Japan (Carlsson et al., 2004; Matsukura and Wada, 2007; Hayes et al., 2008). As a result, considerable research has not only focused on these invasive species from a pest perspective but has also addressed diverse aspects of their basic biology, reviewed by Hayes et al. (2015). For many years the identity of the invasive species was not clearly understood (Cowie et al., 2006), until the molecular analysis of Hayes et al. (2008) demonstrated that there were two species widely distributed in Southeast Asia, *Pomacea canaliculata* and *P. maculata* [then known as *P. insularum* (d'Orbigny, 1835), which was subsequently synonymized with *P. maculata* by Hayes et al., 2012], and two other nonnative but apparently not invasive South American species, *Pomacea scalaris* (d'Orbigny, 1835), known only from Taiwan (Wu et al., 2010), and *P. diffusa* Blume, 1957, known only from Sri Lanka (Hayes et al., 2008).

In addition to the introduction of these species to Southeast Asia, ampullariids have also been introduced to North America, where only one native ampullariid, *Pomacea paludosa* Say, 1829 occurs, in Florida and neighboring states. The identities of these introduced species were also clarified by molecular analysis (Rawlings et al., 2007), with *Pomacea canaliculata* in the southwestern United States and *Pomacea maculata* in the southeastern United States. *Pomacea canaliculata* is now confirmed in Florida (Bernatis et al. 2016). In addition, *Pomacea diffusa* and an unidentified species of *Pomacea* [thought incorrectly to be *P. haustrum* (Reeve, 1856) by Rawlings et al., 2007] were also reported from Florida. *Pomacea maculata* in particular may be causing declines in populations of the native *Pomacea paludosa* (Posch et al., 2013) as well as heavily consuming native wetland plants (Baker et al., 2010). With the exception of *P. diffusa* [often incorrectly identified as *P. bridgesii* (Reeve, 1856)], which was introduced to Florida in the early 1960s (Clench, 1966), these species are thought to be relatively recent introductions to the United States. Thompson (1997) identified specimens from Florida as *P. canaliculata*, but they were almost certainly *P. maculata* (or possibly the unidentified species above). And Neck and Schultz (1992) misidentified specimens of *P. maculata* collected in 1989 in Texas as *P. canaliculata*. *Marisa cornuarietis* (Linnaeus, 1758), with a planispiral rather than globular shell, was introduced to Florida also much earlier, in 1957 (Hunt, 1958), and has since been recorded from other states (Rawlings et al., 2007). Introduction of *P. canaliculata* to the western United States may have been from Asia, either directly or from Hawaii, where it has also been introduced and where it is a major agricultural pest (Tran et al., 2008), as a human food resource (Rawlings et al., 2007). However, introduction of the other species to the United States was probably directly from their native South America and probably associated with the domestic aquarium trade (Rawlings et al., 2007), although *Marisa cornuarietis* may also have been introduced as a biological control agent for aquatic weeds (Cowie, 2002).

Pomacea diffusa has been recorded in the wild in Australia (Hayes, et al., 2008) and *P. maculata* in rice-growing areas in Spain (Horgan et al., 2014a), where unpublished information indicates that *P. canaliculata* is also present. *Pila scutata*, probably from the Philippines, has been reported (as *Pila conica*) from the Pacific islands of Guam, Palau, and Hawaii (Cowie, 2002).

Pomacea canaliculata has also been introduced within South and Central America from its native range (the Uruguay, lower Paraná, and La Plata basins; Hayes et al., 2012) to Chile (Letelier and Soto-Acuña, 2008), Ecuador (Horgan et al., 2014b), and to the Dominican Republic (Rosario and Moquete, 2006). *Pomacea diffusa*, which is native to the Amazon region, now occurs elsewhere in South and Central America and is widely available in the domestic aquarium trade, often incorrectly identified as *P. bridgesii* (Perera and Walls, 1996; Hayes et al., 2008; Cowie and Hayes, 2012).

The nonnative ranges of all introduced ampullariids, as understood at the time, were tabulated by Cowie and Hayes (2012), who summarized key information on the history of introductions outside their native ranges and the reasons for their introduction, their ecology, their behavior and physiology, their importance as pests, and their control and management. A comprehensive review of basic ampullariid biology was published by Hayes et al. (2015), while all aspects of the role of ampullariids as pests, focused on Asia, were reviewed in the book edited by Joshi and Sebastian (2006), both publications expanding on the earlier review of Cowie (2002). A more recent book, edited by Joshi, Cowie, and Sebastian (2017), augments these contributions.

Unlike many other groups of nonmarine mollusks that are suffering high levels of extinction (Lydeard et al., 2004; Régnier et al., 2015a, b; Cowie et al., 2017), to the extent that it is known, most ampullariid species are not threatened or endangered. The IUCN (2017) has assessed 88 ampullariid species. It finds two species to be Critically Endangered [*Lanistes neritoides* Brown & Berthold, 1990 and *Pomacea ocanensis* (Kobelt, 1914)], three Endangered [*Lanistes alexandri* (Bourguignat, 1889), *L. nyassanus* Dohrn, 1865, *L. solidus* Smith, 1878], five Vulnerable [*Lanistes ciliatus* Martens, 1878, *L. farleri* Craven, 1880, *Lanistes grasseti* (Morelet, 1863), *Pomacea palmeri* (Marshall, 1930), and *Pomacea quinindensis* (Miller, 1879)], three Near Threatened [*Afropomus balanoidea* (Gould,

1850), *Lanistes stuhlmanni* Martens, 1897, and *Pomacea expansa* (Miller, 1879)], and one at lower risk or Near Threatened (*Lanistes ellipticus* Martens). Nonetheless, many species were assessed as Data Deficient (i.e., not enough is known about them to make an assessment), and only 88 of the nearly 200 species have been assessed. Evaluation of most ampullariid species is beset with difficulties because of the taxonomic problems that still pervade the family and that preclude definitive identification and assessment of the validity of many species. Despite the ubiquity and abundance of some species, especially in South America, many species have not been collected for many years and may well be declining; for instance, *Forbesopomus atalanta* Bequaert & Clench, 1937 is only known from type material (Cowie, 2015). And there are no doubt numerous cryptic species. Just as for other gastropods (Régnier et al., 2015a; Cowie et al., 2017), the level of extinction, as well as of threats to extant species, may be greater than currently documented.

LITERATURE CITED

Alderson, E.G. 1925. Studies in *Ampullaria*. Heffer, Cambridge, UK. xx + 102 pp., 19 plates.

Baker, P., F. Zimmanck, and S.M. Baker. 2010. Feeding rates of an introduced freshwater gastropod *Pomacea insularum* on native and nonindigenous aquatic plants in Florida. Journal of Molluscan Studies 76: 138-143.

Bernatis, J.L., I.J. Mcgaw, and C.L. Cross. 2016. Abiotic tolerances in different life stages of apple snails *Pomacea canaliculata* and *Pomacea maculata* and the implications for distribution. Journal of Shellfish Research 35 (4): 1013-1025.

Berthold, T. 1991. Vergleichende Anatomie, Phylogenie und historische Biogeographie der Ampullariidae (Mollusca, Gastropoda). Abhandlungen des Naturwissenschaftlichen Vereins in Hamburg (NF) 29: 1-256.

Bouchet, P., J.-P. Rocroi, B. Hausdorf, A. Kaim, Y. Kano, A. Nützel, P. Parkhaev, M. Schrödl, and E. Strong. 2017. Revised classification, nomenclator and typification of gastropod and monoplacophoran families. Malacologia 61: 1-526.

Carlsson, N.O.L., C. Brönmark, and L.-A. Hansson. 2004. Invading herbivory: The golden apple snail alters ecosystem functioning in Asian wetlands. Ecology 85: 1575-1580.

Clench, W.J. 1966. *Pomacea bridgesi* (Reeve) in Florida. Nautilus 79: 105.

Cowie, R.H. 1997. *Pila* Röding, 1798 and *Pomacea* Perry, 1810 (Mollusca, Gastropoda): Proposed placement on the Official List, and Ampullariidae Gray, 1824: Proposed confirmation as the nomenclaturally valid synonym of Pilidae Preston, 1915. Bulletin of Zoological Nomenclature 54: 83-88.

Cowie, R.H. 2002. Apple snails (Ampullariidae) as agricultural pests: Their biology, impacts and management. Pp 145-192 in Molluscs as Crop Pests (G.M. Barker, ed.). CABI, Wallingford, UK.

Cowie, R.H. 2015. The recent apple snails of Africa and Asia (Mollusca: Gastropoda: Ampullariidae: *Afropomus, Forbesopomus, Lanistes, Pila, Saulea*): A nomenclatural and type catalogue. The apple snails of the Americas: Addenda and corrigenda. Zootaxa 3940 (1): 1-92.

Cowie, R.H., A.R. Kabat, and N.L. Evenhuis. 2001. *Ampullaria canaliculata* Lamarck, 1822 (currently *Pomacea canaliculata;* Mollusca, Gastropoda): Proposed conservation of the specific name. Bulletin of Zoological Nomenclature 58 (1): 13-18.

Cowie, R.H., and S.C. Thiengo. 2003. The apple snails of the Americas (Mollusca: Gastropoda: Ampullariidae: *Asolene, Felipponea, Marisa, Pomacea, Pomella*): A nomenclatural and type catalog. Malacologia 45: 41-100.

Cowie, R.H., K.A. Hayes, and S.C. Thiengo. 2006. What are apple snails? Confused taxonomy and some preliminary resolution. Pp. 3-23 in Global Advances in Ecology and Management of Golden Apple Snails (R.C. Joshi and L.C. Sebastian, eds.). Philippine Rice Research Institute, Muñoz, Nueva Ecija, Philippines.

Cowie, R.H., and K.A. Hayes. 2012. Apple snails. Pp. 207-217, plates 18.1-18.4, in A Handbook of Global Freshwater Invasive Species (R.A. Francis, ed.). Earthscan, London.

Cowie, R.H., C. Régnier, B. Fontaine, and P. Bouchet. 2017. Measuring the Sixth Extinction: What do mollusks tell us? *Nautilus* 131 (1): 3-41.

Hayes, K.A. 2009. Evolution, molecular systematics and invasion biology of Ampullariidae. PhD diss., University of Hawaii, Honolulu.

Hayes, K.A., R.C. Joshi, S.C. Thiengo, and R.H. Cowie. 2008. Out of South America: Multiple origins of non-native apple snails in Asia. Diversity and Distributions 14: 701-712.

Hayes, K.A., R.H. Cowie, and S.C. Thiengo. 2009. A global phylogeny of apple snails: Gondwanan origin, generic relationships and the influence of outgroup choice (Caenogastropoda: Ampullariidae). Biological Journal of the Linnean Society 98: 61-76.

Hayes, K.A., R.H. Cowie, S.C. Thiengo, and E.E. Strong. 2012. Comparing apples with apples: Clarifying the identities of two highly invasive Neotropical Ampullariidae (Caenogastropoda). Zoological Journal of the Linnean Society 166: 723-753.

Hayes, K.A., R.L. Burks, A. Castro-Vazquez, P.C. Darby, H. Heras, P.R. Martín, J.-W. Qiu, et al. 2015. Insights from an integrated view of the biology of apple snails (Caenogastropoda: Ampullariidae). Malacologia 58: 245-302.

Horgan, F.G., A.M. Stuart, and E.P. Kudavidanage. 2014a. Impact of invasive apple snails on the functioning and services of natural and managed wetlands. Acta Oecologica 54: 90-100.

Horgan, F.G., M.I. Felix, D.E. Portalanza, L. Sánchez, W.M. Moya Rios, S.E. Farah, J.A. Wither, C.I. Andrade, and E.B. Espin. 2014b. Responses by farmers to the apple snail invasion of Ecuador's rice fields and attitudes toward predatory snail kites. Crop Protection 62: 135-143.

Hunt, B.P. 1958. Introduction of *Marisa* into Florida. Nautilus 72: 53-55.

ICZN (International Commission on Zoological Nomenclature). 1999. Opinion 1913: *Pila* Röding, 1798 and *Pomacea* Perry, 1810 (Mollusca, Gastropoda): Placed on the Official List, and AMPULLARIIDAE Gray, 1824: Confirmed as the nomenclaturally valid synonym of PILIDAE Preston, 1915. Bulletin of Zoological Nomenclature 56: 74-76.

IUCN (International Union for the Conservation of Nature). 2017. The IUCN Red List of Threatened Species. Version 2017-3. www.iucnredlist.org. Accessed 12 June 2018.

Joshi, R.C., and L.S. Sebastian, eds. 2006. Global Advances in Ecology and Management of Golden Apple Snails. Philippine Rice Research Institute, Muñoz, Nueva Ecija, Philippines.

Joshi, R.C., R.H. Cowie, and L.S. Sebastian, eds. 2017. Biology and management of invasive apple snails. Philippine Rice Research Institute, Muñoz, Nueva Ecija, Philippines.

Lamarck, J.B.P.A. de M. de. 1804. Suite des mémoires sur les fossiles des environs de Paris. Annales du Muséum National d'Histoire Naturelle 5 (25): 28-36.

Letelier, S., and Soto-Acuña, S. 2008. Registro de *Pomacea* sp. (Gastropoda: Ampullaridae) en Chile. Amici Molluscarum 16: 6-13.

Lv, S., Y. Zhang, P. Steinmann, G.-J. Yang, K. Yang, X.-N. Zhou, and J. Utzinger, 2011. The emergence of angiostrongyliasis in the People's Republic of China: The interplay between invasive snails, climate change and transmission dynamics. Freshwater Biology 56: 717-734.

Lydeard, C., R.H. Cowie, W.F. Ponder, A.E. Bogan, P. Bouchet, S.A. Clark, K.S. Cummings, et al. 2004. The global decline of nonmarine mollusks. BioScience 54: 321-330.

Matsukura, K., and T. Wada. 2007. Environmental factors affecting the increase of cold hardiness in the apple snail *Pomacea canaliculata* (Gastropoda: Ampullariidae). Applied Entomology and Zoology 42: 533-539.

Mochida, O. 1991. Spread of freshwater *Pomacea* snails (Pilidae, Mollusca) from Argentina to Asia. Micronesica, Supplement 3: 51-62.

Neck, R.W., and J.G. Schultz. 1992. First record of living channeled applesnail *Pomacea canaliculata* (Pilidae) from Texas. Texas Journal of Science 44: 115-116.

Ng, T.H., S.K. Tan, and M.E.Y. Low. 2014. Singapore Mollusca: 7. The family Ampullariidae (Gastropoda: Caenogastropoda: Ampullarioidea). Nature in Singapore 7: 31-47.

Perera, G., and J.G. Walls. 1996. Apple Snails in the Aquarium. T. F. H. Publications, Neptune City, New Jersey.

Perrilliat, M. del C., F.J. Vega, B. Espinosa, and E. Naranjo-Garcia. 2008. Late Cretaceous and Paleogene freshwater gastropods from northeastern Mexico. Journal of Paleontology 82 (2): 255-266.

Posch, H., A.L. Garr, and E. Reynolds. 2013. The presence of an exotic snail, *Pomacea maculata*, inhibits growth of juvenile Florida apple snails, *Pomacea paludosa*. Journal of Molluscan Studies 79: 383-385.

Rawlings, T.A., K.A. Hayes, R.H. Cowie, and T.M. Collins. 2007. The identity, distribution, and impacts of nonnative apple snails in the continental United States. BMC Evolutionary Biology 7: 97.

Régnier, C., G. Achaz, A. Lambert, R.H. Cowie, P. Bouchet, and B. Fontaine. 2015a. Mass extinction in poorly known taxa. Proceedings of the National Academy of Sciences 112 (25): 7761-7766.

Régnier, C., P. Bouchet, K.A. Hayes, N.W. Yeung, C.C. Christensen, D.J.D. Chung, B. Fontaine, and R.H. Cowie. 2015b. Extinction in a hyperdiverse endemic Hawaiian land snail family and implications for the underestimation of invertebrate extinction. Conservation Biology 29: 1715-1723.

Rosario, J., and C. Moquete. 2006. The aquatic snail *Ampullaria canaliculata* L.—plague of irrigated lowland rice in the Dominican Republic. Pp. 514-515 in Global Advances in Ecology and Management of Golden Apple Snails (R.C. Joshi and L.C. Sebastian, eds.). Philippine Rice Research Institute, Muñoz, Nueva Ecija, Philippines.

Thiengo, S.C., K.A. Hayes, A.C. Mattos, M.A. Fernandez, and R.H. Cowie. 2011. A Família Ampullariidae no Brasil: Aspectos morfológicos, biológicos e taxonômicos. Pp. 95-111 in Tópicos em Malacologia: Ecos do XIX Encontro Brasileiro de Malacologia (M.A. Fernandez, S.B. dos Santos, A.D. Pimenta, and S.C. Thiengo, orgs.). Sociedade Brasileira de Malacologia, Rio de Janeiro.

Thompson, F.G. 1997. *Pomacea canaliculata* (Lamarck, 1822)

(Gastropoda, Prosobranchia, Pilidae): A freshwater snail introduced into Florida, U.S.A. Malacological Review 30: 91.

Tran, C.T., K.A. Hayes, and R.H. Cowie. 2008. Lack of mitochondrial DNA diversity in invasive apple snails (Ampullariidae) in Hawaii. Malacologia 50: 351-357.

Wu, J.-Y., Y.-T. Wu, M.-C. Li, Y.-W. Chiu, M.-Y. Liu, and L.-L. Liu. 2011 Reproduction and juvenile growth of the invasive apple snails *Pomacea canaliculata* and *Pomacea scalaris* (Gastropoda: Ampullariidae) in Taiwan. Zoological Studies 50: 61-68.

5 Viviparidae Gray, 1847

BERT VAN BOCXLAER AND ELLEN E. STRONG

Viviparidae (type genus *Viviparus* Montfort, 1810; type species *Viviparus fluviorum* Montfort, 1810, by original designation; = Paludinidae Fitzinger, 1833; invalid [Opinion 573, 1959]) is a family of freshwater gastropods with a global distribution and moderate diversity in the extant fauna (~125-150 valid, described species; Strong et al., 2008). Commonly known as "Mystery Snails" or "River Snails," viviparids are distributed primarily in lakes, rivers, and streams in temperate to tropical regions. Their greatest diversity occurs in tropical and subtropical regions of Asia (Prashad, 1928), where some 60-85 species occur (Strong et al., 2008). North America is home to roughly 29 species, mainly in the eastern and central regions of the continent (Clench, 1962; Strong et al., 2008), whereas Africa and Australia each have about 20-25 species (Brown, 1994; Ponder and Walker, 2003; Strong et al., 2008), and Europe 5 species (Cuttelod et al., 2011). Despite their global distribution, viviparids are absent from large regions of northern Africa, the Middle East, and central Asia and do not occur naturally in South America or Antarctica (Pra-

shad, 1928), although they have been introduced to the former (Ovando and Cuezzo, 2012) among other areas (Jokinen, 1982; Soes et al. 2011).

Viviparids have a rich fossil record (Prashad, 1928) that extends back to the Middle and Late Jurassic of Europe (Bathonian, Tracey et al., 1993; Hudson et al., 1995; Kimmeridgian, Bandel, 1993) and Asia (Bajocian-Bathonian, Pan, 1977). Other Mesozoic finds have been made in Africa (Early Cretaceous, Fischer, 1963; Van Damme et al., 2015), South America (Kimmeridgian/Tithonian, Late Jurassic [identification somewhat uncertain], Perea et al., 2009; Maastrichtian, Late Cretaceous, Prashad, 1928; Ghilardi et al., 2011), North America (Early Cretaceous, possibly Late Jurassic, Henderson, 1935; Yen, 1950), and Australia (Aptian, Early Cretaceous, Kear et al., 2003; Albian, Hamilton-Bruce et al., 2002). It is remarkable that many extinct species are found far outside the family's current geographic range: in South America (Ghilardi et al., 2011), far southwest in North America (Clench 1962), and the Middle East (Sivan et al., 2006; Ashkenazi et al., 2010). The distribution of these early fossils and their

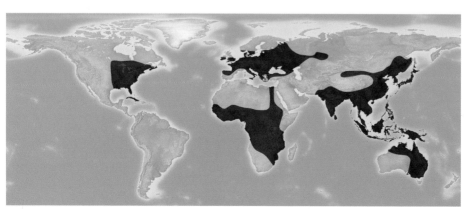

Distribution of Viviparidae.

ages strongly suggest that the family originated on Pangaea (Strong et al., 2008) and subsequently went extinct in South America and Africa, with a recolonization of Africa during the Miocene from Asia via the Middle East (Schultheiß et al., 2014).

As the name implies, the family is characterized by ovoviviparous reproduction, and females develop eggs and juveniles in a large pallial brood pouch (Rohrbach, 1937; Vail, 1977). Another important synapomorphy of the family is suspension feeding, which has resulted in various modifications of the pallial cavity (see below) (Vail, 1977; Simone, 2004, 2011). The family contains three extant subfamilies with largely disjunct distributions: Viviparinae, Bellamyinae, and Lioplacinae (Bouchet and Rocroi, 2005; Van Bocxlaer et al., 2018). Within Viviparinae three extant genera are currently considered valid: *Viviparus* Montfort, 1810 (type genus), *Rivularia* Heude, 1890, and *Tulotoma* Haldeman, 1840; within Bellamyinae 17: *Bellamya* Jousseaume, 1886 (type genus), *Angulyagra* Rao, 1931, *Anularya* Zhang & Chen, 2015, *Anulotaia* Brandt, 1968, *Cipangopaludina* Hannibal, 1912, *Eyriesia* Fischer, 1885, *Filopaludina* Habe, 1964 (with its subgenera *Filopaludina s.s.* and *Siamopaludina* Brandt, 1968), *Idiopoma* Pilsbry, 1901, *Larina* Adams, 1855, *Margarya* Nevill, 1877 (with its subgenera *Margarya s.s.* and *Tchangmargarya* He, 2013), *Mekongia* Crosse & Fischer, 1876, *Neothauma* Smith, 1880, *Notopala* Cotton, 1935, *Sinotaia* Haas, 1939, *Taia* Annandale, 1918, *Torotaia* Haas, 1939, *Trochotaia* Brandt, 1974); and within Lioplacinae two: *Lioplax* Troschel, 1857 (type genus), and *Campeloma* Rafinesque, 1819 (Bouchet, 2011; with modifications from Brandt, 1974; Boss, 1992; Brown, 1994; Hirano et al., 2015; Zhang et al., 2015; Van Bocxlaer et al., 2018). The separation between these subfamilies is based on anatomical differences, mainly in reproductive anatomy (Rohrbach, 1937; Vail, 1977; Van Bocxlaer and Strong, 2016). For example, the testis lies in the pallial cavity in Bellamyinae, but it overlies the digestive gland in the viscera in Viviparinae and Lioplacinae. This difference in arrangements also causes marked differences in the length and elaborations of the vas deferens, with an enlarged proximal pallial vas deferens in the latter two subfamilies. Lioplacinae have a strongly convoluted rather than a straight prostate (as in Viviparinae and Bellamyinae). Molecular studies, thus far

Filopaludina sumatrensis (Dunker, 1852). Thailand: Ban Bu Song near Ban Choen, Phra Nakhon Si Ayutthaya Province. 23 mm. Poppe 959634.

with limited taxon sampling, have not contradicted the monophyly of the Viviparinae or Bellamyinae (Sengupta et al., 2009; Du et al., 2013; Hirano et al., 2015; Wang et al., 2017; Van Bocxlaer et al., 2018). However, the Lioplacinae have not been integrated in these efforts, and consequently the monophyly of the family as a whole has not been assessed molecularly. The relationship of viviparids to other basal caenogastropods, the Campaniloidea in particular, also remains controversial, as is the issue of whether the Architaenioglossa (i.e., Viviparoidea, Ampullarioidea, and Cyclophoroidea) are monophyletic (e.g., Harasewych et al., 1998; Simone, 2004; Colgan et al., 2007; Ponder et al., 2008; Simone, 2011).

The taxonomic history of Viviparidae is complex, with substantial confusion that continues to the present day in the circumscription and extension of many species and genera. As with many other freshwater families, much of the confusion stems from the profusion of names introduced mainly by nineteenth- and early-twentieth-century authors based on minor variations in shell morphology. For example, no fewer than 95 names are available for the 27 species of North American *Campeloma* (Lioplacinae) that are currently considered valid, a state of affairs that impelled Clench (1962: 276) to note that "on a specific level, probably few genera among North American freshwater mollusks are in a more confused state." In Africa, *Bellamya unicolor* and *B. capillata* are both acknowledged species complexes (Brown, 1994; Schultheiß et al., 2011), but both have yet to be formally revised. These examples are far from unique, and the systematics of most species requires comprehensive revision. At the genus level, the systematics is equally deserving of reassessment. For example, Asian bellamyines, especially from China, continue to be erroneously relegated to *Bellamya* (e.g., Soes et al., 2011; Gu et al., 2015; Wang et al., 2017), the type species of which belongs to a deeply divergent clade restricted to Africa (Schultheiß et al., 2014; Van Bocxlaer and Strong, 2016).

As mentioned, viviparids bear anatomical modification for suspension feeding: the gill leaflets have become long and slender, an endostyle is present along the base of the gill that secretes mucus for ensnaring particles, and the mantle cavity floor is modified into a food groove that conveys food from the posterior mantle cavity to the mouth (e.g., Van Bocxlaer and Strong, 2016). Many species still rely, to varying extents, on deposit feeding, for which they possess a taenioglossate radula that may be taxonomically diagnostic (Falniowski et al., 1996b; Anistratenko et al., 2013; but see Rao, 1925). It is small and delicate relative to overall body size, reflecting the decreased reliance on this feeding mode and the soft substrates and small particles on which they graze. Their diet is comprised of algae, including diatoms, bacteria, and rotifers (Stańczykowska et al., 1972; Plinski et al., 1978; Dillon, 2000; Jakubik, 2012), as well as decaying organic material (Annandale and Sewell, 1921; Van Bocxlaer and Strong, 2016). The versatility between feeding modes may influence the adaptability of viviparids—those with a greater capacity for deposit feeding, which include the invasive taxa, are more easily maintained in captivity (Van Bocxlaer, pers. obs.).

Most viviparids are gonochoristic with internal fertilization, but some *Campeloma* species reproduce parthenogenetically (Mattox, 1938; Johnson and Bragg, 1999). In males, the right cephalic tentacle is modified as a copulatory organ. The brood pouch, formed by the distal pallial oviduct, contains encapsulated eggs and often hatchlings at various stages of development. The size and number of eggs and juveniles in the brood pouch at any given time vary greatly between species. As larger females can accommodate more eggs and juveniles, fecundity typically increases with body size and hence age until senescence (Stańczykowska et al., 1971; Browne, 1978; Buckley, 1986; Van Bocxlaer and Strong, 2016). Ovoviviparity is perhaps one of the causes for the modest sexual dimorphism in shell shape observed in many species (e.g., Minton and Wang, 2011), but as females of some species have faster growth rates (Khan and Chaudhuri, 1984) or longer life spans (Stańczykowska et al., 1971) than males, allometric changes may also contribute to sexual dimorphism. Ovoviviparity likely contributes to the female-biased sex ratios in many species (Annandale and Sewell, 1921; Stańczykowska et al., 1971; Berry, 1974; De Bernardi et al., 1976; Brown et al., 1989), but in some species this bias may be caused primarily by increased male mortality (Annandale and Sewell, 1921; Jakubik, 2012).

Most species appear to have successive or prolonged periods of reproduction (Stańczykowska et al., 1971; Berry, 1974; Browne, 1978; Jakubik, 2012), although some populations have been reported to be semelparous (Jokinen et al., 1982; Dillon, 2000: 161). Gestation times are on the order of 1–3 months in tropical and subtropical taxa (Brown et al., 1989; Ma et al., 2010), but at least some temperate species overwinter juveniles, leading to gestation times of up to 9 months (Browne, 1978). After being released from the brood pouch, juveniles may mature within as few as 3 months, but more typically within ~1 year (Stańczykowska et al., 1971; Khan and Chaudhuri, 1984; Brown, 1994: 516). Generation times may therefore range from 6 to 24 months and are typically shorter for smaller and/or tropical species. In temperate regions, strong seasonal differences are observed in reproduction (Stańczykowska et al., 1971; Fretter and Graham, 1994), but temporal variation in the state of the ovaries and the contents of the brood pouch are also observed in tropical species, likely related to rainfall and food availability (Berry, 1974; Brown, 1994). Estimates of fecundity are challenging in viviparids and are most often obtained by examining the contents of the brood pouch. However, given seasonal and ontogenetic variation in juvenile production and interspecific differences in gestation period, estimates are difficult to compare at face value.

Life expectancies typically range from 1 to 5 years, with a few species surviving up to a maximum of 11 years; in many species males have a substantially shorter life expectancy than females and tropical species tend to have shorter life spans than those in temperate zones (Van der Spoel, 1958; Stańczykowska et al., 1971; Khan and Chaudhuri, 1984; Brown et al., 1989; Heller, 1990; Brown, 1994). Although growth is indeterminate, adults grow much more slowly than juveniles (Khan and Chaudhuri, 1984). Viviparids mature at a shell size of 1.5 to 3.0 cm, with sizes at adulthood typically ranging from 2.0 to 5.0 cm, although some can grow up to 10.0 cm (Stańczykowska et al., 1971; Khan and Chaudhuri, 1984; Brown, 1994; Ying et al., 2013). Upon reaching adulthood, shells may show modifications such as downward deflections of the aperture. Shells are typically smooth, although periostracal ornamentation and hairs on juvenile shells are common and taxonomically diagnostic (Jokinen, 1984; Brown, 1994; Falniowski et al., 1996a), but some extant and extinct viviparids have remarkably sculptured shells, with nodules, carinae, or spines (e.g., Annandale, 1924; Ying et al., 2013). Viviparids are dextrally coiled, but in some of the parthenogenetic *Campeloma* species, elevated percentages of sinistral specimens have been observed, which also possess a reversed arrangement of the internal organs (Savage, 1938). Sinistral individuals are much rarer in other viviparids, where it would hamper copulation, but in parthenogenetic taxa the sinistral condition is much less disadvantageous.

Viviparids are intermediate and definitive hosts to several parasites, but as most of these interactions do not affect human health, they remain poorly studied (e.g., Annandale and Sewell, 1921; Jokinen, 1982; Jezewski, 2004; Jakubik, 2012). Two possible exceptions are echinostomes (digenean trematodes, Chung and Jung, 1999) and the nematode *Angiostrongylus* (Chang et al., 1968), both parasites with low host specificity that cause food-borne diseases (echinostomiasis and encephalitis, respectively) in various mammals, including humans. As viviparids are almost exclusively consumed by humans in Asia (Köhler et al., 2012), it is unlikely that they are important transmitters of human parasitic diseases elsewhere.

At least partly as a consequence of their life history

traits, most viviparids are poor dispersers, resulting in patchy distributions and regional endemism. Long-lived lakes are notable for their numbers of endemic species, many of which are characterized by moderate to great phenotypic disparity, e.g., in Philippine Lake Lanao (Bartsch, 1910), in the Chinese Yunnan lakes and paleolakes (Kira, 1999; Ying et al., 2013), and in the Congolese-Ugandese Palaeolake Obweruka (Van Damme and Pickford, 1999; Salzburger et al., 2014). Clench (1962: 277) observed that North American viviparids display morphological patterns that vary by river system, whereas molecular studies on the African *Bellamya* and Australian *Notopala* suggest that species are typically confined to single drainages (Carini and Hughes, 2006; Schultheiß et al., 2014). However, much more work on the circumscription and distribution of species is required before biogeographic generalizations can be made.

Viviparids prefer clean waters, tolerate low to moderate levels of pollution, and are typically absent from ephemeral waters (Burch and Jung, 1987; Brown, 1994; Jakubik, 2012). They are an important component in many benthic freshwater ecosystems (Browne, 1978; Jakubik, 2012) and occur in densities from just a few individuals up to several hundred per m^2 (Stańczykowska et al., 1971; Browne, 1978; Khan and Chaudhuri, 1984; Jakubik, 2012). Seasonally they can reach exceptionally high biomass in some locations, especially in the (sub)tropics (1500+ individuals per m^2) (Brown et al., 1989; Jakubik, 2012). Despite the often high local abundance of viviparids, many species are, as mentioned, highly geographically restricted. This combination has led to the description of an Australian *Notopala* species as an "endangered pest": nowadays it occurs almost exclusively in irrigation pipelines, where it can reach great population densities and cause economic losses (Sheldon and Walker, 1993; Walker, 1996). Whereas rather limited information is available on the ecological sensitivity of viviparids, substrate composition, dissolved oxygen concentration, and water hardness are considered to be the most important determinants of habitat suitability (Jakubik, 2012).

In North America ~25% of viviparids are considered imperiled (Johnson et al., 2013), whereas in Africa at least seven *Bellamya* species are Endangered or Critically Endangered, and two more are considered Near Threatened, accounting for a rate of imperilment of nearly 50% of the species currently considered valid (IUCN, 2017). In Asia, one species of *Anulotaia,* two of *Cipangopaludina* (one of which represents *Heterogen,* which is likely a junior synonym of *Cipangopaludina* [Hirano et al., 2015]), and five *Margarya* species are currently considered Endangered or Critically Endangered (Kira, 1999; Köhler et al., 2012; IUCN, 2017). In Europe no viviparids are considered to be imperiled (Cuttelod et al., 2011), although the ranges of many are shrinking. A similar situation exists in Australia, where two *Notopala* species are considered Endangered, mainly because of flow modifications that have impacted the biofilms on which they feed (Ponder and Walker, 2003). As for other freshwater gastropods, the increased level of threat for viviparids is caused by habitat degradation, fragmentation and loss, e.g., via pollution, eutrophication, increased sedimentation, structural alterations such as damming, warming, and introduction of non-native species (Graf et al., 2011). Several Asian species, including *Margarya,* face the additional stressor that they are collected in great numbers and overexploited for human consumption (Köhler et al., 2012). Although some widespread and invasive taxa may be resilient in the face of these threats (e.g., *Sinotaia* and *Cipangopaludina;* Gu et al., 2015; Jokinen, 1982), many narrow range endemics are especially vulnerable (e.g., *Tulotoma* and *Notopala;* Ponder and Walker, 2003; Johnson et al., 2013). However, most viviparids are Data Deficient or have not been evaluated, posing significant challenges for accurately assessing their conservation status (Graf et al., 2011; Köhler et al., 2012). Consequently, the recognized rate of imperilment is almost certainly underestimated, with many other species warranting protection. Further investigations into the systematics, genetic diversity, ecology, biology, and distribution are urgently needed to demystify the natural history of mystery snails and to facilitate their worldwide conservation.

LITERATURE CITED

Anistratenko, V.V., Y.S. Ryabceva, and E.V. Degtyarenko. 2013. Morphological traits of the radula in Viviparidae (Mollusca, Caenogastropoda) as a master key to discrimination of closely related species. Vestnik zoologii 47: 40-51.

Annandale, N. 1924. The evolution of shell-sculpture in fresh-water snails of the Family Viviparidae. Proceedings of the Royal Society of London, Series B, 96: 60-76.

Annandale, N., and R.B.S. Sewell. 1921. The banded pond-snail of India (*Vivipara bengalensis*). Records of the Indian Museum 22: 215-299.

Ashkenazi, S., K. Klass, H. K. Mienis, B. Spiro, and R. Abel. 2010. Fossil embryos and adult Viviparidae from the Early-Middle Pleistocene of Gesher Benot Ya'aqov, Israel: Ecology, longevity and fecundity. Lethaia 43: 116-127.

Bandel, K. 1993. Caenogastropoda during Mesozoic times. Scripta Geologica Special Issue 2: 7-56.

Bartsch, P. 1910. Notes on the Philippine pond snails of the Genus *Vivipara,* with descriptions of new species. Proceedings of the US National Museum 37: 365-367.

Berry, A.J. 1974. Reproductive condition in two Malayan freshwater viviparid gastropods. Journal of Zoology 174: 357-367.

Boss, K.J. 1992. Edited and annotated translation of 'Starobogatov, Y. I. 1985. Generic composition of the family Viviparidae (Gastropoda Pectinibranchia Vivipariformes). Transactions of the Zoological Institute, Academy of Sciences USSR 135: 26-32, figs. 1-18.' Special Occasional Publications of the Harvard University Museum of Comparative Zoology 10: 1-27.

Bouchet, P. 2011. Viviparidae Gray, 1847. MolluscaBase (2016), World Register of Marine Species. www.molluscabase.org.

Bouchet, P., and J.-P. Rocroi. 2005. Classification and nomenclator of gastropod families. Malacologia 47: 1-397.

Brandt, R.A.M. 1974. The non-marine aquatic Mollusca of Thailand. Archiv für Molluskenkunde 105: i-iv, 1-423.

Brown, D.S. 1994. Freshwater Snails of Africa and Their Medical Importance. 2nd ed. Taylor & Francis, London.

Brown, K.M., D. Varza, and T.D. Richardson. 1989. Life histories and population dynamics of two subtropical snails (Prosobranchia: Viviparidae). Journal of the North American Benthological Society 8: 222-228.

Browne, R.A. 1978. Growth, mortality, fecundity, biomass and productivity of four lake populations of the prosobranch snail, *Viviparus georgianus.* Ecology 59: 742-750.

Buckley, D.E. 1986. Bioenergetics of age-related versus size-related reproductive tactics in female *Viviparus georgianus.* Biological Journal of the Linnean Society 27: 293-309.

Burch, J.B., and Y. Jung. 1987. A review of the classification, distribution and habitats of the freshwater gastropods of the North American Great Lakes. Walkerana 2: 233-291.

Carini, G., and J.M. Hughes. 2006. Subdivided population structure and phylogeography of an endangered freshwater snail, *Notopala sublineata* (Conrad, 1850) (Gastropoda: Viviparidae), in Western Queensland, Australia. Biological Journal of the Linnean Society 88: 1-16.

Chang, P.K., J.H. Cross, and S.S.S. Chen. 1968. Aquatic snails as intermediate hosts for *Angiostrongylus cantonensis* on Taiwan. Journal of Parasitology 54: 182-183.

Chung, P.R., and Y. Jung. 1999. *Cipangopaludina chinensis malleata* (Gastropoda: Viviparidae): A new second molluscan intermediate host of a human intestinal fluke *Echinostoma cenetorchis* (Trematoda: Echinostomatidae) in Korea. Journal of Parasitology 85: 963-964.

Clench, W.J. 1962. A catalogue of the Viviparidae of North America with notes on the distribution of *Viviparus georgianus* Lea. Occasional Papers on Mollusks 2: 261-287.

Colgan, D.J., W.F. Ponder, E. Beacham, and J. Macaranas. 2007. Molecular phylogenetics of Caenogastropoda (Gastropoda: Mollusca). Molecular Phylogenetics and Evolution 42: 717-737.

Cuttelod, A., M. Seddon, and E. Neubert. 2011. European Red List of Non-Marine Molluscs. Publications Office of the European Union, Luxembourg.

De Bernardi, R., O. Ravera, and B. Oregioni. 1976. Demographic structure and biometric characteristics of *Viviparus ater* Cristofori and Jan (Gasteropoda: Prosobranchia) from Lake Alserio (Northern Italy). Journal of Molluscan Studies 42: 310-318.

Dillon, R.T., Jr. 2000. The Ecology of Freshwater Molluscs. Cambridge University Press, Cambridge, UK.

Du, L.N., J.X. Yang, T. von Rintelen, X.Y. Chen, and D. Aldridge. 2013. Molecular phylogenetic evidence that the Chinese viviparid genus *Margarya* (Gastropoda: Viviparidae) is polyphyletic. Chinese Science Bulletin 58: 2154-2162.

Falniowski, A., K. Mazan, and M. Szarowska. 1996a. Embryonic shells of *Viviparus*—what they may tell us about taxonomy and phylogeny? (Gastropoda: Architaenioglossa: Viviparidae). Malakologische Abhandlungen 18: 35-42.

Falniowski, A., K. Mazan, and M. Szarowska. 1996b. Tracing the viviparid evolution: Radular characters (Gastropoda: Architaenioglossa: Viviparidae). Malakologische Abhandlungen 18: 43-52.

Fischer, J.-C. 1963. Les gastéropodes du 'Continental Intercalaire' du Sahara. In Les mollusques du 'Continental Intercalaire' (Mésozoïque) du Sahara Central. Mémoires de la Société Géologique de France 96: 41-50.

Fretter, V., and A. Graham. 1994. British prosobranch molluscs—their functional anatomy and ecology. Dorset Press, Dorchester, UK.

Ghilardi, R.P., S.C. Rodrigues, L.R.L. Simone, F.A. Carbonaro, and W.R. Nava. 2011. Moluscos fósseis do Grupo Bauru. Pp. 239-250 in Paleontologia: Cenários de Vida, vol. 4 (I.D. Souza Carvalho, N. Kumar Srivastava,

O. Strohschoen Jr., and C. Cunha Lana, eds.). Interciência, Rio de Janeiro, Brazil.

Graf, D.L., A. Jørgensen, D. Van Damme, and T.K. Kristensen. 2011. The status and distribution of freshwater molluscs (Mollusca). Pp. 48-61 in The Status and Distribution of Freshwater Biodiversity in Central Africa. (E.G.E. Brooks, D.J. Allen, and W.R.T. Darwall, eds.), IUCN, Cambridge, UK.

Gu, Q.H., M. Husemann, B. Ding, Z. Luo, and B.X. Xiong. 2015. Population genetic structure of *Bellamya aeruginosa* (Mollusca: Gastropoda: Viviparidae) in China: Weak divergence across large geographic distances. Ecology and Evolution 5: 4906-4919.

Hamilton-Bruce, R.J., B.J. Smith, and K.L. Gowlett-Holmes. 2002. Descriptions of a new genus and two new species of viviparid snails (Mollusca: Gastropoda: Viviparidae) from the Early Cretaceous (middle-late Albian) Griman Creek Formation of Lightning Ridge, northern New South Wales. Records of the South Australian Museum 35: 193-203.

Harasewych, M.G., S.L. Adamkewicz, M. Plassmeyer, and P.M. Gillevet. 1998. Phylogenetic relationships of the lower Caenogastropoda (Mollusca, Gastropoda, Architaenioglossa, Campaniloidea, Cerithioidea) as determined by partial 18S rDNA sequences. Zoologica Scripta 4: 361-272.

Heller, J. 1990. Longevity in molluscs. Malacologia 31: 259-295.

Henderson, J. 1935. Fossil non-marine mollusca of North America. Geological Society of North America, Special Papers 3: 1-313.

Hirano, T., T. Saito, and S. Chiba. 2015. Phylogeny of freshwater viviparid snails in Japan. Journal of Molluscan Studies 81: 435-441.

Hudson, J.D., R.G. Clements, J.B. Riding, M.I. Wakefield, and W. Walton. 1995. Jurassic paleosalinities and brackish-water communities—A case study. Palaios 10: 392-407.

IUCN Red List of Threatened Species. 2017. Version 2017-3. www.iucnredlist.org.

Jakubik, B. 2012. Life strategies of Viviparidae (Gastropoda: Caenogastropoda: Architaenioglossa) in various aquatic habitats: *Viviparus viviparus* (Linaeus, 1758) and *V. contectus* (Millet, 1813). Folia Malacologica 20: 145-179.

Jezewski, W. 2004. Occurrence of Digenea (Trematoda) in two *Viviparus* species from lakes, rivers and a dam reservoir. Helminthologia 41: 147-150.

Johnson, P.D., A.E. Bogan, K.M. Brown, N.M. Burkhead, J.R. Cordeiro, J.T. Garner, P.D. Hartfield, et al. 2013. Conservation status of freshwater gastropods of Canada and the United States. Fisheries 38: 247-282.

Johnson, S.G., and E. Bragg. 1999. Age and polyphyletic origins of hybrid and spontaneous parthenogenetic *Cam-*

peloma (Gastropoda: Viviparidae) from the Southeastern United States. Evolution 53: 1769-1781.

Jokinen, E.H. 1982. *Cipangopaludina chinensis* (Gastropoda: Viviparidae) in North America, review and update. Nautilus 96: 89-95.

Jokinen, E.H. 1984. Periostracal morphology of viviparid snail shells. Transactions of the American Microscopical Society 103: 312-316.

Jokinen, E.H., J. Guerette, and R.W. Kortmann. 1982. The natural history of an ovoviviparous snail, *Viviparus georgianus* (Lea), in a soft-water eutrophic lake. Freshwater Invertebrate Biology 1: 2-17.

Kear, B.P., R.J. Hamilton-Bruce, B.J. Smith, and K.L. Gowlett-Holmes. 2003. Reassessment of Australia's oldest freshwater snail, *Viviparus* (?) *albascopularis* Etheridge, 1902 (Mollusca: Gastropoda: Viviparidae), from the Lower Cretaceous (Aptian, Wallumbilla Formation) of White Cliffs. New South Wales Molluscan Research 23: 149-158.

Khan, R.A., and S. Chaudhuri. 1984. The population and production ecology of a freshwater snail *Bellamya bengalensis* (Lamarck) (Gastropoda: Viviparidae) in an artificial lake of Calcutta, India. Bulletin of the Zoological Survey of India 5: 59-76.

Kira, T. 1999. Prehistoric shell-mounds and current situation of pond snails (*Margarya* spp.) around ancient lakes of Yunnan Province, China. Pp. 123-133 in Ancient Lakes: Their Cultural and Biological Diversity (H. Kawanabe, G.W. Coulter, and A.C. Roosevelt, eds.). Kenobi Productions, Belgium.

Köhler, F., M. Seddon, A.E. Bogan, D.V. Tu, P. Sri-Aroon, and D. Allen. 2012. The status and distribution of freshwater molluscs of the Indo-Burma region. Pp. 67-88 in The Status and Distribution of Freshwater Biodiversity in Indo-Burma (D.J. Allen, K.G. Smith, and W.R.T. Darwall, eds.). IUCN, Cambridge, UK.

Ma, T., S. Gong, K. Zhou, C. Zhu, K. Deng, Q. Luo, and Z. Wang. 2010. Laboratory culture of the freshwater benthic gastropod *Bellamya aeruginosa* (Reeve) and its utility as a test species for sediment toxicity. Journal of Environmental Sciences 22: 304-313.

Mattox, N.T. 1938. Morphology of *Campeloma rufum*, a parthenogenetic snail. Journal of Morphology 62: 243-261.

Minton, R.L., and L.L. Wang. 2011. Evidence of sexual shape dimorphism in *Viviparus* (Gastropoda: Viviparidae). Journal of Molluscan Studies 77: 315-317.

Opinion 573. 1959. Determination under the plenary powers of a lectotype for the nominal species *Helix vivipara* Linnaeus, 1758, and addition to the Official List of the generic name *Viviparus* Montfort, 1810, and the family-group name Viviparidae Gray, 1847 (Class Gastropoda). Bulletin of Zoological Nomenclature 17: 117-131.

Ovando, X.M.C., and M.G. Cuezzo. 2012. Discovery of an established population of a non-native species of Viviparidae (Caenogastropoda) in Argentina. Molluscan Research 32: 121-131.

Pan, H.Z. 1977. Mesozoic and Cenozoic gastropods from Yunnan. Pp. 83-152 in Mesozoic Fossils from Yunnan fasc. 2. Science Press, Peking.

Perea, D., M. Soto, G. Veroslavsky, S. Martínez, and M. Ubilla. 2009. A Late Jurassic fossil assemblage in Gondwana: Biostratigraphy and correlations of the Tacuarembó Formation, Parana Basin, Uruguay. Journal of South American Earth Sciences 28: 168-179.

Plinski, M., W. Lawacz, A. Stańczykowska, and E. Magnin. 1978. Etude quantitative et qualitative de la nourriture des *Viviparus malleatus* (Reeve) (Gastropoda, Prosobranchia) dans deux lacs de la région de Montréal. Canadian Journal of Zoology 56: 272-279.

Ponder, W.F., and K.F. Walker. 2003. From mound springs to mighty rivers: The conservation status of freshwater molluscs in Australia. Aquatic Ecosystem Health and Management 6: 19-28.

Ponder, W.F., D.J. Colgan, J.M. Healy, A. Nützel, L.R.L. Simone, and E.E. Strong. 2008. Caenogastropoda. Pp. 331-383 in Phylogeny and Evolution of the Mollusca (W.F. Ponder, and D.R. Lindberg, eds.). University of California Press, Berkeley.

Prashad, B. 1928. Recent and fossil Viviparidae: A study in distribution, evolution and palaeogeography. Memoirs of the Indian Museum 8: 153-252.

Rao, H.S. 1925. On the comparative anatomy of Oriental Viviparidae. Records of the Indian Museum 27: 129-135.

Rohrbach, F. 1937. Ökologische und morphologische Untersuchungen an *Viviparus* (*Bellamya*) *capillatus* Frauenfeld und *Viviparus* (*Bellamya*) *unicolor* Olivier, unter Berücksichtigung anderer tropischer Formen und im Hinblick auf phyletische Beziehungen. Archiv für Molluskenkunde 69: 177-218.

Salzburger, W., B. Van Bocxlaer, and A.S. Cohen. 2014. Ecology and evolution of the African Great Lakes and their faunas. Annual Review of Ecology, Evolution and Systematics 45: 519-545.

Savage, A.E. 1938. A comparison of the nervous system in normal and sinistral snails of the species *Campeloma rufum*. American Naturalist 72: 160-169.

Schultheiß, R., T. Wilke, A. Jørgensen, and C. Albrecht. 2011. The birth of an endemic species flock: Demographic history of the *Bellamya* group (Gastropoda, Viviparidae) in Lake Malawi. Biological Journal of the Linnean Society 102: 130-143.

Schultheiß, R., B. Van Bocxlaer, F. Riedel, T. von Rintelen, and C. Albrecht. 2014. Disjunct distributions of freshwater snails testify to a central role of the Congo system in shaping biogeographical patterns in Africa. BMC Evolutionary Biology 14: e42, 1-12.

Sengupta, M.E., T.K. Kristensen, H. Madsen, and A. Jørgensen. 2009. Molecular phylogenetic investigations of the Viviparidae (Gastropoda: Caenogastropoda) in the lakes of the Rift Valley area of Africa. Molecular Phylogenetics and Evolution 52: 797-805.

Sheldon, F., and K.F. Walker. 1993. Pipelines as a refuge for freshwater snails. Regulated Rivers: Research & Management 8: 295-299.

Simone, L.R.L. 2004. Comparative morphology and phylogeny of representatives of the superfamilies of architaenioglossans and the Annulariidae (Mollusca, Caenogastropoda). Arquivos do Museu Nacional 62: 387-504.

Simone, L.R.L. 2011. Phylogeny of the Caenogastropoda (Mollusca), based on comparative morphology. Arquivos de Zoologia 42: 161-323.

Sivan, N., J. Heller, and D. Van Damme. 2006. Fossil Viviparidae (Mollusca: Gastropoda) of the Levant. Journal of Conchology 39: 207-219.

Soes, D.M., G.D. Majoor, and S.M.A. Keulen. 2011. *Bellamya chinensis* (Gray, 1834) (Gastropoda: Viviparidae), a new alien snail species for the European fauna. Aquatic Invasions 6: 97-102.

Stańczykowska, A., E. Magnin, and A. Dumouchel. 1971. Etude de trois populations de *Viviparus malleatus* (Reeve) (Gastropoda, Prosobranchia) de la région de Montréal. I. Croissance, fécondité, biomasse et production annuelle. Canadian Journal of Zoology 49: 1431-1441.

Stańczykowska, A.M., M. Plinski, and E. Magnin. 1972. Etude de trois populations de *Viviparus malleatus* (Reeve) (Gastropoda, Prosobranchia) de la région de Montréal. II. Etude qualitative et quantitative de la nourriture. Canadian Journal of Zoology 50: 1617-1624.

Strong, E.E., O. Gargominy, W.F. Ponder, and P. Bouchet. 2008. Global diversity of gastropods (Gastropoda: Mollusca) in freshwater. Hydrobiologia 595: 149-166.

Tracey, S., J.A. Todd, and D.H. Erwin. 1993. Mollusca: Gastropoda. Pp. 131-167 in The Fossil Record 2 (M.J. Benton, ed.). Chapman and Hall, London.

Vail, V.A. 1977. Comparative reproductive anatomy of 3 viviparid gastropods. Malacologia 16: 519-540.

Van Bocxlaer, B., and E.E. Strong. 2016. Anatomy, functional morphology, evolutionary ecology and systematics of the invasive gastropod *Cipangopaludina japonica* (Viviparidae: Bellamyinae). Contributions to Zoology 85: 235-263.

Van Bocxlaer, B., E.E. Strong, R. Richter, B. Stelbrink, and T. von Rintelen. 2018. Anatomical and genetic data reveal that *Rivularia* Heude, 1980 belongs to Viviparinae (Gastropoda: Viviparidae). Zoological Journal of the Linnean Society 182: 1-23.

Van Damme, D., and M. Pickford. 1999. The late Cenozoic Viviparidae (Mollusca, Gastropoda) of the Albertine Rift Valley (Uganda-Congo). Hydrobiologia 390: 171-217.

Van Damme, D., A.E. Bogan, and M. Dierick. 2015. A revision of the Mesozoic naiads (Unionoida) of Africa and the biogeographic implications. Earth-Science Reviews 147: 141-200.

Van der Spoel, S. 1958. Groei en ouderdom bij *Viviparus contectus* (Millet, 1813) en *Viviparus viviparus* (Linné, 1758). Basteria 22: 77-90.

Walker, K.F. 1996. The river snail *Notopala hanleyi*: An endangered pest. Xanthopus 14: 5-7.

Wang, J.-G., D. Zhang, I. Jakovlić, and W.-M. Wang. 2017. Sequencing of the complete mitochondrial genomes of eight freshwater snail species exposes pervasive paraphyly within the Viviparidae family (Caenogastropoda). Plos One 12: e0181699, 1-17.

Yen, T.-C. 1950. Fresh-water mollusks of Cretaceous Age from Montana and Wyoming. Geological Survey Professional Paper 233-A. US Government Printing Office, Washington.

Ying, T., F.T. Fürsich, and S. Schneider. 2013. Giant Viviparidae (Gastropoda: Architaenioglossa) from the Oligocene of the Nanning Basin (Guangxi, SE China). Neues Jahrbuch für Geologie und Paläontologie, Abhandlungen 267: 75-87.

Zhang, L.-J., S.-C. Chen, L.-T. Yang, L. Jin, and F. Köhler. 2015. Systematic revision of the freshwater snail *Margarya* Nevill, 1877 (Mollusca: Viviparidae) endemic to the ancient lakes of Yunnan, China, with description of new taxa. Zoological Journal of the Linnean Society 174: 760-800.

6 Hemisinidae Fischer & Crosse, 1891

MATTHIAS GLAUBRECHT AND MARCO T. NEIBER

Hemisinidae are medium-sized to large neotropical freshwater snails with an elongate-ovate to ovate-conical shell that is often ornamented with distinctly incised spiral lines and sometimes also with sinuous ribs or nodules at the shoulder of the whorls. The aperture is pointed above and basally well rounded or with a slight columellar notch. The operculum is corneous and paucispiral, with a subbasal nucleus (Morrison, 1954; Gomez et al., 2011). The mantle edge is fringed. *Hemisinus lineolatus* (Wood, 1828), representing the type genus of the family, possesses a large midgut with a textured accessory pad and two short, prominent caecal folds as well as fused typhlosoles; the renal oviduct is branching, and a cephalic brood pouch is present in females and a middorsal spermatophore organ in males (Gomez et al., 2011). Usually only few, large juveniles are present in the brood pouch (Morrison, 1954; Gomez et al., 2011).

Although limnic Cerithoideans form a significant component of Neotropical freshwater biota, those taxa are relatively poorly studied to date (Glaubrecht, 1996; Gomez et al., 2011). Next to representatives of the Pachychilidae, such as *Pachychilus* H.C. Lea & I. Lea, 1851, and *Doryssa* Swainson, 1840, other Neotropical Cerithioidea comprise species hitherto considered as representatives of the Thiaridae. On the basis of recent molecular genetic studies (Glaubrecht et al., unpublished data; see also Gimnich, 2015; Strong et al., 2011), it could be shown that, aside from Pachychilidae, another, independent evolutionary lineage of cerithioidean freshwater gastropods is present in the Neotropics that is here suggested to be considered a family, Hemisinidae Fischer & Crosse, 1891, on its own.

Fischer and Crosse (1891: 312) introduced Semisinusinae as a subfamily of "Melaniidae," which has later been corrected to Hemisinuseae (Thiele, 1928) because the emendation of *Hemisinus* Swainson, 1840, to *Semisinus* by Fischer (1885) in Fischer (1880-1887) or Fischer and Crosse (1891) was unjustified. Recently the name was again corrected to Hemisininae because of an incorrect stem formation, which cannot be preserved as in prevailing usage (Bouchet and Rocroi, 2005; Bouchet et al., 2005).

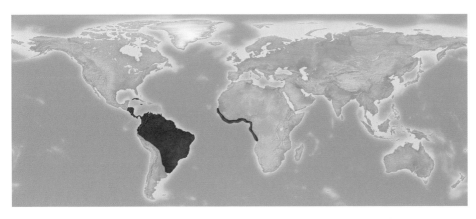

Distribution of Hemisinidae.

The type genus of the family, *Hemisinus,* has been included in the "Melaniidae," or Thiaridae sensu lato (see Glaubrecht, 2011), by most authors so far (c.g., Thiele, 1928; Pilsbry and Olsson, 1935; Wenz, 1938–1944; Morrison, 1952, 1954; Ponder and Warén, 1988; Nuttall, 1990; Glaubrecht, 1996). First descriptions of the anatomy and reproductive biology of Neotropical "thiarids" are given in Fischer and Crosse (1891), Thiele (1928, 1929-1931), and Hylton Scott (1953), whereas more detailed accounts of a few species only became available in the past two decades (Glaubrecht, 1996; Simone, 2001; Gómez, 2009; Gomez et al., 2011; Vogler, 2013a, b; Vogler et al., 2014). Phylogenetic analyses based on morphological (Glaubrecht, 1996; Simone, 2001, 2011) and/or molecular markers (Lydeard et al., 2002; Glaubrecht and Rintelen, 2008; Gómez, 2009; Strong et al., 2011; Glaubrecht, 2011; Gimnich, 2015) recovered a sister group relationship of Thiaridae and Hemisinidae. These later studies suggest a Gondwanan origin of a clade including, aside from Thiaridae and Hemisinidae, the African/Asian Paludomidae.

The genus *Hemisinus* shares many characteristics with Thiaridae, such as the structure of the operculum, mantle edge, rachidian teeth (shape, basal denticles), midgut (large, textured accessory pad; two short, prominent caecal folds; fused typhlosoles), renal oviduct (branching), and cephalic brood pouch. It also shares some anatomical similarities with paludomids (statoconia, spermatophore organ), especially basal paludomids (mantle papillae, right dialyneury) (Gomez et al., 2011). The presence of a partially open female gonoduct can be regarded as an autapomorphy of Hemisinidae (Glaubrecht, 1996). This character state, along with a very similar radula (Binder, 1957; Glaubrecht, 1996), has led to the hypothesis that the western African *Pachymelania* E. A. Smith, 1893, species that live in brackish environments and can tolerate strong fluctuations in salinity (Brown, 1994; Bandel and Kowalke, 1999), are the sister group of the Neotropical genus *Hemisinus* (Glaubrecht, 1996), suggesting a possible amphi-Atlantic distribution of the family. This distribution pattern was recently confirmed by molecular analyses (Glaubrecht and Rintelen, 2008; Glaubrecht et al., unpublished data).

Generic names such as *Aylacostoma* Spix, 1827 (for nomenclatorial comments, see Glaubrecht, 1996;

Hemisinus guayaquilensis Reeve, 1859. Ecuador: Balsas, El Oro Province. 17.4 mm. Poppe 846719.

Cowie et al., 2004), *Hemisinus,* and *Cubaedomus* Thiele, 1928, together with the subgenera *Verena* Adams & Adams, 1854, *Longiverena* Pilsbry & Olsson, 1935, and *Basistoma* Lea, 1852, have been assigned to recent taxa inhabiting Central and/or South America that are here united in the family Hemisinidae (Fischer, 1885, in Fischer 1880-1887; Thiele, 1928; Pilsbry & Olsson, 1935; Morrison, 1952, 1954; Jaeckel, 1969; Nuttall, 1990; Simone, 2006). The limits among these taxa still remain to be established, though. As a consequence, the distributional range of *Hemisinus,* which was originally thought to be restricted to Jamaica and Cuba only (Gomez et al., 2011), was progressively expanded since the middle of the nineteenth century to include Central America, Peru, Venezuela, Surinam, Brazil, Argentina, and Paraguay (Reeve, 1860; Brot, 1862; Martens, 1873; Fischer, 1885, in Fischer 1880-1887; Ihering, 1901, 1909; Vernhout, 1914; Hylton Scott, 1953, 1954; Simone, 2001, 2006; Pointier et al., 2005; Vogler, 2012, 2013a, b; Vogler et al., 2012, 2014; Pointier and Noya Alarcón, 2015). On the other hand, species from Jamaica, Cuba, and Central America have been included by Simone (2006) in *Aylacostoma,* which was originally described from Brazil by J. B. von Spix (Wagner, 1827). The same applies to fossil taxa that have been assigned to the group because of conchological resemblance (e.g., Nuttall, 1990).

Brot (1874) lists 36 morphospecies under the generic name *Hemisinus,* of which 29 are recorded from the New World, while the remaining 5 taxa could be shown to belong to the Melanopsidae (Glaubrecht, 1996). Simone (2006) lists 32 species under the name *Aylacostoma,* including the type species of *Aylacostoma* and *Hemisinus.* The total diversity within the family amounts to approximately 40 recent taxa including several species from Mesoamerica and the Caribbean region not listed by Simone (2006) and the western African *Pachymelania* species that comprise a total of 3 currently recognized taxa distributed from Senegal to Angola along the coast. However, the family is in urgent need of a thorough monographic revision in an evolutionary systematics context (see Glaubrecht, 2010).

Relatively few species belonging to the Hemisinidae have been assessed by the International Union for Conservation of Nature (IUCN) so far. The three *Pachymelania* species have been assessed as Least Concern, while three of the four species of *Aylacostoma* that have been considered, *A. guaraniticum* Hylton Scott, 1953, *A. chloroticum* Hylton Scott, 1954, and *A. stigmaticum* Hylton Scott, 1954, have been assessed as Extinct in the Wild as a consequence of the construction of the Yacyretá Dam at the Paraná River (Argentina and Paraguay), where the species were endemic on hard substrates in regions with fast-flowing water (IUCN, 2017). *Aylacostoma stigmaticum* and *A. guaraniticum* are considered globally extinct (Peso et al., 2013a, b; Vogler, 2013b). However, an ongoing ex situ conservation program was developed in the 1990s for *A. chloroticum,* and recently a new morphotype was discovered (Vogler, 2012, 2013b). Moreover, two populations of *A. chloroticum* were recently reported to have survived in the wild (Vogler, 2012; Quintana and Ostrowski de Núñez, 2016), though they are seriously under threat from the alteration of their natural habitat.

There are few species in the Hemisinidae that have been surveyed for their importance as vectors for parasitic trematodes, and reports on human pathogenic parasites are lacking. Ostrowski de Núñez and Quintana (2008) and Quintana and Ostrowski de Núñez (2014, 2016) report, however, on a total of five different cercariae from *A. chloroticum.* The life cycle of three of the trematode species detected were stud-ied by these authors in detail, and it could be shown that the adult worms were parasites of different poeciliid and tetragonopterid fish species.

LITERATURE CITED

Bandel, K., and T. Kowalke. 1999. Gastropod fauna of the Cameroonian coast. Helgoland Marine Research 53: 129-140.

Binder, E. 1957. Mollusques aquatiques de Côte d'Ivoire. I. Gastéropodes. Bulletin de l'Institut Fondamental d'Afrique Noire, Series A, 19: 97-125.

Bouchet, P., and J.-P. Rocroi. 2005. Part 1. Nomenclator of gastropod family-group names. In Classification and Nomenclator of Gastropod Families (P. Bouchet and J.-P. Rocroi, eds.). Malacologia 47: 5-239.

Bouchet, P., J. Frýda, B. Hausdorf, W.F. Ponder, Á. Valdés, and A. Warén. 2005. Part 2. Working Classification of the Gastropoda. In Classification and Nomenclator of Gastropod Families (P. Bouchet and J.-P. Rocroi, eds.). Malacologia 47: 239-283.

Brot, A. 1862. Materiaux pour servir a l'Etude de la Famille des Mélaniens: Catalogue systématique des Espèces qui composent la Famille des Mélaniens. J.-G. Flick, Geneva.

Brot, A. 1874. Die Melaniaceen (Melanidae) in Abbildungen nach der Natur mit Beschreibungen. Systematisches Conchylien-Cabinet Martini Chemnitz 1 (24): 1-488, plates 1-49.

Brown, D.S. 1994. Freshwater Snails of Africa and Their Medical Importance. 2nd ed. Taylor & Francis, London.

Cowie, R.H., N.J. Cazzaniga, and M. Glaubrecht. 2004. The South American Mollusca of Johann Baptist Ritter von Spix and their publication by Johann Andreas Wagner. Nautilus 118: 71-87.

Fischer, P. 1880-1887. Manuel de conchyliologie et de paléontologie conchyliologique ou histoire naturelle des mollusques vivants et fossiles. F. Savy, Paris.

Fischer, P., and H. Crosse. 1891. Etudes sur les Mollusques terrestres et fluviatiles du Mexique et du Guatemala: Mission scientifique au Mexique et dans l'Amérique Centrale. Recherches zoologiques. Vol. 2, part 7, issue 12. Imprimerie National, Paris.

Gimnich, F. 2015. Molecular approaches to the assessment of biodiversity in limnic gastropods (Cerithioidea, Thiaridae) with perspectives on a Gondwanian origin. PhD diss., Humboldt Universität zu Berlin.

Glaubrecht, M. 1996. Evolutionsökologie und Systematik am Beispiel von Süß- und Brackwasserschnecken (Mollusca: Caenogastropoda: Cerithioidea): Ontogenese-Strategien, paläontologische Befunde und Historische Zoogeographie. Backhuys, Leiden.

Glaubrecht, M. 2010. Evolutionssystematik limnischer Gastropoden. Habilitation thesis, Humboldt Universität zu Berlin.

Glaubrecht, M. 2011. Towards solving Darwin's 'mystery': Speciation and radiation in lacustrine and riverine freshwater gastropods. American Malacological Bulletin 29: 187-216.

Glaubrecht, M., and T. von Rintelen. 2008. The first comprehensive molecular phylogeny reveals ancient vicariance, parallel evolution of viviparity and extensive convergence in shell morphology of thiarid freshwater gastropods. In Systematics 2008, Programme and Abstracts, 10th Annual Meeting of Gesellschaft für Biologische Systematik, Göttingen, 7-11 April 2008. 18th International Symposium, 'Biodiversity and Evolutionary Biology,' of the German Botanical Society, p. 211 (S.R. Gradstein, S. Klatt, F. Normann, P. Weigelt, R. Willmann, and R. Wilson, eds.). Universitätsverlag Göttingen, Göttingen.

Gómez, M.I. 2009. Systematics, phylogeny and biogeography of Mesoamerican and Caribbean freshwater gastropods (Cerithioidea: Thiaridae and Pachychilidae). PhD diss., Humboldt University, Berlin.

Gomez, M.I., E.E. Strong, and M. Glaubrecht. 2011. Redescription and anatomy of the viviparous freshwater gastropod Hemisinus lineolatus (W. Wood, 1828) from Jamaica (Cerithioidea, Thiaridae). Malacologia 53: 229-250.

Hylton Scott, M.I. 1953. El género Hemisinus (Melaniidae) en la costa fluvial argentina (Mol. Prosobr.). Physis 20: 438-443.

Hylton Scott, M.I. 1954. Dos nuevos melánidos del Alto Paraná (Mol. Prosobr.). Neotropica 1: 45-48.

Ihering, H. von. 1901. As melanias do Brasil. Revista do Museu Paulista 5: 653-681.

Ihering, H. von. 1909. Les mélaniidés Américains. Journal de Conchyliologie 57: 289-316.

IUCN Red List of Threatened Species. 2017. Version 2017-1. www.iucnredlist.org. Accessed 25 May 2017.

Jaeckel, S.G. 1969. Die Mollusken Südamerikas. Pp. 794-827 in Biogeography and Ecology in South America, vol. 2 (E.J. Fittkau, J. Illies, H. Klinge, G.H. Schwabe, and H. Sioli, eds.). Junk, The Hague.

Lydeard, C., W.E. Holznagel, M. Glaubrecht, and W.F. Ponder. 2002. Molecular phylogeny of a circum-global, diverse gastropod superfamily (Cerithioidea: Mollusca: Caenogastropoda): Pushing the deepest phylogenetic limits of ditochondrial LSU rDNA sequences. Molecular Phylogenetics and Evolution 22: 399-406.

Martens, E. von. 1873. Die Binnenmollusken Venezuela's. Pp. 157-225 in Festschrift zur Feier des hundertjährigen Bestehens der Gesellschaft naturforschender Freunde zu Berlin (K.B. Reichert, ed.). F. Dümmlers Verlagsbuchhandlung, Berlin.

Morrison, J.P.E. 1952. World relations of the melanians (an abstract). American Malacological Union News Bulletin and Annual Report 1951: 6-9.

Morrison, J.P.E. 1954. The relationships of Old and New World melanians. Proceedings of the United States National Museum 103: 357-394.

Nuttall, C.P. 1990. A review of the Tertiary non-marine molluscan faunas of the Pebasian and other inland basins of north-western South America. Bulletin of the British Museum (Natural History) Geology Series 45: 165-371.

Ostrowski de Núñez, M., and M.G. Quintana. 2008. The life cycle of Stephanoprora aylacostoma n. sp. (Digenea: Echinostomatidae), parasite of the threatened snail Aylacostoma chloroticum (Prosobranchia, Thiaridae), in Argentina. Parasitology Research 102: 647-655.

Peso, J.G., C. Costigliolo Rojas, and M.J. Molina. 2013a. Aylacostoma stigmaticum Hylton Scott, 1954: Antecedentes de la especie. Amici Molluscarum 21: 43-46.

Peso, J.G., M.J. Molina, and C. Costigliolo Rojas. 2013b. Aylacostoma guaraniticum (Hylton Scott, 1953): Antecedentes de la especie. Amici Molluscarum 21: 39-42.

Pilsbry, H.A., and A.A. Olsson. 1935. Tertiary fresh-water mollusks of the Magdalena embayment, Colombia. Proceedings of the Academy of Natural Sciences of Philadelphia 7: 7-39.

Pointier, J.-P., and O. Noya Alarcón. 2015. Family Thiaridae. Pp. 84-94 in Freshwater Molluscs of Venezuela and Their Medical and Veterinary Importance (J.-P. Pointier, ed.). ConchBooks: Harxheim, Germany.

Pointier, J.-P., M. Yong, and A. Gutiérrez, A. 2005. Guide to the Freshwater Molluscs of Cuba. ConchBooks, Hackenheim, Germany.

Ponder, W.F., and A. Warén. 1988. Classification of the Caenogastropoda and Heterostropha—a list of the family-group names and higher taxa. In Prosobranch phylogeny: Proceedings of a Symposium Held at the 9th International Malacological Congress, Edinburgh, 1986 (W.F. Ponder, ed.). Malacological Review Supplement 4: 288-328.

Quintana, M.G., and M. Ostrowski de Núñez. 2014. The life cycle of Pseudosellacotyla lutzi (Digenea: Cryptogonimidae), in Aylacostoma chloroticum (Prosobranchia: Thiaridae), and Hoplias malabaricus (Characiformes: Erythrinidae), in Argentina. Journal of Parasitology 100: 805-811.

Quintana, M.G., and M. Ostrowski de Núñez. 2016. The life cycle of Neocladocystis intestinalis (Vaz, 1932) (Digenea: Cryptogonimidae), in Aylacostoma chloroticum (Prosobranchia: Thiaridae), and Salminus brasiliensis (Characiformes:

Characidae), in Argentina. Parasitology Research 115: 2589-2595.

Reeve, L.A. 1860. Monograph of the Genus *Hemisinus:* Conchologia Iconica: Or Illustrations of the Shells of Molluscous Animals, vol. 12, plates 1-6. London.

Simone, L.R.L. 2001. Phylogenetic analyses of Cerithioidea (Mollusca, Caenogastropoda) based on comparative morphology. Arquivos de Zoologia 36: 147-263.

Simone, L.R.L. 2006. Land and Freshwater Molluscs of Brazil: An Illustrated Inventory of the Brazilian Malacofauna, Including Neighboring Regions of South America, Respect to the Terrestrial and Freshwater Ecosystems. EGB, São Paulo.

Simone, L.R.L. 2011. Phylogeny of the Caenogastropoda (Mollusca), based on comparative morphology. Arquivos de Zoologia 42: 161-323.

Strong, E.E., D.J. Colgan, J.M. Healy, C. Lydeard, W.F. Ponder, and M. Glaubrecht. 2011. Phylogeny of the gastropod superfamily Cerithioidea using morphology and molecules. Zoological Journal of the Linnean Society 162: 43-89.

Thiele, J. 1928. Revision des Systems der Hydrobiiden und Melaniiden. Zoologische Jahrbücher, Abteilung für Systematik, Ökologie und Geographie der Tiere 55: 351-402.

Thiele, J. 1929-1931. Handbuch der Systematischen Wiechtierkunde, vol. 1. Gustav Fischer, Jena, Germany.

Vernhout, J.H. 1914. The non-marine molluscs of Surinam. Notes from the Leyden Museum 36: 1-46.

Vogler, R.E. 2012. *Aylacostoma chloroticum* Hylton Scott, 1954: Antedecedentes de la especie. Amici Molluscarum 20: 43-46.

Vogler, R.E. 2013a. Inferencia filogeográfica aplicada a la conservación de hembras partenogenéticas del género *Aylacostoma* Spix, 1827: Especies amenazadas del río Paraná. PhD diss., Universidad Nacional de La Plata.

Vogler, R.E. 2013b. The radula of the extinct freshwater snail *Aylacostoma stigmaticum* (Caenogastropoda: Thiaridae) from Argentina and Paraguay. Malacologia 56: 329-332.

Vogler, R.E., A.A. Beltramino, D.E. Gutiérrez-Gregoric, J.G. Peso, M. Griffin, and A. Rumi. 2012. Threatened Neotropical mollusks: Analysis of shape differences in three endemic snails from high Paraná River by geometric morphometrics. Revista Mexicana de Biodiversidad 83: 1045-1052.

Vogler, R.E., A.A. Beltramino, J.G. Peso, and A. Rumi. 2014. Threatened gastropods under the evolutionary genetic species concept: Redescription and new species of the genus *Aylacostoma* (Gastropoda: Thiaridae) from High Paraná River (Argentina-Paraguay). Zoological Journal of the Linnean Society 172: 501-520.

Wagner, J.A. 1827. Testacea fluviatilia quae in itinere per Brasiliam annis MDCCCXVII-MDCCCXX jussu et auspiciis Maximiliani Josephi I. Bavariae Regis augustissimi suscepto, collegit et pingenda curavit Dr. J. B. de Spix. C. Wolf, Monachii, Germany.

Wenz, W. 1938-1944. Gastropoda. Part 1. Allgemeiner Teil und Prosobranchia. Pp. i-viii, 1-1639 in Handbuch der Paläozoologie, vol 6, part 1 (O.H. Schindewolf, ed.). Gebrüder Borntraeger, Berlin.

7 Melanopsidae H. Adams & A. Adams, 1854

MARCO T. NEIBER AND MATTHIAS GLAUBRECHT

Melanopsidae are medium-sized freshwater snails, with ovate to elongate conical shells that either have a smooth surface or may be ornamented with often nodulose ribs. The aperture is usually narrow, pointed above and usually with a truncated columella. The operculum is corneous and paucispiral, with a basal nucleus. The posterior oviduct is open in melanopsids, the cerebralpedal connectives are short, and the osphradium is less than half the ctenidial length (Houbrick, 1988; Glaubrecht, 1996). The mantle edge is smooth. Both sexes occur, females with large grooved ovipositor in a pit on the right side of the foot (Morrison, 1954; Glaubrecht, 1996). Eggs are laid by the female, from which veliger larvae or juveniles hatch (Morrison, 1954; Glaubrecht, 1996).

The taxa that are currently united in the family Melanopsidae have almost unequivocally been classified as "Melaniidae" and later as Thiaridae by nineteenth- and twentieth-century malacologists (Brot, 1874; Bouvier, 1887; Sturany and Wagner, 1914; Ankel, 1928; Thiele, 1928, 1929-1931; Sunderbrinck, 1929; Wenz, 1938-1944; Jaeckel et al., 1957; Starmühlner and Edlauer, 1957; Willmann and Pieper, 1978; Fretter, 1984; Houbrick, 1987; Falkner, 1990; Starmühlner, 1991). Even though significant anatomical differences (e.g., in the nervous system) of Thiaridae and Melanopsidae were known for a long time (see Bouvier, 1887; Simroth, 1896-1907), only a few authors have regarded Melanopsidae as a family distinct from Thiaridae (Simroth, 1896-1907; Moore, 1899; Morrison, 1954; Glaubrecht, 1996). However, the phylogenetic analyses by Strong et al. (2011) have shown that Melanopsidae is not closely related to Thiaridae and is here considered to be valid.

The Melanopsidae, including the extant genera *Melanopsis* Férussac, 1807, in Férussac and Férussac (1807), *Esperiana* Bourguignat, 1877, *Holandriana* Bourguignat, 1884, *Microcolpia* Bourguignat, 1884, and *Zemelanopsis* Finlay, 1926, are characterized by the following apomorphic character states within Cerithioidea: (1) a zygoneurous nervous system, (2) lamellar pockets on both sides of the oesophagus, (3) a reduced number of whorls (5-10) of the shell, and (4) a posterior parietal callus, although second-

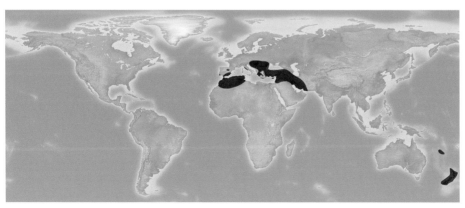

Distribution of Melanopsidae.

Melanopsis magnifica (Bourguignat, 1884). Morocco: Ain Chkef, Moulay Yacoub Province. 35.2 mm. Poppe 740302.

arily absent in recent species of *Esperiana, Holandriana,* and *Microcolpia* (Glaubrecht, 1996).

The group has an extraordinarily rich fossil record that dates back to the Cretaceous (Glaubrecht, 1996; Bandel, 2000; Neubauer et al., 2016). Taking this into account along with an exceptional morphological variability of the shell, it is hardly surprising that more than 1300 species group names have been introduced over the past 250 years (Neubauer, 2016), many of which certainly do not represent biological entities (Glaubrecht, 1993, 1996, 2011; Altaba, 1998; Neubauer et al., 2016). Glaubrecht (1993, 1996) provided comprehensive data on the morphology, anatomy, distribution, biogeography, evolutionary history, ecology, mode of reproduction, and ontogeny as well as an overview of the fossil record of the family and especially data on the *Melanopsis* species in the Mediterranean basin. In the past decades a number of studies, on both recent and fossil representatives, focused on the underlying evolutionary processes that may have contributed to the observed morphological plasticity in the family and contributed to the understanding of the paleobiogeography and historical biogeography of the family on different geographical scales (Papp, 1955; Geary, 1990, 1992; Glaubrecht, 1993, 1996; Heller et al., 1999, 2005; Bandel, 2000; Geary et al., 2002, 2012; Heller and Sivan, 2001, 2002a, b; Elkarmi and

Ismail, 2006; Bandel et al., 2007; Neubauer et al., 2013, 2014a, b, 2016).

Early representatives of the family were apparently pure brackish-water dwellers, while, from the early Miocene on, several lineages started to adapt to freshwater environments (Glaubrecht, 1996; Neubauer et al., 2016; Neubauer, 2016). This change in life style coincided with a series of radiations in the middle and late Miocene, e.g., in the Dinaride lake system and especially in Lake Pannon, and resulted in a diversity maximum in the late Miocene and a subsequent decline during the Pliocene and Quaternary (Neubauer et al., 2016).

The four currently widely accepted recent Palaearctic genera, with the exception of *Melanopsis,* usually contain only one to two species: *Esperiana* from southeastern Europe and northwestern Anatolia (one to two species), *Holandriana* from southeastern central to southeastern Europe (one species), and *Microcolpia* also from southeastern central to southeastern Europe (one to two species with a few subspecies). Depending on the authority, *Melanopsis* contains only a few polymorphic species (Glaubrecht, 1996) or up to an estimated number of 25–50 species (Strong et al., 2008; Neubauer et al., 2016; Heller et al., 2005; García et al., 2010). In any case the taxonomic redundancy in this particular genus, with a total of at least several hundred recent and fossil named taxa, appears to be very high and close to that found, for example, in thiarids (Glaubrecht, 2011).

Melanopsis has a distribution range including southern and eastern Spain and parts of Morocco, Algeria, and Tunisia (e.g., Glaubrecht, 1993, 1996; Pallary, 1924, 1926, 1939; Pujante and Gallardo, 1990), a few thermal springs in the Italian Tuscany region (Glaubrecht, 1993, 1996; Altaba, 1998; Cianfanelli, 2010), mainland Greece and many of the Aegean islands, the coastal regions of Turkey, including a few records from inland localities in that country as well as Cyprus (e.g., Schütt, 1965; Schütt and Bilgin, 1974; Paget, 1976; Glaubrecht, 1993, 1996; Frank, 1997; Çabuk et al., 2004; Odabaşi and Arslan, 2015), and the Near and Middle East from northeastern Egypt and eastern Saudi Arabia in the south through Israel, Jordan, Lebanon, Palestine, and Syria to Iraq and Iran in the east (e.g., Olenev, 1987; Ismail, 1989; Glaubrecht, 1993, 1996; Neubert, 1998; Heller et al., 1999, 2002, 2005;

Heller and Sivan, 2000, 2001, 2002a, b; Neubauer et al., 2014b, 2016). With an enormous distribution gap (Brot, 1874; Glaubrecht, 1996), the family is also reported from New Caledonia, with two endemic species currently assigned to the genus *Melanopsis,* and from New Zealand, with a single recent species assigned to the genus *Zemelanopsis* (see Glaubrecht, 1996).

Surprisingly, the family has received relatively little attention in morphology-based or molecular phylogenetic studies. Glaubrecht (1996) discussed melanopsids as a monophyletic group with *Faunus ater* (Linnaeus, 1758) as its sister taxon. However, he and coauthors have subsequently shown that *Faunus* is not closely related to the Melanopsidae but belongs to the Pachychilidae (Lydeard et al., 2002; Strong et al., 2011). The latter studies found a close relationship of the genera *Melanopsis* and *Holandriana* with representatives of the Pleuroceridae and Semisulcospiridae, with the monophyly of Melanopsidae recovered in one of the combined analyses of morphological characters and molecular markers but neither on the basis of morphological characters or molecular markers alone (Strong et al., 2011). Smoleń and Falniowski (2009) recovered the monophyly of Melanopsidae using 18S rDNA sequences but not on the basis of *cox1* sequences. These authors, however, did not include representatives of the closely related Pleuroceridae and Semisulcospiridae in their analyses and hence were unable to support the monophyly of the Melanopsidae with regard to these families. Thus, two important questions with regard to the phylogeny of the family currently remain unanswered: (1) Are the Melanopsidae, including the genera *Melanopsis, Holandriana, Microcolpia, Esperiana,* and *Zemelanopsis,* a monophylum? And (2) Are the species from the Pacific region more closely related to each other and not, as current taxonomies imply, with the New Caledonian species more closely related to the western Palearctic taxa? Another open question is the actual number of recent species in the genus *Melanopsis.* Investigations using molecular genetic techniques (Glaubrecht et al., unpublished data) could lead to new insights here and to a better understanding of the phylogeny and phylogeography of this group in the Mediterranean basin.

A total of 36 species assigned to the family *Melan-*

opsidae are currently included in the IUCN Red List, 10 as Critically Endangered, 6 as Endangered, 2 as Vulnerable, 1 as Near Threatened, 8 as Least Concern, and 6 as Data Deficient, implying that approximately half of the taxa in the family are threatened with a comparatively high number of taxa with very high extinction risk.

As in other cerithioidean freshwater families, melanopsids may be important intermediate hosts for trematode species. Cichy et al. (2011) list nine different species for European Melanopsidae. Various studies have reported different types of cercariae from the Near and Middle East as well as the Caucasus region (e.g., Ullman, 1954; Lengy and Stark, 1971; Galaktionov, 1980; Olenev, 1987; Ismail, 1989; Farahnak et al., 2006; Bdir and Adwan, 2011, 2012; Manafov, 2011, 2012, 2013), among them potentially human pathogen cercariae of eye flukes belonging to the genus *Philophthalmus* Looss, 1899 (Gold et al., 1993).

LITERATURE CITED

Altaba, C.R. 1998. Testing vicariance: Melanopsid snails and Neogene tectonics in the western Mediterranean. Journal of Biogeography 25: 541-551.

Ankel, W.E. 1928. Beobachtungen über Eiablage und Entwicklung von *Fagotia esperi* (Férussac). Archiv für Molluskenkunde 60: 251-256.

Bandel, K. 2000. Speciation among the Melanopsidae (Caenogastropoda): Special emphasis to the Melanopsidae of the Pannonian Lake at Pontian time (Late Miocene) and the Pleistocene and Recent of Jordan. Mitteilungen aus dem Geologisch-Paläontologischen Institut der Universität Hamburg 84: 131-208.

Bandel, K., N. Sivan, and J. Heller. 2007. *Melanopsis* from Al-Qarn, Jordan Valley (Gastropoda: Cerithioidea). Paläontologische Zeitschrift 81: 304-315.

Bdir, S., and G. Adwan. 2011. Larval stages of digenetic trematodes of *Melanopsis praemorsa* snails from freshwater bodies in Palestine. Asian Pacific Journal of Tropical Biomedicine 1: 200-204.

Bdir, S., and G. Adwan. 2012. Three new species of cercariae from *Melanopsis praemorsa* (L. 1758, *Buccinum*) snails in Al-Bathan fresh water body, Palestine. Asian Pacific Journal of Tropical Biomedicine 2: 1064-1069.

Bouvier, E.-L. 1887. Système nerveux, morphologie générale et classicication des gastéropodes prosobranches. Annales des Sciences Naturelles. Zoologie 3: 1-510.

Brot, A. 1874. Die Melaniaceen (Melanidae) in Abbildungen nach der Natur mit Beschreibungen. Systematisches

Conchylien-Cabinet Martini Chemnitz 1 (24): 1-488, plates 1-49.

Çabuk, Y., N. Arslan, and V. Yilmaz. 2004. Species composition and seasonal variations, of the Gastropoda in Upper Sakarya River System (Turkey) in relation to water quality. Acta Hydrochimica et Hydrobiologica 6: 393-400.

Cianfanelli, S. 2010. *Melanopsis etrusca*. The IUCN Red List of Threatened Species 2010: e.T40077A10300265.

Cichy, A., A. Faltýnková, and E. Żbikowska. 2011. Cercariae (Trematoda, Digenea) in European freshwater snails—a checklist of records from over one hundred years. Folia Malacologica 19: 165-189.

Elkarmi, A.Z., and N.S. Ismail. 2006. Allometry of the gastropod *Melanopsis praemorsa* (Thiaridae: Prosobranchia) from Azraq Oasis, Jordan. Pakistan Journal of Biological Sciences 9: 1359-1363.

Falkner, G. 1990. Binnenmollusken. Pp. 112-280 in Weichtiere: Europäische Meeres- und Binnenmollusken. Steinbach's Naturführer, vol. 10 (R. Fechter and G. Falkner, eds.). Mosaik Verlag, Munich.

Farahnak, A., R. Vafaie-Darian, and I. Mobedi. 2006. A faunistic survey of cercariae from fresh water snails: *Melanopsis* spp. and their role in disease transmission. Iranian Journal of Public Health 35: 70-74.

Férussac, J.B.L. d'Audebard de, and A.E.J.P.J.F. d'Audebard Férussac. 1807. Essai d'une méthode conchyliologique appliquée aux mollusques fluviatiles et terrestres d'après la considération de l'animal et de son test. Delance, Paris.

Frank, C. 1997. Die Molluskenfauna der Insel Rhodos, 2. Teil. Stapfia 48: 1-179.

Fretter, V. 1984. Prosobranchs. Pp. 1-45 in The Mollusca, vol. 7, Reproduction (A.S. Tompa, N.H. Verdonk, and J.A.M. van den Biggelaar, eds.). Academic Press, Orlando, Florida.

Galaktionov, K.V. 1980. Two species of cyathocotylid cercariae from the fresh water mollusk *Melanopsis praemorsa*. Parasitologyia 14: 299-307.

García, N., A. Cuttelod, and D. Abdul Malak. 2010. The Status and Distribution of Freshwater Biodiversity in Northern Africa. IUCN, Gland, Switzerland.

Geary, D.H. 1990. Patterns of evolutionary tempo and mode in the radiation of *Melanopsis* (Gastropoda: Melanopsidae). Paleobiology 16: 492-511.

Geary, D.H. 1992. An unusual pattern of divergence between two fossil gastropods: Ecophenotypy, dimorphism, or hybridization. Paleobiology 18: 93-109.

Geary, D.H., A.W. Staley, P. Müller, and I. Magyar. 2002. Iterative changes in Lake Pannon *Melanopsis* reflect a recurrent theme in gastropod morphological evolution. Paleobiology 28: 208-221.

Geary, D.H., E. Hoffmann, I. Magyar, J. Freiheit, and

D. Padilla. 2012. Body size, longevity, and growth rate in lake Pannon melanopsid gastropods and their predecessors. Paleobiology 38: 554-568.

Glaubrecht, M. 1993. Mapping the diversity: Geographical distribution of the freshwater snail *Melanopsis* (Gastropoda: ?Cerithioidea: Melanopsidae) with focus on its systematics in the Mediterranean Basin. Mitteilungen aus dem Hamburgischen Zoologischen Museum und Institute 90: 41-97.

Glaubrecht, M. 1996. Evolutionsökologie und Systematik am Beispiel von Süß- und Brackwasserschnecken (Mollusca: Caenogastropoda: Cerithioidea): Ontogenese-Strategien, paläontologische Befunde und Historische Zoogeographie. Backhuys, Leiden.

Glaubrecht, M. 2011. Towards solving Darwin's 'mystery': Speciation and radiation in lacustrine and riverine freshwater gastropods. American Malacological Bulletin 29: 187-216.

Gold, D., Y. Lang, and J. Lengy. 1993. *Philophthalmus* species, probably *P. palpebrarum,* in Israel: Description of the eye fluke from experimental infection. Parasitology Research 79: 372-377.

Heller, J., and N. Sivan. 2000. A new species of *Melanopsis* from the Golan Heights, Southern Levant (Gastropoda: Melanopsidae). Journal of Conchology 37: 1-5.

Heller, J., and N. Sivan. 2001. *Melanopsis* from the Mid-Pleistocene site of Gesher Benot Ya'aqov (Gastropoda: Cerithioidea). Journal of Conchology 37: 127-147.

Heller, J., and N. Sivan. 2002a. *Melanopsis* from the Pleistocene site of 'Ubeidiya, Jordan Valley: Direct evidence of early hybridization (Gastropoda: Cerithioidea). Biological Journal of the Linnean Society 75: 39-57.

Heller, J., and N. Sivan. 2002b. *Melanopsis* from the Pliocene site of 'Erq el-Ahmar, Jordan Valley (Gastropoda: Cerithioidea). Journal of Conchology 37: 607-625.

Heller, J., N. Sivan, and U. Motro. 1999. Systematics, distribution and hybridization of *Melanopsis* from the Jordan Valley (Gastropoda: Prosobranchia). Journal of Conchology 36: 49-81.

Heller, J., N. Sivan, and F. Ben-Ami. 2002. Systematics of *Melanopsis* from the coastal plain of Israel (Gastropoda: Cerithioidea). Journal of Conchology 37: 589-606.

Heller, J., P. Mordan, F. Ben-Ami, and N. Sivan. 2005. Conchometrics, systematics and distribution of *Melanopsis* (Mollusca: Gastropoda) in the Levant. Zoological Journal of the Linnean Society 144: 229-260.

Houbrick, R.S. 1987. Anatomy, reproductive biology, and phylogeny of the Planaxidae (Cerithiacea: Prosobranchia). Smithsonian Contributions to Zoology 445: 1-57.

Houbrick, R.S. 1988. Cerithioidean phylogeny. In Prosobranch Phylogeny: Proceedings of a Symposium Held at the 9th International Malacological Congress, Edinburgh, 1986 (W.F. Ponder, ed.). Malacological Review Supplement 4: 88-128.

Ismail, N.S. 1989. Two new furcocercariae from *Melanopsis praemorsa* in Jordan. Helminthologia 261: 15-20.

Jaeckel, S.G., W. Klemm, and W. Meise. 1957. Die Land- und Süßwasser-Mollusken der nördlichen Balkanhalbinsel. Abhandlungen und Berichte des Staatlichen Museums für Tierkunde Dresden 23: 141-205.

Lengy, J., and A. Stark. 1971. Studies on larval stages of digenetic trematodes in aquatic molluscs of Israel 2. On three cercariae encountered in the freshwater snail *Melanopsis praemorsa* L. Israel Journal of Zoology 20: 41-51.

Lydeard, C., W.E. Holznagel, M. Glaubrecht, and W.F. Ponder. 2002. Molecular phylogeny of a circum-global, diverse gastropod superfamily (Cerithioidea: Mollusca: Caenogastropoda): Pushing the deepest phylogenetic limits of mitochondrial LSU rDNA sequences. Molecular Phylogenetics and Evolution 22: 399-406.

Manafov, A.A. 2011. New virgulid cercaria (Trematoda, Lecithodendroidea) from the mollusk *Melanopsis praemorsa* (Melanopsidae) from Azerbaijan water bodies: Morphology and chaetotaxy of *Cercaria agstaphensis* 11. Vestnik Zoologii 45: 105-111.

Manafov, A.A. 2012. Cercaria of trematodes in freshwater mollusc *Melanopsis praemorsa* from Azerbaijan. 1. Morphology and chaetotaxy of three new species of stylet cercaria (Trematoda, Plagiorchiida, Lecithodendroidea). Zoologicheskiĭ Zhurnal 91: 1443-1456.

Manafov, A.A. 2013. Cercaria of trematodes of freshwater mollusc *Melanopsis praemorsa* from Azerbaijan. 2. Morphology and chaetotaxy of two new virgulid cercaria (Plagiorchiida, Lecithodendroidea). Zoologicheskiĭ Zhurnal 92: 389-398.

Moore, J.E.S. 1899. On the divergent forms at present incorporated in the family Melaniidae. Proceedings of the Malacological Society, London, 3: 65-79.

Morrison, J.P.E. 1954. The relationships of Old and New World melanians. Proceedings of the United States National Museum 103: 357-394.

Neubauer, T.A. 2016. A nomenclator of extant and fossil taxa of the Melanopsidae (Gastropoda, Cerithioidea). ZooKeys 602: 1-358.

Neubauer, T.A., M. Harzhauser, and A. Kroh. 2013. Phenotypic evolution in a fossil gastropod species lineage: Evidence for adaptive radiation? Palaeogeography, Palaeoclimatology, Palaeoecology 370: 117-126.

Neubauer, T.A., O. Mandic, and M. Harzhauser. 2014a. A new melanopsid species from the Middle Miocene Kupres Basin (Bosnia and Herzegovina). Nautilus 128: 51-54.

Neubauer, T.A., M. Harzhauser, E. Georgopoulou, and C. Wrozyna. 2014b. Population bottleneck triggering millennial-scale morphospace shifts in endemic thermal-spring melanopsids. Palaeogeography, Palaeoclimatology, Palaeoecology 414: 116-128.

Neubauer, T.A., M. Harzhauser, O. Mandic, E. Georgopoulou, and A. Kroh. 2016. Paleobiogeography and historical biogeography of the non-marine caenogastropod family Melanopsidae. Palaeogeography, Palaeoclimatology, Palaeoecology 444: 124-143.

Neubert, E. 1998. Annotated checklist of the terrestrial and freshwater molluscs of the Arabian Peninsula with descriptions of new species. Fauna of Arabia 17: 333-461.

Odabaşi, D.A., and N. Arslan. 2015. Description of a new subterranean nerite: *Theodoxus gloeri* n. sp. with some data on the freshwater gastropod fauna of Balikdami Wetland (Sakarya River, Turkey). Ecologica Montenegrina 2: 327-333.

Olenev, A.V. 1987. Cercarial fauna in the freshwater mollusk, *Melanopsis praemorsa* in western Georgia Part 1. Ekologi Eksperiment Parazitology 1: 73-96.

Paget, O. 1976. Die Molluskenausbeute der Insel Rhodos. Annalen des Naturhistorischen Museums in Wien 80: 681-780.

Pallary, P. 1924. Révision des *Melanopsis* de l'Espagne. Bulletin de la Société d'histoire naturelle d'Afrique du Nord 15: 240-255.

Pallary, P. 1926. Répertoire des *Melanopsis* fossiles et vivantes connus en 1925. Bulletin de la Société d'histoire naturelle d'Afrique du Nord 17: 126-136.

Pallary, P. 1939. Deuxième addition à la faune malacologique de la Syrie. Mémoires de l'Institut d'Egypte 39: 1-141.

Papp, A. 1955. Brack- u. Süsswasserfaunen Griechenlands. V. Bemerkungen über Melanopsiden der Untergattung *Melanosteira* Oppenheim, 1891. Annales Géologiques des Pays Helléniques 6: 122-132.

Pujante, A., and A. Gallardo. 1990. Distribución del género *Melanopsis* Férussac, 1807 en algunos ríos de Andalucia Occidental (España). Iberus 9: 439-447.

Schütt, H. 1965. Zur Systematik und Ökologie türkischer Süßwasserprosobranchier. Zoologische Mededelingen, Leiden 41: 43-72.

Schütt, H., and F.H. Bilgin. 1974. Recent Melanopsines of the Aegean. Archiv für Molluskenkunde 104: 59-64.

Simroth, H. 1896-1907. Dr. H. G. Bronn's Klassen und Ordnungen des Tier-Reichs, wissenschaftlich dargestellt in Wort und Bild. 3. Band. Mollusca. II. Abteilung: Gastropoda Prosobranchia, Winter, Leipzig.

Smoleń, M., and A. Falniowski. 2009. Molecular phylogeny and estimated time of divergence in the Central European Melanopsidae: *Melanopsis, Fagotia* and *Holandriana*

(Mollusca: Gastropoda: Cerithioidea). Folia Malacologica 17: 1-9.

Starmühlner, F. 1991. Die Gasteropoden der Berg-Fliessgewässer isolierter kontinentaler und ozeanischer Inseln des Indopazifik und der Karibik. Proceedings of the 10th International Malacological Congress, Tübingen, 1989: 403-416.

Starmühlner, F., and A. Edlauer. 1957. Ergebnisse der Österreichischen Iran Expedition 1949/50 (Mit Berücksichtigung der Ausbeute der Österreichischen Iran-Expedition 1956). Osterreichische Akademie der Wissenschaften Mathematisch-naturwissenschaftlichen Klasse Sitzungsberichte (I), 166: 435-494.

Strong, E.E., O. Gargominy, W.F. Ponder, and P. Bouchet. 2008. Global diversity of gastropods (Gastropoda: Mollusca) in freshwater. Hydrobiologia 595: 149-166.

Strong, E.E., D.J. Colgan, J.M. Healy, C. Lydeard, W.F. Ponder, and M. Glaubrecht. 2011. Phylogeny of the gastropod superfamily Cerithioidea using morphology and molecules. Zoological Journal of the Linnean Society 162: 43-89.

Sturany, R., and A.J. Wagner. 1914. Über schalentragende Landmollusken aus Albanien und Nachbargebieten. Denkschriften der Kaiserlichen Akademie der Wissenschaften. Mathematisch-Naturwissenschaftliche Classe 91: 19-138.

Sunderbrinck, O. 1929. Zur Frage der Verwandschaft zwischen Melaniiden und Cerithiiden. Zeitschrift für Morphologie und Ökologie der Tiere 14: 261-337.

Thiele, J. 1928. Revision des Systems der Hydrobiiden und Melaniiden. Zoologisches Jahrbücher, Abteilung für Systematik, Ökologie und Geographie der Tiere 55: 351-402.

Thiele, J. 1929-1931. Handbuch der Systematischen Wiechtierkunde, vol. 1. Gustav Fischer, Jena, Germany.

Ullman, H. 1954. Observations on a new cercaria developing in *Melanopsis praemorsa* in Israel. Parasitology 44: 1-15.

Wenz, W. 1938-1944. Gastropoda. Teil 1. Allgemeiner Teil und Prosobranchia. Pp. i-viii, 1-1639 in Handbuch der Paläozoologie, vol 6, part 1 (O.H. Schindewolf, ed.). Gebrüder Borntraeger, Berlin.

Willmann, R., and H. Pieper. 1978. Gastropoda. Pp. 118-134 in Limnofauna Europaea (J. Illies, ed.), Fischer, Stuttgart, Germany.

8 Pachychilidae Fischer and Crosse, 1892

MARCO T. NEIBER AND MATTHIAS GLAUBRECHT

As a result of detailed morphological and/or molecular investigations on the family, genus, and species levels, Pachychilidae Fischer and Crosse, 1892, are currently widely recognized as an independent family of mostly freshwater inhabiting species within the diverse and primarily marine gastropod superfamily Cerithioidea Férussac, 1819 (Glaubrecht, 1996, 1999; Simone, 2001; Köhler and Glaubrecht, 2001, 2002a, 2003, 2007, 2010; Lydeard et al., 2002; Glaubrecht and Rintelen, 2003, 2008; Glaubrecht and Köhler, 2004; Köhler et al., 2004; Rintelen et al., 2004; Rintelen and Glaubrecht, 2005; Gómez, 2009; Köhler and Dames, 2009; Köhler and Deein, 2010; Rintelen et al., 2010, 2014; Strong, 2011; Strong et al., 2011).

More traditional views on the systematics of the group have placed genera that are now united in the family either in a large family "Melaniidae" = Thiaridae Gill, 1871 (1823) (e.g., Thiele, 1928, 1929; Rensch, 1934; Morrison, 1954; Benthem-Jutting, 1956; Solem, 1966; Brandt, 1968, 1974; Dudgeon, 1982, 1989; Houbrick, 1988), which is now known to represent a polyphyletic assemblage (Lydeard et al., 2002; Strong et al., 2011), or on the basis of distribution either in the Pleuroceridae (e.g., Ponder and Warén, 1988; Vaught, 1989) or a less broadly defined Thiaridae.

Pachychilidae as here delimited have a truly circumtropical distribution with the freshwater inhabiting *Pachychilus* Lea & Lea, 1851 (ca. 60 species), and *Doryssa* Swainson, 1840 (ca. 40 species), from South and Central America (Simone, 2006; Gómez, 2009; Thompson, 2011; Gomez-Berning et al., 2012; Pointier and Noya Alarcón, 2015), *Potadoma* Swainson, 1840 (19 species), from tropical western Africa and the Congo River drainage system (Brown, 1994), and *Madagasikara* Köhler & Glaubrecht, 2010 (five species), from Madagascar (Köhler and Glaubrecht, 2010). Asian taxa include *Paracrostoma* Cossmann, 1900 (three species), from India (Köhler and Glaubrecht, 2007), *Sulcospira* Troschel, 1858 (ca. 13 species), and *Brotia* Adams, 1866 (ca. 35 species), from Southeast Asia (Köhler and Glaubrecht, 2001, 2002a, b, 2005, 2006, 2007; Köhler, 2003; Glaubrecht and Köhler, 2004; Köhler et al., 2008; Köhler and Dames, 2009; Köhler and Deein, 2010), *Tylomelania* Sarasin &

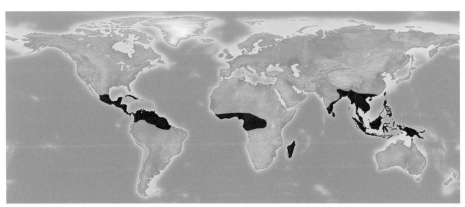

Distribution of Pachychilidae.

Pachychilus immanis (Morelet, 1851). Guatemala: Lake Izabal, Izabal Department. 75 mm. INHS 56018.

Sarasin, 1897 (44 described plus ca. 32 undescribed species), from the Indonesian island Sulawesi (Rintelen and Glaubrecht, 1999, 2003, 2005, 2008; Rintelen, 2003; Rintelen et al., 2004, 2007, 2010, 2012, 2014; Glaubrecht and Rintelen, 2008; Glaubrecht, 2011), *Jagora* Köhler & Glaubrecht, 2003 (two species), from the Philippines (Köhler and Glaubrecht, 2003), and *Pseudopotamis* Martens, 1894 (two species), which is restricted to the Torres Strait Islands between the Cape York Peninsula in the northeastern-most part of Australia and New Guinea (Glaubrecht and Rintelen, 2003). Interestingly, the only taxon inhabiting brackish environments, viz. *Faunus* Montfort, 1810 (one species), is restricted to coastal regions from eastern India to New Guinea, the region of the highest generic diversity of the family. *Faunus* was first recognized to be closely related to pachychilids by Strong and Glaubrecht (2000) on the basis of morphology, a result which has been corroborated by subsequent molecular analyses that resolved *F. ater* as sister group of all other confamiliar taxa (Lydeard et al., 2002; Köhler et al., 2004; Strong et al., 2011).

Morphologically, Pachychilidae can be characterized by features of the midgut anatomy, i.e., a pointed, crescent-shaped sorting area and the appearance of crescentic pads bordering the sorting area at the left, and which are separated from one another by a broad, shallow trough (Strong and Glaubrecht, 2000; Strong, 2011). It is an exceptionally diverse group with respect to realized reproductive strategies. Oviparity is the symplesiomorphic trait

among Pachychilidae, which is found in the Asian *Faunus* (cf. Houbrick, 1991), the Neotropical genera *Pachychilus* and *Doryssa* (cf. Simone, 2001), the African *Potadoma* (cf. Glaubrecht, 1996), and most species in the Malagasy *Madagasikara* (Köhler and Glaubrecht, 2010). Among Asian Pachychilidae different forms of parental care involving the retention of eggs and development of offspring in incubatory structures evolved (Köhler and Glaubrecht, 2010). The Asian mainland taxa *Brotia, Sulcospira,* and *Paracrostoma,* which were recovered as a monophyletic group (Köhler and Glaubrecht, 2007, 2010), possess a sub-haemocoelic brood pouch (Köhler and Glaubrecht, 2001, 2002a, b, 2005, 2006), whereas the adelphotaxa *Tylomelania* from Sulawesi and *Pseudopotamis* from the Torres Strait Islands brood in a uterine pouch (Glaubrecht and Rintelen, 2003; Rintelen and Glaubrecht, 2003, 2005) and the Philippine *Jagora* species brood in the mantle cavity (Köhler and Glaubrecht, 2003). Similarly to *Jagora, Madagasikara vivipara* Köhler & Glaubrecht, 2003, which was placed in a derived position within the otherwise oviparous *Madagasikara,* also broods in the mantle cavity (Köhler and Glaubrecht, 2010).

All endemic pachychilids from the island of Sulawesi are now suggested to be classified in *Tylomelania,* with currently 23 fluviatile and 53 lacustrine species including unnamed taxa (see Glaubrecht and Rintelen, 2008; Glaubrecht, 2011). The species flocks from the ancient lakes on that island, i.e., the Malili lake system and the isolated Lake Poso, offer an instructive model case for speciation mechanisms and truly adaptive radiations testing allopatry, escalation (coevolution with crabs), ecology (e.g., habitat preference), and trophic specialization (exhibited by radula morphology) to soft or hard substrates, as documented and discussed in Rintelen et al. (2004, 2007, 2010, 2014), Rintelen and Glaubrecht (2005), and Glaubrecht and Rintelen (2008). Results suggest four independent colonizations of the lakes, of which three involve species in the Malili lakes plus one in Lake Poso, with riverine taxa found as sister groups to three lacustrine clades (Rintelen et al., 2004, 2010). Aside from ecological and geographic factors as driving forces for diversification, there is also some evidence for the involvement of introgressive hybridization in speciation processes for many *Tylomelania* species (Glaubrecht and Rintelen, 2008; Glaubrecht, 2011) that need to be further investigated.

Riverine radiations have not attracted the same attention as lacustrine radiations, and details of the phylogeny, phylogeography, and evolution of these radiations is largely unknown (Glaubrecht, 2011). For species belonging to *Brotia,* Davis (1982) noted that usually only one or at most two species co-occur in rivers in Southeast Asia. A notable exception to this pattern is a small radiation comprising, according to our most recent understanding, seven morphologically distinct species in the Kaek River drainage system in north central Thailand that have been shown to form a monophyletic group originating from a Mekong River ancestor and that belong to a larger clade of *Brotia* species from Thailand (Glaubrecht and Köhler, 2004; Köhler and Glaubrecht, 2006; Köhler et al., 2010). Rather low levels of genetic variation and the genetic "nonmonophyly" of most recognized species suggest a relatively recent origin of this intrariverine radiation and rapid morphological divergence possibly driven by adaptation to different habitats and trophic specialization (Glaubrecht and Köhler, 2004; Köhler et al., 2010; Glaubrecht, 2011).

The circumtropical distribution of the family has led to the hypothesis of a Gondwanan origin of pachychilids (Glaubrecht, 1996, 2000). In the context of Gondwanan vicariance, the biotic ferry model (Briggs, 2003) became the standard explanation for the evolution of distributional and genealogical patterns of groups with recent distributions in Africa, Madagascar, and India. As an extension of this model, it has been suggested that Gondwanan elements have colonized Southeast Asia out of India (McKenna, 1973). However, the "out of India" scenario could not be corroborated for the Indian endemic *Paracrostoma,* which has been found to be nested within a clade including Southeast Asian *Brotia* and *Sulcospira* species by Köhler and Glaubrecht (2007). Rather, an "out of Asia and into India" scenario is supported, hinting at a more complex origin of Indian biota than the standard Mesozoic vicariance model suggests. Reviewing the complex palaeogeographical data available, vicariance events were discussed as most likely for the largely disjunct adelphotaxon *Tylomelania* on Sulawesi and *Pseudopotamis* on the Torres Strait Islands by Glaubrecht and Rintelen (2003). In

contrast, the tree topology resulting from the phylogenetic analysis and the timing of events obtained via a molecular clock approach in Köhler and Glaubrecht (2010) for the whole family is not easily reconcilable with a simple Mesozoic vicariance scenario within the framework of Gondwanan fragmentation, nor with any other subsequent palaeogeographical events such as, e.g., the break-off of Madagascar from continental Africa. Alternatively, Köhler and Glaubrecht (2010) discussed transoceanic dispersal playing a role in shaping current distribution patterns, albeit in the striking absence of factual evidence for a free-swimming veliger larvae in all limnic pachychilids.

A total of 74 species assigned to the Pachychilidae are currently included in the IUCN Red List, 4 as Critically Endangered, 10 as Endangered, 10 as Vulnerable, 3 as Near Threatened, 29 as Least Concern, and 18 as Data Deficient (IUCN, 2017). However, assessment coverage is very heterogeneous; i.e., while *Pseudopotamis* and *Jagora* have not been assessed at all, only one of the 44 described species of *Tylomelania,* one of the ca. 60 species and subspecies of *Pachychilus,* and one of the ca. 40 species of *Doryssa* have been assessed so far. Only for *Faunus, Brotia, Paracrostoma, Sulcospira,* and *Potadoma* assessment coverage is complete or nearly complete. Taking *Brotia* (about one-third threatened species) and *Potadoma* (about half of the species are threatened) as representative cases for pachychilids, respectively, about 40% of the total diversity within the family or even more may currently be threatened, thus suggesting a high need of conservation measures for many confamiliar taxa.

Like other freshwater cerithioideans, Pachychilidae are intermediate hosts of trematode cercariae. In Southeast Asia species belonging to *Brotia* and *Jagora* have been shown to host the human pathogen lung fluke *Paragonimus westermanni* (Kerbert, 1878) (Yokogawa, 1965; Attwood, 2010). Dechruska et al. (2007) have found two pachychilid species to be susceptible to trematode infections in the Kaek River in Thailand. However, infection rates were very low when compared to those observed in species belonging to the Thiaridae, especially *Melanoides tuberculata* (Müller, 1774) and *Tarebia granifera* (Lamarck, 1822). Species in *Potadoma* have been suspected to be intermediate hosts of the African lung fluke *Paragonimus africanus* Voelker & Vogel, 1965 (Brown, 1994), and vectors of schistosomiasis (Agbolade et al., 2004) in Africa. Relatively little is known on parasites of South American pachychilids; however, Nasir et al. (1966) report cercarial infections in a species belonging to *Pachychilus*.

LITERATURE CITED

Agbolade, O.M., D.O. Akinboye, O.T. Fajebe, O.M. Abolade, and A.A. Adebambo. 2004. Human urinary schistosomiasis transmission foci and period in an endemic town of Ijebu North, Southwest Nigeria. Tropical Biomedicine 21: 15-22.

Attwood, S.W. 2010. Studies on the parasitology, phylogeography and the evolution of host-parasite interactions for snail intermediate hosts of medically important trematode genera in Southeast Asia. Pp. 405-440 in Important Helminth Infections in Southeast Asia: Diversity and Potential for Control and Elimination, Part B (X.-N. Zhou, R. Bergquist, R. Olveda, and J. Utzinger, eds.). Advances in Parasitology 73. Elsevier, London.

Benthem-Jutting, W.S.S. van. 1956. Systematic studies on the non-marine Mollusca of the Indo-Australian archipelago. 5. Critical revision of the Javanese freshwater gastropods. Treubia 23: 259-477.

Brandt, R.A.M. 1968. Description of new non-marine mollusks from Asia. Archiv für Molluskenkunde 98: 213-289.

Brandt, R.A.M. 1974. The non-marine aquatic Mollusca of Thailand. Archiv für Molluskenkunde 105: i-iv, 1-423.

Briggs, J.C. 2003. Fishes and birds: Gondwana life rafts reconsidered. Systematic Biology 52: 548-553.

Brown, D.S. 1994. Freshwater Snails of Africa and Their Medical Importance. 2nd ed. Taylor & Francis, London.

Davis, G.M. 1982. Historical and ecological factors in the evolution, adaptive radiation, and biogeography of freshwater mollusks. American Zoologist 22: 375-395.

Dechruska, W., D. Krailas, S. Ukong, W. Inkapatanakul, and T. Koonchornboon. 2007. Trematode infections of the freshwater snail family Thiaridae in the Khek River, Thailand. Southeast Asian Journal of Tropical Medicine and Public Health 38: 1016-1028.

Dudgeon, D. 1982. The life history of *Brotia hainanensis* (Brot 1872) (Gastropoda: Prosobranchia: Thiaridae) in a tropical forest stream. Zoological Journal of the Linnean Society 76: 141-154.

Dudgeon, D. 1989. Ecological strategies of Hong Kong Thiaridae (Gastropoda: Prosobranchia). Malacological Review 22: 39-53.

Glaubrecht, M. 1996. Evolutionsökologie und Systematik am Beispiel von Süß- und Brackwasserschnecken (Mollusca: Caenogastropoda: Cerithioidea): Ontogenese-Strategien, Paläontologische Befunde und Zoogeographie. Backhuys, Leiden.

Glaubrecht, M. 1999. Systematics and evolution of viviparity in tropical freshwater gastropods (Cerithioidea: Thiaridae sensu lato)—an overview. Courier Forschungsinstitut Senckenberg 125: 91-96.

Glaubrecht, M. 2000. A look back in time: Towards a historical biogeography as a synthesis of systematic and geological patterns outlined with freshwater gastropods. Zoology 102: 127-147.

Glaubrecht, M. 2006. Independent evolution of reproductive modes in viviparous freshwater Cerithioidea (Gastropoda, Sorbeoconcha)—a brief review. Basteria 69, Supplement 3: 23-28.

Glaubrecht, M. 2011. Towards solving Darwin's 'mystery': Speciation and radiation in lacustrine and riverine freshwater gastropods. American Malacological Bulletin 29: 187-216.

Glaubrecht, M., and F. Köhler. 2004. Radiating in a river: Systematics, molecular genetics and morphological differentiation of viviparous freshwater gastropods endemic to the Kaek River, central Thailand (Cerithioidea, Pachychilidae). Biological Journal of the Linnean Society 82: 275-311.

Glaubrecht, M., and T. von Rintelen. 2003. Systematics, molecular genetics and historical zoogeography of the viviparous freshwater gastropod *Pseudopotamis* Martens, 1894 (Cerithioidea: Pachychilidae): A relic on the Torres Strait Islands, Australia. Zoologica Scripta 32: 415-435.

Glaubrecht, M., and T. von Rintelen. 2008. The species flocks of lacustrine gastropods: *Tylomelania* on Sulawesi as models in speciation and adaptive radiation. Hydrobiologia 615: 181-199.

Gómez, M.I. 2009. Systematics, phylogeny and biogeography of Mesoamerican and Caribbean freshwater gastropods (Cerithioidea: Thiaridae and Pachychilidae). PhD diss., Humboldt University, Berlin.

Gomez-Berning, M., F. Köhler, and M. Glaubrecht. 2012. Catalogue of the nominal taxa of Mesoamerican Pachychilidae (Mollusca: Caenogastropoda). Zootaxa 3381: 1-44.

Houbrick, R.S. 1988. Cerithiodean phylogeny. In Prosobranch Phylogeny: Proceedings of a Symposium Held at the 9th International Malacological Congress, Edinburgh, 1986 (W.F. Ponder, ed.). Malacological Review Supplement 4: 88-128.

Houbrick, R.S. 1991. Anatomy and systematic placement of *Faunus* Montfort, 1810 (Prosobranchia: Melanopsidae). Malacological Review 24: 35-54.

IUCN Red List of Threatened Species. 2017. Version 2017-1. www.iucnredlist.org. Accessed 16 June 2017.

Köhler, F. 2003. *Brotia* in space and time: Phylogeny and

evolution of Southeast Asian freshwater gastropods of the family Pachychilidae (Caenogastropoda, Cerithioidea). PhD diss., Humboldt University, Berlin.

Köhler, F., and C. Dames. 2009. Phylogeny and systematics of Pachychilidae of mainland South-East Asia—novel insights from morphology and mitochondrial DNA (Mollusca, Caenogastropoda, Cerithioidea). Zoological Journal of the Linnean Society 157: 697-699.

Köhler, F., and G. Deein. 2010. Hybridization as potential source of incongruence in the morphological and mitochondrial diversity of a Thai freshwater gastropod (Pachychilidae, *Brotia* H. Adams, 1866). Zoosystematics and Evolution 86: 301-314.

Köhler, F., and M. Glaubrecht. 2001. Toward a systematic revision of the Southeast Asian freshwater gastropod *Brotia* H. Adams, 1866 (Cerithioidea: Pachychilidae): An account of species from around the South China Sea. Journal of Molluscan Studies 67: 281-318.

Köhler, F., and M. Glaubrecht. 2002a. Annotated catalogue of the nominal taxa of Southeast Asian freshwater gastropods, family Pachychilidae Troschel, 1857 (Mollusca: Caenogastropoda: Cerithioidea), with an evaluation of the types. Mitteilungen aus dem Zoologischen Museum Berlin, Zoologische Reihe, 78: 121-156.

Köhler, F., and M. Glaubrecht. 2002b. A new species of *Brotia* H. Adams, 1866 (Caenogastropoda, Cerithioidea, Pachychilidae). Journal of Molluscan Studies 68: 353-357.

Köhler, F., and M. Glaubrecht. 2003. Morphology, reproductive biology and molecular genetics of ovoviviparous freshwater gastropods (Cerithioidea: Pachychilidae) from the Philippines, with description of a new genus *Jagora*. Zoologica Scripta 32: 35-59.

Köhler, F., and M. Glaubrecht. 2005. Fallen into oblivion—the systematic affinities of the enigmatic *Sulcospira* Troschel, 1857 (Cerithioidea: Pachychilidae), a genus of viviparous freshwater gastropods from Java. Nautilus 11: 15-27.

Köhler, F., and M. Glaubrecht. 2006. A systematic revision of the Southeast Asian freshwater gastropod *Brotia* (Cerithioidea: Pachychilidae). Malacologia 48: 159-251.

Köhler, F., and M. Glaubrecht. 2007. Out of Asia and into India: On the molecular phylogeny and biogeography of the endemic freshwater gastropod *Paracrostoma* Cossmann, 1900 (Caenogastropoda: Pachychilidae). Biological Journal of the Linnean Society 91: 627-651.

Köhler, F., and M. Glaubrecht. 2010. Uncovering an overlooked radiation: Molecular phylogeny and biogeography of Madagascar's endemic river snails (Caenogastropoda: Pachychilidae: *Madagasikara* gen. nov.). Biological Journal of the Linnean Society 99: 867-894.

Köhler, F., T. von Rintelen, A. Meyer, and M. Glaubrecht. 2004. Multiple origin of viviparity in Southeast Asian gas-

tropods (Cerithioidea: Pachychilidae) and its evolutionary implications. Evolution 58: 2215-2226.

Köhler, F., N. Brinkmann, and M. Glaubrecht. 2008. Convergence caused confusion: On the systematics of the freshwater gastropod *Sulcospira pisum* (Brot, 1868) (Cerithioidea, Pachychilidae). Malacologia 50: 331-339.

Köhler, F., S. Panha, and M. Glaubrecht. 2010. Speciation and radiation in a river: Assessing the morphological and genetic differentiation in a species flock of viviparous gastropods (Cerithioidea: Pachychilidae). Pp. 513-550 in Evolution in Action: Case Studies in Adaptive Radiation, Speciation and the Origin of Biodiversity (M. Glaubrecht, ed.). Springer, Berlin.

Lydeard, C., W.E. Holznagel, M. Glaubrecht, and W.F. Ponder. 2002. Molecular phylogeny of a circum-global, diverse gastropod superfamily (Cerithioidea: Mollusca: Caenogastropoda): Pushing the deepest phylogenetic limits of mitochondrial LSU rDNA sequences. Molecular Phylogenetics and Evolution 22: 399-406.

McKenna, M.C.C. 1973. Sweepstakes, filters, corridors, Noah's arks, and beached Viking funerals ships in palaeogeography. Pp. 291-304 in Implications of Continental Drift to the Earth Sciences (D.H. Tarling and S.K. Runcorn, eds.). Academic Press, London.

Morrison, J.P.E. 1954. The relationships of Old and New World melanians. Proceedings of the United States National Museum 103: 357-394.

Nasir, P., C.A. Acuna, and C.S. Guevarra. 1966. Studies on freshwater larval trematodes. Part XII. Two new species of cercariae infecting *Pachychilus laevissimus* Sowerby in Venezuela. Zoologischer Anzeiger 177: 133-138.

Pointier, J.-P., and O. Noya Alarcón. 2015. Family Pachychilidae. Pp. 76-77 in Freshwater Molluscs of Venezuela and their Medical and Veterinary Importance (J.-P. Pointier, ed.). ConchBooks, Harxheim, Germany.

Ponder, W.F., and A. Warén. 1988. Classification of the Caenogastropoda and Heterostropha—a list of the family-group names and higher taxa. In Prosobranch phylogeny: Proceedings of a Symposium Held at the 9th International Malacological Congress, Edinburgh, 1986 (W.F. Ponder, ed.). Malacological Review Supplement 4: 288-328.

Rensch, B. 1934. Süßwasser-Mollusken der deutschen limnologischen Sunda-Expedition. Archiv für Hydrobiologie Supplement 13: 3-53.

Rintelen, T. von. 2003. Phylogenetic analysis and systematic revision of a species flock of viviparous freshwater gastropods in the ancient lakes on Sulawesi (Indonesia)—a model case of adaptive radiation? PhD diss., Humboldt University, Berlin.

Rintelen, T. von, and M. Glaubrecht. 1999. On the reproductive anatomy of freshwater gastropods of the genera *Brotia* H. Adams, 1866 and *Tylomelania* Sarasin & Sarasin, 1897 in the central lakes on Sulawesi, Indonesia (Cerithioidea: Melanatriidae). Courier Forschungsinstitut Senckenberg 125: 163-170.

Rintelen, T. von, and M. Glaubrecht. 2003. New discoveries in old lakes: Three new species of *Tylomelania* Sarasin & Sarasin, 1897 from the Malili lake system on Sulawesi, Indonesia. Journal of Molluscan Studies 69: 3-18.

Rintelen, T. von, and M. Glaubrecht. 2005. The anatomy of an adaptive radiation: A unique reproductive strategy in the endemic freshwater gastropod *Tylomelania* (Cerithioidea: Pachychilidae) on Sulawesi, Indonesia, and its biogeographic implications. Biological Journal of the Linnean Society 86: 513-542.

Rintelen, T. von, and M. Glaubrecht. 2008. Three new species of the freshwater snail genus *Tylomelania* (Caenogastropoda: Cerithioidea) from the Malili lake system, Sulawesi, Indonesia. Zootaxa 1852: 37-49.

Rintelen, T. von, A.B. Wilson, A. Meyer, and M. Glaubrecht. 2004. Escalation and trophic specialization drive adaptive radiation of freshwater gastropods in ancient lakes on Sulawesi, Indonesia. Proceedings of the Royal Society London B 271: 2541-2549.

Rintelen, T. von, P. Bouchet, and M. Glaubrecht. 2007. Ancient lakes as hotspots of diversity: A morphological review of an endemic species flock of *Tylomelania* (Gastropoda: Cerithioidea: Pachychilidae) in the Malili lake system on Sulawesi, Indonesia. Hydrobiologia 592: 11-94.

Rintelen, T. von, K. von Rintelen, and M. Glaubrecht. 2010. The species flocks of the viviparous freshwater gastropod *Tylomelania* (Mollusca: Cerithioidea: Pachychilidae) in the ancient lakes of Sulawesi, Indonesia: The role of geography, trophic morphology and color as driving forces in adaptive radiation. Pp. 485-512 in Evolution in Action: Case Studies in Adaptive Radiation, Speciation and the Origin of Biodiversity (M. Glaubrecht, ed.). Springer, Berlin.

Rintelen, T. von, K. von Rintelen, M. Glaubrecht, C.D. Schubart, and F. Herder. 2012. Aquatic biodiversity hotspots in Wallacea: The species flocks in the ancient lakes of Sulawesi, Indonesia. Pp. 290-315 in Biotic Evolution and Environmental Change in Southeast Asia (D.J. Gower, K.G. Johnson, J.E. Richardson, B.R. Rosen, L. Rüber, and S.T. Williams, eds.). Cambridge University Press, New York.

Rintelen, T. von, B. Stelbrink, R.M. Marwoto, and M. Glaubrecht. 2014. A snail perspective on the biogeography of Sulawesi, Indonesia: Origin and intra-island dispersal of the viviparous freshwater gastropod *Tylomelania*. PLoS One 9: e98917.

Simone, L.R.L. 2001. Phylogenetic analyses of Cerithioidea (Mollusca, Caenogastropoda) based on comparative morphology. Arquivos de Zoologia 36: 147-263.

Simone, L.R.L. 2006. Land and Freshwater Molluscs of Brazil: An Illustrated Inventory of the Brazilian Malacofauna, Including Neighboring Regions of South America, Respect to the Terrestrial and Freshwater Ecosystems. EGB, São Paulo.

Solem, A. 1966. Some non-marine mollusks from Thailand, with notes on classification of the Helicarionidae. Spolia zoologica Musei Hauniensis 24: 1-110, plates 1-3.

Strong, E.E. 2011. More than a gut feeling: Utility of midgut anatomy in phylogeny of the Cerithioidea (Mollusca: Caenogastropoda). Zoological Journal of the Linnean Society 162: 585-630.

Strong, E.E., and M. Glaubrecht. 2000. On the systematics of the limnic Pachychilidae: New evidence for the placement of the enigmatic *Faunus* (Caenogastropoda: Cerithioidea). In Abstracts of the 66th American Malacological Society Congress, 7-12 July, San Francisco.

Strong, E.E., D.J. Colgan, J.M. Healy, C. Lydeard, W.F. Ponder, and M. Glaubrecht. 2011. Phylogeny of the gastropod superfamily Cerithioidea using morphology and molecules. Zoological Journal of the Linnean Society 162: 43-89.

Thiele, J. 1928. Revision des Systems der Hydrobiiden und Melaniiden. Zoologische Jahrbücher, Abteilung für Systematik, Ökologie und Geographie der Tiere 55: 351-402.

Thiele, J. 1929. Handbuch der systematischen Weichtierkunde. Teil 1 (Loricata; Gastropoda: Prosobranchia). Fischer, Jena, Germany.

Thompson, F.G. 2011. An annotated checklist and bibliography of the land and freshwater snails of Mexico and Central America. Bulletin of the Florida Museum of Natural History 50: 1-299.

Vaught, K.C. 1989. A Classification of the Living Mollusca. American Malacologists, Melbourne, Florida.

Yokogawa, M. 1965. *Paragonimus* and Paragonimiasis. Pp. 99-158 in Advances in Parasitology, vol. 3 (B. Dawes, ed.). Academic Press, London.

9   Paludomidae Stoliczka, 1868

MARCO T. NEIBER AND MATTHIAS GLAUBRECHT

The genera and species suggested to be included in the Paludomidae (see Glaubrecht, 2011) have hitherto been classified as Thiaridae, especially the endemic thalassoid species from Lake Tanganyika (Michel et al., 1992, 2003; Brown, 1994; Michel, 1994, 2000; West and Michel, 2000), or have been assigned to distantly or even very distantly related groups such as Pleuroceridae or Rissooidea by Bandel (1998) following the outdated systematics of Wenz (1938) and/or Morrison (1954). Strong et al. (2011) recovered the African genera *Cleopatra* Troschel, 1857, *Tiphobia* Smith, 1880, *Lavigeria* Bourguignat, 1888, and *Tanganyicia* Crosse, 1881, and the Asian genus *Paludomus* Swainson, 1840, as the sister group of Thiaridae and Hemisinidae on the basis of a combined analysis of morphological characters and 16S and 28S rDNA sequences, with *Paludomus* as the sister group of the African paludomids (Glaubrecht and Strong, 2004). Synapomorphies of the family are the presence of an anterior spermatophore forming organ that is a hollow glandular tube extending dorsally from the gonoductal groove, a kidney bladder with a divided chamber and anatomical details of the accessory marginal fold in the midgut (Strong et al., 2011). A sister group relationship of *Paludomus* and African paludomid genera *Anceya* Bourguignat, 1885, *Syrnolopsis* Smith, 1880, *Martelia* Dautzenberg, 1908, *Tanganyicia* Crosse, 1881, *Paramelania* Smith, 1881, *Tiphobia*, *Limnotrochus* Smith, 1880, *Bridouxia* Bourguignat, 1885, *Stormsia* Leloup, 1953, *Spekia* Bourguignat, 1879, *Reymondia* Bourguignat, 1885, *Cleopatra* Troschel, 1857, and *Lavigeria* Bourguignat, 1888, was also recovered in the study of Wilson et al. (2004), suggesting a possible division of the family into an Asian subfamily Paludominae and an African subfamily, for which the oldest available name would be Hauttecoeriinae Bourguignat, 1885. However, such a subdivision, or an even further subdivision of the family into tribes as summarized in the working classification of Gastropoda by Bouchet et al. (2005), is premature until better resolved phylogenies, including representatives of all genera, are available.

The family has a general distribution range including most of tropical sub-Saharan Africa, Madagas-

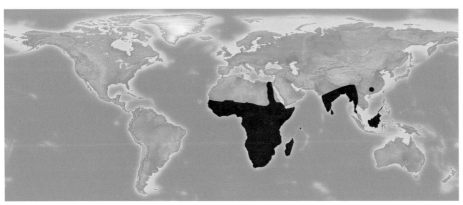

Distribution of
Paludomidae.

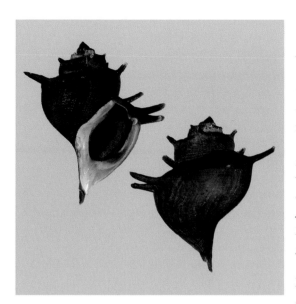

Tiphobia horei E. A. Smith, 1880. Burundi: Lake Tanganyika, south of Bujumbura, Rumonge Province,. 45.2 mm. Poppe 1025298.

car, and South and Southeast Asia. Four paludomid genera are recorded from Asia: *Tanalia* Gray, 1847, and *Philopotamis* Layard, 1855, with a center of diversity in Sri Lanka (Starmühlner, 1974, 1977, 1979; Subba Rao, 1989; Amarasinghe and Krishnarajah, 2009) and possibly a single species each in India (Subba Rao, 1989), the monotypic genus *Stomatodon* Benson, 1856, which is endemic to the Western Ghats in India (Subba Rao, 1989), and *Paludomus* Swainson, 1840, the type genus of the family, with a single species from the Seychelles Islands (Brown and Gerlach, 1991), numerous nominal species from India, Sri Lanka, and Mainland Southeast Asia (Mendis and Fernando, 1962; Brandt, 1974; Starmühlner, 1974, 1977, 1979; Subba Rao, 1989; Krailas et al., 2003a, b, 2012), two to six species from Southwest China (Gredler, 1890; Liu et al., 1994), and several described species from Borneo and Palawan (Brot, 1891; Martens, 1908; Thiele, 1928), as well as, interestingly, a single record from Lombok east of the Wallace Line (Glaubrecht, unpublished data).

Generic diversity of African paludomids is concentrated in the Lake Tanganyika basin and adjacent water bodies, with only two genera, *Cleopatra* and *Pseudocleopatra* Thiele, 1928, recorded from outside of this region. The genus *Cleopatra* is widespread in rivers, lakes, and even temporary water bodies of sub-Saharan Africa reaching North Africa through the Nile River system and in Madagascar (Starmühlner, 1969; Mandahl-Barth, 1954, 1974; Brown, 1994; El-Kady et al., 2000; Glaubrecht, 2010; Köhler and Glaubrecht, 2010), while *Pseudocleopatra* is reported from Ghana and the Congo River basin (Thiele, 1928; Mandahl-Barth, 1973, 1974; Brown, 1994). The morphological disparity in shell form of the endemic thalassoid, i.e., resembling marine gastropods, species flock in Lake Tanganyika including the genera *Anceya* Bourguignat, 1885 (two species), *Bathanalia* Moore, 1898 (one to two species), *Bridouxia* Bourguignat, 1885 (at least four species), *Chytra* Moore, 1898 (monotypic), *Lavigeria* Bourguignat, 1888 (at least two species, probably more), *Limnotrochus* Smith, 1880 (monotypic), *Martelia* Dautzenberg, 1908 (monotypic), *Mysorelloides* Leloup, 1953 (monotypic), *Paramelania* Smith, 1881 (three species), *Reymondia* Bourguignat, 1885 (two species), *Spekia* Bourguignat, 1879 (monotypic), *Stanleya* Bourguignat, 1885 (monotypic), *Stormsia* Leloup, 1953 (monotypic), *Syrnolopsis* Smith, 1880 (about 3 species), *Tanganyicia* Crosse, 1881 (probably a single, variable species), *Tiphobia* Smith, 1880 (monotypic), *Vinundu* Michel, 2004 (two species) (Leloup, 1953; Brown, 1994; Strong and Glaubrecht, 2002, 2003, 2007, 2008; Glaubrecht and Strong, 2004, 2007; Glaubrecht, 2008; Michel, 2004) has long been held as a striking example of an in situ radiation (Boss, 1978; Johnston and Cohen, 1987; Coulter, 1991; Michel et al., 1992; Michel, 1994, 2000; West and Cohen, 1996; Glaubrecht, 1996; Martens, 1997; West and Michel, 2000). However, more recently evidence has been amassing that Lake Tanganyika served as an evolutionary reservoir, with major evolutionary lineages possibly predating lake formation and at least two independent colonizations of the lake from riverine habitats (Wilson et al., 2004; Glaubrecht and Strong, 2007; Glaubrecht, 2008). Glaubrecht and Strong (2007), on the basis of an analysis of morphological characters, found a close relationship of the endemic Lake Tanganyika genera *Lavigeria* with the riverine genus *Potadomoides* Leloup, 1953, which is known from the Malagarasi Delta that flows into Lake Tanganyika from the east (one species) and the upper reaches of the Congo River (three species) west of the lake. Based on an analysis of mitochondrial cytochrome oxidase subunit 1 (*cox1*) sequences, Michel (2004) found *Cleopatra* nested within a clade, albeit without statistical support, including the Lake Tanganyika genera *Bathanalia, Chytra, Limnotrochus, Tiphobia, Paramelania, Reymondia, Stanleya,* and *Tanganyicia,* while Wilson et al. (2004) found a moderately supported sister group relationship of *Cleopatra* and a clade including *Anceya, Syrnolopsis, Martelia, Tanganyicia, Paramelania, Tiphobia, Limnotrochus, Bridouxia, Stormsia, Spekia,* and *Reymondia* on the basis of an analysis of *cox1* and 16S rDNA sequences. This clade in turn was found to be the well-supported sister group of *Lavigeria* by these authors, corroborating the systematization suggested in Glaubrecht (2008).

The Paludomidae comprises about 24 genera. Approximately 420 nominal species and subspecies have been described, with an overwhelming majority of names having been introduced for taxa endemic to eastern Africa's Lake Tanganyika by J.-R. Bourguignat (1885, 1886, 1888, 1889, 1890), who is well known—and has been harshly criticized for it—as an excessive splitter (e.g., Dance, 1970). Strong et al. (2008) give a very tentative estimate of 100 valid species, a value that has not been substantiated yet. Pending comprehensive revisions on the basis of morphological and molecular data for most genera currently render an accurate estimation of actual species diversity in the family difficult, especially when taking into account the recent finding that the taxonomic redundancy in freshwater Mollusca in general and Cerithioidea in particular might be as high as 70% and more (Glaubrecht 2009; Glaubrecht et al. 2009).

Freshwater ecosystems may well be the most imperiled ecosystems in the world, with major, often interacting, threats including water pollution, overexploitation, flow modification, habitat degradation, species invasion, and climate change (Dudgeon et al., 2006; Woodward et al., 2010). Knowledge of the total diversity of freshwater species is woefully incomplete on a global scale, and especially in tropical latitudes. In the case of Paludomidae, the incomplete state of our knowledge is exemplified by the fact that of the 99 species included in the IUCN Red List, 34 are listed as Data Deficient, while 6 species are listed as Near Threatened, 4 species as Vulnerable, 10 species as Endangered, and 2 species as Critically Endangered

(IUCN, 2015). Threats contributing to the demise of paludomid species include pollution, habitat decline through human disturbance, increasing sedimentation, dam construction, and increases in the frequency of severe droughts.

Species in the family Paludomidae seem to be less important as vectors of parasitic infections than other freshwater snail groups. *Cleopatra bulimoides* (Olivier, 1804) has been reported as the first intermediate host of the digenean *Prohemistomum vivax* (Sonsino, 1893), which is rarely reported to infect humans (Nasr, 1941), and the intestinal parasite *Gastrodiscus aegyptiacus* (Cobbold, 1876) (Lotfy and Lotfy, 2015). Furthermore, *C. bulimoides* is a host of *Angiostrongylus cantonensis* (Chen, 1935), which causes eosinophilic meningitis in humans (El-Shazly et al., 2002), and a host of a species of *Philophthalamus* Loos, 1899, which causes conjunctivitis in birds (Lotfy and Abo El-Hadid, 2005). Relatively little is known about parasites of Asian paludomids; however, Krailas et al. (2003a) detected four types of cercariae in *Paludomus petrosas* (Gould, 1843) from Thailand.

LITERATURE CITED

Amarasinghe, A.A.T., and S.R. Krishnarajah. 2009. Distribution patterns of the genus *Paludomus* (Gastropoda: Thiaridae: Paludominae) in Mahaweli, Kelani, Kalu, Gin and Maha-Oya River basins of Sri Lanka. Taprobanica 1: 130-134.

Bandel, K. 1998. Evolutionary history of East African fresh water gastropods interpreted from the fauna of Lake Tanganyika and Lake Malawi. Zentralblatt für Geologie und Paläontologie. Teil I 1997 (1-2): 233-292.

Boss, K.J. 1978. On the evolution of gastropods in ancient lakes. Pp. 385-428 in Pulmonates, vol. 2a, Systematics, Evolution and Ecology (V. Fretter and J. Peake, eds.). Academic Press, London.

Bouchet, P., J. Frýda, B. Hausdorf, W.F. Ponder, Á. Valdés, and A. Warén. 2005. Part 2. Working Classification of the Gastropoda. In Classification and Nomenclator of Gastropod Families (P. Bouchet and J.-P. Rocroi, eds.). Malacologia 47: 239-283.

Bourguignat, J.R. 1885. Notice prodromique sur les mollusques terrestres et fluviatiles recueillis dans la région méridionale du Lac Tanganika. V. Tremblay, Paris.

Bourguignat, J.R. 1886. Des tiphobies du Lac Tanganika. Bulletin de la Société Malacologique de France 3: 141-149, plate 6.

Bourguignat, J.R. 1888. Iconographie malacologique des animaux mollusques fluviatiles du Lac Tanganika. Crété, Corbeil.

Bourguignat, J.R. 1889. Mollusques de l'Afrique Equatoriale de Moguedouchou a Bagamoyo et de Bagamoyo au Tanganika. D. Dumoulin, Paris.

Bourguignat, J.R. 1890. Histoire malacologique du Lac Tanganika (Afrique Équatoriale). Annales des Sciences Naturelles Zoologie et Paleontologie 10: 1-267, plates 1-12.

Brandt, R.A.M. 1974. The non-marine aquatic Mollusca of Thailand. Archiv für Molluskenkunde 105: i-iv, 1-423.

Brot, A. 1891. *Paludomus palawanicus,* n. sp. Nautilus 5: 17-18.

Brown, D.S. 1994. Freshwater Snails of Africa and Their Medical Importance. 2nd ed. Taylor & Francis, London.

Brown, D.S., and J. Gerlach. 1991. On *Paludomus* and *Cleopatra* (Thiaridae) in Africa and the Seychelles Islands. Journal of Molluscan Studies 57: 471-479.

Coulter, G.W. 1991. Lake Tanganyika and Its Life. Oxford University Press, Oxford.

Dance, S.P. 1970. 'Le Fanatisme du nobis': A study of J.-R. Bourguignat and the 'Nouvelle Ecole.' Journal of Conchology 27: 65-86.

Dudgeon, D., A.H. Arthington, M.O. Gessner, Z.-I. Kawabata, D.J. Knowler, C. Lévêque, R.J. Naiman, et al. 2006. Freshwater biodiversity: Importance, threats, status and conservation challenges. Biological Reviews of the Cambridge Philosophical Society 81: 163-182.

El-Kady, G.A., A. Shoukry, L.A. Reda, and Y.S. El-Badri. 2000. Survey and population dynamics of freshwater snails in newly settled areas of the Sinai Peninsula. Egyptian Journal of Biology 2: 42-48.

El-Shazly, A.M., E.M. El-Hamshary, K.M. El-Shewy, M.M. Rifaat, and I.M. El-Sharkawy. 2002. Incidence of *Parastrongylus cantonensis* larvae in different fresh water snails in Dakahlia Governorate. Journal of the Egyptian Society of Parasitology 32: 579-588.

Glaubrecht, M. 1996. Evolutionsökologie und Systematik am Beispiel von Süß- und Brackwasserschnecken (Mollusca: Caenogastropoda: Cerithioidea): Ontogenese-Strategien, paläontologische Befunde und Historische Zoogeographie. Backhuys, Leiden.

Glaubrecht, M. 2008. Adaptive radiation of thalassoid gastropods in Lake Tanganyika, East Africa: Morphology and systematization of a paludomid species flock in an ancient lake. Zoosystematics and Evolution 84: 71-122.

Glaubrecht, M. 2009. On 'Darwinian Mysteries' or molluscs as models in evolutionary biology: From local speciation to global radiation. American Malacological Bulletin 27: 3-23.

Glaubrecht, M. 2010. The enigmatic *Cleopatra broecki*

Putzeys, 1899 of the Congo River system in Africa—re-transfer from *Potadomoides* Leloup, 1953 (Caenogastropoda, Cerithioidea, Paludomidae). Zoosystematics and Evolution 86: 283-293.

Glaubrecht, M. 2011. Towards solving Darwin's 'mystery': Speciation and radiation in lacustrine and riverine freshwater gastropods. American Malacological Bulletin 29: 187-216.

Glaubrecht, M., and E.E. Strong. 2004. Spermatophores of thalassoid gastropods (Paludomidae) in Lake Tanganyika, East Africa, with a survey of their occurrence in Cerithioidea: Functional and phylogenetic implications. Invertebrate Biology 123: 218-236.

Glaubrecht, M., and E.E. Strong. 2007. Ancestry to an endemic radiation in Lake Tanganyika? Evolution of the viviparous gastropod *Potadomoides* Leloup, 1953 in the Congo River system (Caenogastropoda, Cerithioidea, Paludomidae). Biological Journal of the Linnean Society. 92: 367-401.

Glaubrecht, M., N. Brinkmann, and J. Pöppe. 2009. Diversity and disparity 'down under': Systematics, biogeography and reproductive modes of the 'marsupial' freshwater Thiaaridae (Caenogastropoda, Cerithiodea) in Australia. Zoosystematics and Evolution 85: 199-275.

Gredler, P.V. 1890. Zur Conchylien-Fauna von China. XVI. Stück. Nachrichtsblatt der Deutschen Malakozoologischen Gesellschaft 22: 145-153.

IUCN Red List of Threatened Species. 2015. Version 2015-1. www.iucnredlist.org. Accessed 24 July 2017.

Johnston, M.R., and A.S. Cohen. 1987. Morphological divergence in endemic gastropods from Lake Tanganyika: Implications for models of species flock formation. Palaios 2: 413-425.

Köhler, F., and M. Glaubrecht. 2010. Uncovering an overlooked radiation: Molecular phylogeny and biogeography of Madagascar's endemic river snails (Caenogastropoda: Pachychilidae: *Madagasikara* gen. nov.). Biological Journal of the Linnean Society 99: 867-894.

Krailas, D., W. Dechruksa, S. Ukong, and T. Janecharut. 2003a. Cercarial infection in *Paludomus petrosus,* freshwater snail in Pa La-U waterfall. Southeast Asian Journal of Tropical Medicine and Public Health 34: 286-290.

Krailas, D., T. Janecharat, S. Ukong, N. Notesiri, and P. Ratanathai. 2003b. Preliminary report on freshwater snails in Toa Dum Forest, Saiyok District, Kanchanaburi. Silpakorn University Journal of Social Sciences, Humanities and Arts 3: 194-205.

Krailas, D., S. Chotesaengsri, W. Dechruksa, S. Namchote, C. Chuanprasit, N. Veeravechsukij, D. Boonmekam, and T. Koonchornboon. 2012. Species diversity of aquatic mollusks and their cercarial infections; Khao Yai National

Park, Thailand. Journal of Tropical Medicine and Parasitology 35: 37-47.

Leloup, E. 1953. Exploration hydrobiologique du Lac Tanganika (1946-1947). Résultats scientifiques, vol. 3, Gastéropodes. Institut Royal des Sciences Naturelles de Belgique, Brussells.

Liu, Y.-Y., W.-Z. Zhang, Y.-X. Wang, and Y.-H. Duan. 1994. Eight new species of freshwater molluscs in Southwest China (Gastropoda: Bivalvia). Acta Zootaxonomica Sinica 19: 25-36.

Lotfy, H.S., and S.M. Abo El-Hadid. 2005. Life cycle of *Philophthalamus* species for the first time in Egypt. Beni-Suef Veterinary Medicine Journal 15: 244-246.

Lotfy, W.M., and L.M. Lotfy. 2015. Synopsis of the Egyptian freshwater snail fauna. Folia Malacologica 23: 19-40.

Mandahl-Barth, G. 1954. The freshwater molluscs of Uganda and adjacent territories. Annales du Musée Royal du Congo Belge, Série in 8°, Sciences Zoologiques 32: 1-206.

Mandahl-Barth, G. 1973. Description of new species of African freshwater molluscs. Proceedings of the Malacological Society, London, 40: 277-286.

Mandahl-Barth, G. 1974. New or little known species of freshwater Mollusca from Zaire and Angola, with remarks on the genus *Sierraia* Connolly. Revue de Zoologie Africaine 88: 352-362.

Martens, E. von. 1908. Beschreibung einiger im östlichen Borneo von Dr. Martin Schmidt gesammelten Land- und Süßwasser-Conchylien. Mitteilungen aus dem Zoologischen Museum, Berlin, 4: 249-291, plates 5-6.

Martens, K. 1997. Speciation in ancient lakes. Trends in Ecology and Evolution 12: 177-182.

Mendis, A.S., and C.H. Fernando. 1962. A guide to the freshwater fauna of Ceylon. Bulletin of the Fisheries Research Station, Ceylon, 12: 1-162.

Michel, E. 1994. Why snails radiate: A review of gastropod evolution in long-lived lakes, both Recent and fossil. Archiv für Hydrobiologie, Ergebnisse der Limnologie 44: 285-317.

Michel, E. 2000. Phylogeny of a gastropod species flock: Exploring speciation in Lake Tanganyika in a molecular framework. Advances in Ecological Research 31: 275-302.

Michel, E. 2004. *Vinundu,* a new genus of gastropod (Cerithioidea: Thiaridae) with two species from Lake Tanganyika, East Africa, and its molecular phylogenetic relationships. Journal of Molluscan Studies 70: 1-19.

Michel, E., A.S. Cohen, K. West, M.R. Johnston, and P.W. Kat. 1992. Large African lakes as natural laboratories for evolution: Examples from the endemic gastropod fauna of Lake Tanganyika. International Association of Theoretical and Applied Limnology 23: 85-99.

Michel, E., J.A. Todd, F.R. Cleary, I. Kingma, A.S. Cohen, and M. Genner. 2003. Scales of endemism: Challenges for conservation and incentives for evolutionary studies in a gastropod species flock from Lake Tanganyika. Journal of Conchology, Special Publication 3: 155-172.

Morrison, J.P.E. 1954. The relationships of Old and New World melanians. Proceedings of the United States National Museum 103: 357-394.

Nasr, M. 1941. The occurrence of *Prohemistomum vivax* infections in man with redescription of the parasite. Laboratory Medical Progress 2: 135-149.

Starmühlner, F. 1969. Die Gastropoden der madagassischen Binnengewässer. Malacologia 8: 1-434.

Starmühlner, F. 1974. The freshwater gastropods of Ceylon. Bulletin of the Fisheries Research Station, Ceylon, 25: 97-181.

Starmühlner, F. 1977. The genus *Paludomus* in Ceylon. Malacologia 16: 261-264.

Starmühlner, F. 1979. Distribution of freshwater mollusks in mountain streams of tropical Indo-Pacific islands (Madagascar, Ceylon, New Caledonia). Malacologia 18: 245-255.

Strong, E.E., and M. Glaubrecht. 2002. Evidence for convergent evolution of brooding in a unique gastropod from Lake Tanganyika: Anatomy and affinity of *Tanganyicia rufofilosa* (Caenogastropoda, Cerithioidea, Paludomidae). Zoologica Scripta 31: 167-184.

Strong, E.E., and M. Glaubrecht. 2003. Anatomy and systematic affinity of *Stanleya neritinoides* (Smith, 1880), an enigmatic member of the thalassoid gastropod species flock from Lake Tanganyika, East Africa (Cerithioidea: Paludomidae). Acta Zoologica (Stockholm) 84: 249-265.

Strong, E.E., and M. Glaubrecht. 2007. The morphology and independent origin of ovoviviparity in *Tiphobia* and *Lavigeria* (Caenogastropoda: Cerithioidea: Paludomidae) from Lake Tanganyika. Organisms, Diversity & Evolution 7: 81-105.

Strong, E.E., and M. Glaubrecht. 2008. Anatomy and systematics of the minute syrnolopsine gastropods from Lake Tanganyika (Caenogastropoda, Cerithioidea, Paludomidae). Acta Zoologica (Stockholm) 89: 289-310.

Strong, E.E., O. Gargominy, W.F. Ponder, and P. Bouchet. 2008. Global diversity of gastropods (Gastropoda: Mollusca) in freshwater. Hydrobiologia 595: 149-166.

Strong, E.E., D.J. Colgan, J.M. Healy, C. Lydeard, W.F. Ponder, and M. Glaubrecht. 2011. Phylogeny of the gastropod superfamily Cerithioidea using morphology and molecules. Zoological Journal of the Linnean Society 162: 43-89.

Subba Rao, N.V. 1989. Handbook Freshwater Molluscs of India. Zoological Survey of India, Calcutta.

Thiele, J. 1928. Revision des Systems der Hydrobiiden und Melaniiden. Zoologische Jahrbücher, Abteilung Systematik, Ökologie und Geographie der Tiere 55: 351-402.

Wenz, W. 1938. Gastropoda. Teil 1. Allgemeiner Teil und Prosobranchia. Pp. i-viii, 1-1639 in Handbuch der Paläozoologie, vol 6, part 1 (O.H. Schindewolf, ed.). Gebrüder Borntraeger, Berlin.

West, K., and A. Cohen. 1996. Shell microstructure of gastropods from Lake Tanganyika, Africa: Adaptation, convergent evolution, and escalation. Evolution 50: 672-681.

West, K., and E. Michel. 2000. The dynamics of endemic diversification; molecular phylogeny suggests an explosive origin of the thiarid gastropods of Lake Tanganyika. Advances in Ecological Research 31: 275-302.

Wilson, A.B., M. Glaubrecht, and A. Meyer. 2004. Ancient lakes as evolutionary reservoirs: Evidence from the thalassoid gastropods of Lake Tanganyika. Proceedings of the Royal Society B: Biological Sciences 271: 529-536.

Woodward, G., D.M. Perkins, and L.E. Brown. 2010. Climate change and freshwater ecosystems: Impacts across multiple levels. Philosophical Transactions of the Royal Society of London Series B, Biological Sciences 365: 2093-2106.

10 Pleuroceridae P. Fischer, 1885

ELLEN E. STRONG AND CHARLES LYDEARD

The Pleuroceridae (type genus *Pleurocera* Rafinesque, 1818; type species *Pleurocera acuta* Blainville, 1824, by subsequent designation, Opinion 1195 [1981: 259]) is the second most speciose family of freshwater gastropods in North America after the Hydrobiidae, and is the most imperiled. Commonly known as freshwater periwinkles, pleurocerids are distributed mainly in temperate to subtropical regions of North America east of the Rocky Mountains, with four species that just extend into the tropics of northern Mexico (Burch and Tottenham, 1980; Thompson, 2011; Johnson et al., 2013). Pleurocerids inhabit primarily lakes, rivers, and streams and attain their greatest diversity in the southeastern United States, a globally recognized hotspot of aquatic diversity (Lydeard and Mayden, 1995; Neves et al., 1997; Strong et al., 2008).

The fossil record of the family extends at least to the Cretaceous, with undisputed pleurocerid fossils known from the Albian (see Strong and Köhler, 2009). Today, the family is comprised of 167 species currently considered valid, distributed among seven Recent genera: seven genera from eastern North America north of Mexico including *Pleurocera, Athearnia* Morrison, 1971, *Elimia* H. Adams & A. Adams, 1854, *Gyrotoma* Shuttleworth, 1845, *Io* I. Lea, 1831, *Leptoxis* Rafinesque, 1819, and *Lithasia* Haldeman, 1840 (Burch and Tottenham, 1980; Burch, 1982, 1988, 1989), and one genus, *Lithasiopsis* Pilsbry, 1910, restricted to the Rio Panuco and Rio Guayalejo systems of northeastern Mexico (Thompson, 2011). The IUCN (2017) has assessed 102 pleurocerid species and of these, concluded 31 species to be Extinct, 7 Critically Endangered, 4 Endangered, 31 Vulnerable, 9 Near Threatened, 5 Data Deficient, and 15 of Least Concern. A similarly bleak picture emerged from a comprehensive conservation assessment of North American freshwater gastropods, which determined that, of the 163 species that occur north of Mexico, 33 are extinct and 100 are critically imperiled (G1), imperiled (G2), or vulnerable (G3), yielding an imperilment rate of ~80% for the family (Johnson et al., 2013). This is the highest rate of imperilment among North American freshwater gastropods, with the family as a whole accounting for more than half of all

Distribution of Pleuroceridae.

Pleurocera canaliculata (Say, 1821). USA: Center Hill Reservoir, Pate's Ford, White County, Tennessee. 28 mm. Poppe 718640.

freshwater gastropod extinctions in North America (Johnson et al., 2013).

The Pleuroceridae is one of seven freshwater families in the basal caenogastropod superfamily Cerithioidea, and is the second most diverse after the Pachychilidae (Lydeard et al., 2002; Strong et al., 2008, 2011; Bouchet et al., 2017). Circumscription and taxonomy of limnic cerithioidean families has had a complex and confusing history (e.g., Glaubrecht, 1996; Houbrick, 1988; Ponder, 1991). Until recently, Pleuroceridae *s.l.* included two subfamilies, with species from eastern North America united with those from western North America (*Juga* H. & A. Adams, 1854) in the Pleurocerinae, and species from eastern Asia (*Semisulcospira* Boettger, 1886, *Biwamelania* Matsuoka & Nakamura, 1981, *Hua* Chen, 1943, *Koreanomelania* Burch & Jung, 1988, *Koreoleptoxis* Burch & Jung, 1988, and "*Parajuga*," a nomenclaturally unavailable name) united in the Semisulcospirinae (Bouchet and Rocroi, 2005). However, based on morphological and molecular data, Strong and Köhler (2009) restricted the Pleuroceridae *s.s.* to eastern North American forms and elevated the Semisulcospirinae to the rank of family, and redefined it to include *Juga*. Pleurocerids differ from semisulcospirids in lacking a seminal receptacle and in features of the prostate, gonoductal groove, albumen gland, and kidney (Strong, 2005; Strong

and Frest, 2007; Strong and Köhler, 2009). Reciprocal monophyly of the two families has been supported in an expanded phylogenetic analysis of the superfamily based on morphological and molecular data (Strong et al., 2011).

Pleurocerids and semisulcospirids are closely related to the freshwater Melanopsidae, which is largely distributed in northern Africa and southern and eastern Europe, with a disjunct distribution in New Zealand and New Caledonia (Strong et al., 2008, 2011; Strong and Köhler, 2009; p. 56, this volume), but relationships between the three families are still uncertain. Parsimony analysis of a combined morphological and molecular data set supported a monophyletic Melanopsidae as sister to the Semisulcospiridae, while Bayesian analysis resulted in a paraphyletic Melanopsidae at the base of a clade uniting Pleuroceridae plus Semisulcospiridae (Strong et al., 2011). Regardless of how the relationships are resolved, it is likely that the common ancestor of the three families was widespread across Laurasia after its separation from Pangaea and that they diverged rapidly during the Cretaceous following the separation of Laurasia into Asiamerica and Euramerica approximately 100 mya caused by the flooding of epicontinental seas (Strong and Köhler, 2009).

At the species level, the documentation and description of pleurocerid diversity has had a similarly complex and confusing history owing to the more than 800 available species-group names (Graf, 2001; Garner, Bogan, Johnson, and Strong, unpublished data). The impact of this confusion is evident in a case of mistaken identity of the federally threatened Painted Rocksnail, known by the incorrect name and its historical distribution misinterpreted for more than 100 years (Whelan et al., 2017). The astounding number of nominal taxa is largely due to the efforts of Isaac Lea and J. G. Anthony, who described roughly 440 and 120 species, respectively, in the early 1800s based on minor differences in shell morphology among and within populations. In a series of papers from 1917 to 1950, Goodrich examined the family from a population perspective by drainage basin and synonymized many taxa (e.g., Goodrich, 1920, 1922, 1936, 1941). Goodrich's influential works had a significant impact on the widely used classification of North American gastropods

(Burch and Tottenham, 1980; Burch, 1982, 1988, 1989), which recognized 144 pleurocerid species north of Mexico. These works form the cornerstone of the classification currently in use (Johnson et al., 2013), but are based almost entirely on traditional notions of morphological shell variation.

Thus far, molecular systematic studies of the group have relied almost exclusively on the COI and 16S mitochondrial genes. Early analyses were based typically on only one or two representatives per species and did not support the monophyly of many of the traditionally conceived genera (Lydeard et al., 1997; Holznagel and Lydeard, 2000). With increased sampling, the incongruence between topology and taxonomy extended to species, with significant levels of para- or polyphyly of species-level taxa which were attributed to the presence of cryptic species (Minton and Lydeard, 2003; Minton et al., 2005; Sides, 2005) and the retention of ancestral polymorphisms and/or historical introgression (Dillon and Frankis, 2004; Dillon and Robinson, 2009). Pronounced within-population mitochondrial heterogeneity has been recovered within semisulcospirids as well, similarly producing nonmonophyletic putative species (Kim et al., 2010) and mismatch with nuclear gene trees (Lee et al., 2007). Dense population-level sampling suggested that the pattern could be the result of among-drainage migration and possibly also introgression (Lee et al., 2007; Miura et al., 2013; Köhler, 2016). Whelan and Strong (2016) completed a population-level analysis of four species from the Cahaba River and Paint Rock River drainages, and they assembled and annotated a mitochondrial genome for *Leptoxis ampla,* to explore if similar causal mechanisms could explain the pattern in pleurocerids. Cryptic diversity, paralogous nuclear copies of mitochondrial genes (NUMTs), and doubly uniparental inheritance were rejected. Despite the dense population-level sampling, the pattern could not be unambiguously attributed to migration and/or introgression, although these may have been contributing factors. It is possible that a combination of processes is responsible, including endosymbiont infection, which has not been adequately explored. A clear signature of balancing selection seems to be acting to maintain this mitochondrial diversity (Whelan and Strong, 2016). Recently, Köhler (2017) concluded that

nonneutral mitochondrial evolution could be a driver among semisulcospirids as well. Regardless of the source and the mechanisms of its maintenance, this heterogeneity has effectively suppressed meaningful molecular approaches to pleurocerid systematics based on mitochondrial markers. Consequently, modern, synthetic approaches to the systematics are still lacking, and validity of the included species remains untested with molecular data.

As in other freshwater caenogastropods, slow growth, long maturation times, and narrow ecological tolerances make pleurocerids highly sensitive to human mediated impacts (Brown and Johnson, 2004; Brown et al., 2008; Strong et al., 2008). Factors that have contributed to the demise of pleurocerids include habitat destruction from dams, channel modification, siltation, water-quality degradation from point and nonpoint pollution, and the introduction of nonindigenous species. The vast majority of extinctions, including of the entire genus *Gyrotoma,* occurred among the extraordinary radiation of species endemic to the Coosa River in Alabama after it was impounded (Lydeard and Mayden, 1995; Neves et al., 1997; Johnson et al., 2013; Whelan et al., 2017). Despite the high rate of imperilment, this has rarely translated into formal recognition and protection under the Endangered Species Act; only seven species are listed as endangered or threatened by the US Fish and Wildlife Service, although more species are currently under consideration for listing. While few, conservation success stories demonstrate there is hope, where improvements to water quality and habitat restoration have allowed some populations to rebound and have led to the successful reintroduction of captive bred populations (Johnson et al., 2013). Recent surveys have revealed the persistence of several species presumed extinct (Minton et al., 2003; Whelan et al., 2012b).

Despite their diversity and acknowledged level of threat, remarkably few studies have been completed on aspects of basic biology, including ecology, life history, and anatomy, and few species have been studied in detail. *Elimia livescens* and *Pleurocera acuta* were the subject of an intensive study by Dazo (1965), and they remain two of the better-known species (see also Strong, 2005). Pleurocerids are microherbivorous grazers (see Brown et al., 2008), and in habitats with high periphyton biomass, densities can exceed 1500/

m^2 (Johnson and Brown, 1997). Conversely, a link has been observed between eutrophication and higher prevalence of infection by trematode parasites, which can reach 49% in some populations (Ciparis et al., 2013). The average life span ranges from 2 to 6 years (Huryn et al., 1994; Johnson et al., 2013) but may be as long as 10 years or more in some species (Richardson et al., 1988). They reach a maximum height of ~1 to 5 cm at adulthood (Burch 1989; Johnson et al., 2013) and, like other freshwater gastropods, show extensive levels of intra- and interspecific variation in shell morphology, making it difficult to diagnose species based on shell characters alone. The source of this variation may have both an ecological (i.e., ecophenotypic; Minton et al., 2008) and genetic component (Whelan et al., 2012a, 2015).

Sexes are separate, populations often showing significant female biased sex ratios (Ciparis et al., 2012), and may be iteroparous or semelparous (Richardson et al., 1988; Miller-Way and Way, 1989; Whelan et al., 2015). Males are aphallate and transfer sperm via spermatophores (e.g., Dazo, 1965; Glaubrecht and Strong, 2004; Strong, 2005). As mentioned, pleurocerids lack a discrete seminal receptacle, a structure found in both semisulcospirids and melanopsids. Rather, females possess only a single sperm-storing pocket homologous to the spermatophore bursa of other cerithioideans (Strong, 2005; Strong et al., 2011), but which functions both as a bursa for the short-term storage of unorientated sperm and as a seminal receptacle for longer storage of orientated sperm (Whelan and Strong, 2014). A ciliated egg groove on the right side of the foot conveys eggs to the ovipositor (Houbrick, 1988; Strong, 2005), through which the eggs pass before they are deposited on solid substrates by the foot sole. Oviposition occurs between January and July, depending on the species, and egg capsules are deposited singly or in clutches (e.g., Dazo, 1965; Whelan and Strong, 2014; Whelan et al., 2015). Gametogenic activity is reduced or ceases and the volume of the gonads shrinks during nonreproductive periods; accessory reproductive structures also atrophy significantly outside the reproductive season (Whelan and Strong, 2014). Females are not capable of overwintering sperm, requiring that mating occurs before the seasonal onset of oviposition in the spring (Whelan and Strong, 2014).

The family is in urgent need of studies on basic biology, life history, physiology, biogeography, and ecology in concert with comprehensive systematic revision (Brown et al., 2008; Lysne et al., 2008; Strong et al., 2008; Johnson et al., 2013; Whelan et al., 2017). The presence of pronounced mitochondrial heterogeneity within pleurocerid species has prevented meaningful systematic revision based on mitochondrial data. Interim taxonomic rearrangements using mitochondrial data or based on small subsets of taxa and/or data (e.g., Dillon, 2011) are not in the best interest of stability and are likely to be overturned. Among candidate nuclear DNA barcoding loci such as ITS, 28S, and Histone H3, comparative data for pleurocerid species have been generated only for H3; these data show that H3 is of limited utility at the species level within the family (Whelan and Strong, 2016); further work is needed to assess the utility of 28S and ITS. Genomics remains the most promising way forward to identify additional nuclear loci and to generate the comparative molecular framework necessary to engage in meaningful systematic studies of the family (Whelan and Strong, 2016).

LITERATURE CITED

Bouchet, P., and J.-P. Rocroi. 2005. Classification and nomenclator of gastropod families. Malacologia 47: 1–397.

Bouchet, P., J.-P. Rocroi, B. Hausdorf, A. Kaim, Y. Kano, A. Nützel, P. Parkhaev, M. Schrödl, and E.E. Strong. 2017. Revised classification, nomenclator and typification of gastropod and monoplacophoran families. Malacologia 61: 1–526.

Brown, K.M., and P.D. Johnson. 2004. Comparative conservation ecology of pleurocerid and pulmonate gastropods of the United States. American Malacological Bulletin 19: 57–62.

Brown, K.M., B. Lang, and K.E. Perez. 2008. The conservation ecology of North American pleurocerid and hydrobiid gastropods. Journal of the North American Benthological Society 27: 484–495.

Burch, J.B. 1982. North American freshwater snails: Identification keys, generic synonymy, supplemental notes, glossary, references, index. Walkerana 1: 217–365.

Burch, J.B. 1988. North American freshwater snails. Introduction, systematics, nomenclature, identification, morphology, habitats, distribution. Walkerana 2: 1–80.

Burch, J.B. 1989. North American Freshwater Snails. Malacological Publications, Hamburg, Michigan.

Burch, J.B., and J.L. Tottenham. 1980. North American freshwater snails: Species list, ranges, and illustrations. Walkerana 1: 81 215.

Ciparis, S., W.F. Henley, and J.R. Voshell. 2012. Population sex ratios of pleurocerid snails (*Leptoxis* spp.): Variability and relationships with environmental contaminants and conditions. American Malacological Bulletin 30: 287-298.

Ciparis, S., D.D. Iwanowicz, and J.R. Voshell. 2013. Relationships between nutrient enrichment, pleurocerid snail density and trematode infection rate in streams. Freshwater Biology 58: 1392-1404.

Dazo, B.C. 1965. The morphology and natural history of *Pleurocera acuta* and *Goniobasis livescens* (Gastropoda: Cerithiacea: Pleuroceridae). Malacologia 3: 1-80.

Dillon, R.T., Jr. 2011. Robust shell phenotype is a local response to stream size in the genus *Pleurocera* (Rafinesque, 1818). Malacologia 53: 265-277.

Dillon, R.T., Jr., and R.C. Frankis Jr. 2004. High levels of mitochondrial DNA sequence divergence in isolated populations of freshwater snails of the genus *Goniobasis* Lea, 1862. American Malacological Bulletin 19: 69-77.

Dillon, R.T., Jr., and J.D. Robinson. 2009. The snails the dinosaurs saw: Are the pleurocerid populations of the Older Appalachians a relict of the Paleozoic era? Journal of the North American Benthological Society 28: 1-11.

Glaubrecht, M. 1996. Evolutionsökologie und Systematik am Beispiel von Süß- und Brackwasserschnecken (Mollusca: Caenogastropoda: Cerithioidea): Ontogenese-Strategien, paläontologische Befunde und Historische Zoogeographie. Backhuys, Leiden.

Glaubrecht, M., and E.E. Strong. 2004. Spermatophores of thalassoid gastropods (Paludomidae) in Lake Tanganyika, East Africa, with a survey of their occurrence in Cerithioidea: Functional and phylogenetic implications. Invertebrate Biology 123: 218-236.

Goodrich, C. 1920. Goniobases of Ohio. Nautilus 33: 73-74.

Goodrich, C. 1922. The Anculosae of the Alabama River drainage. Miscellaneous Publications, Museum of Zoology, University of Michigan, 7: 1-57.

Goodrich, C. 1936. *Goniobasis* of the Coosa River, Alabama. Miscellaneous Publications, Museum of Zoology, University of Michigan, 31: 1-60.

Goodrich, C. 1941. Pleuroceridae of the small streams of the Alabama River system. Occasional Papers of the Museum of Zoology, University of Michigan, 427: 1-10.

Graf, D.L. 2001. The cleansing of the Augean stables, or a lexicon of the nominal species of the Pleuroceridae (Gastropoda: Prosobranchia) of recent North America, north of Mexico. Walkerana 12: 1-124.

Holznagel, W.E., and C. Lydeard. 2000. A molecular phylogeny of North American Pleuroceridae (Gastropoda: Cerithioidea) based on mitochondrial 16S rDNA sequences. Journal of Molluscan Studies 66: 233-257.

Houbrick, R.S. 1988. Cerithioidean phylogeny. In Prosobranch Phylogeny: Proceedings of a Symposium Held at the 9th International Malacological Congress, Edinburgh, 1986 (W.F. Ponder, ed.). Malacological Review Supplement 4: 88-128.

Huryn, A.D., J. Koebel, and A.C. Benke. 1994. Life history and longevity of the pleurocerid snail *Elimia:* A comparative study of eight populations. Journal of the North American Benthological Society 13: 540-556.

IUCN Red List of Threatened Species. 2017. Version 2017-3. www.iucnredlist.org.

Johnson, P.D., and K.M. Brown. 1997. The role of current and light in explaining the habitat distribution of the lotic snail *Elimia semicarinata* (Say). Journal of the North American Benthological Society 16: 545-561.

Johnson, P.D., A.E. Bogan, K.M. Brown, N.M. Burkhead, J.R. Cordeiro, J.T. Garner, P.D. Hartfield, et al. 2013. Conservation status of freshwater gastropods of Canada and the United States. Fisheries 38: 247-282.

Kim, W.-J., D.-H. Kim, J.S. Lee, I.-C. Bang, W.-O. Lee, and H. Jung. 2010. Systematic relationships of Korean freshwater snails of *Semisulcospira, Koreanomelania,* and *Koreoleptoxis* (Cerithioidea: Pleuroceridae) revealed by mitochondrial cytochrome oxidase I sequences. Korean Journal of Malacology 26: 275-283.

Köhler, F. 2016. Rampant taxonomic incongruence in a mitochondrial phylogeny of *Semisulcospira* freshwater snails from Japan (Cerithioidea, Semisulcospiridae). Journal of Molluscan Studies 82: 268-281.

Köhler, F. 2017. Against the odds of unusual mtDNA inheritance, introgressive hybridisation and phenotypic plasticity: Systematic revision of Korean freshwater gastropods (Semisulcospiridae, Cerithioidea). Invertebrate Systematics 31: 249-268.

Lee, T., H.C. Hong, J.J. Kim, and D. Ó Foighil. 2007. Phylogenetic and taxonomic incongruence involving nuclear and mitochondrial markers in Korean populations of the freshwater snail genus *Semisulcospira* (Cerithioidea: Pleuroceridae). Molecular Phylogenetics and Evolution 43: 386-397.

Lydeard, C., and R.L. Mayden. 1995. A diverse and endangered aquatic ecosystem of the Southeast United States. Conservation Biology 9: 800-805.

Lydeard, C., W.E. Holznagel, J. Garner, P. Hartfield, and M. Pierson. 1997. A molecular phylogeny of Mobile River drainage pleurocerid snails (Caenogastropoda: Cerithioidea). Molecular Phylogenetics and Evolution 7: 117-128.

Lydeard, C., W.E. Holznagel, M. Glaubrecht, and W.F. Ponder. 2002. Molecular phylogeny of a circum-global, diverse gastropod superfamily (Cerithioidea: Mollusca: Caenogastropoda): Pushing the deepest phylogenetic limits of mitochondrial LSU rDNA sequences. Molecular Phylogenetics and Evolution 22: 399-406.

Lysne, S.J., K.E. Perez, K.M. Brown, R.L. Minton, and J.D. Sides. 2008. A review of freshwater gastropod conservation: Challenges and opportunities. Journal of the North American Benthological Society 27: 463-470.

Miller-Way, C.A., and C.M. Way. 1989. The life history of *Leptoxis dilatata* (Conrad) (Prosobranchia: Pleuroceridae) from the Laurel Fork River, West Virginia. American Midland Naturalist 122: 193-198.

Minton, R.L., and C. Lydeard. 2003. Phylogeny, taxonomy, genetics and global heritage ranks of an imperiled, freshwater snail genus *Lithasia* (Pleuroceridae). Molecular Ecology 12: 75-87.

Minton, R.L., J.T. Garner, and C. Lydeard. 2003. Rediscovery, systematic position, and re-description of '*Leptoxis*' *melanoides* (Conrad, 1834) (Mollusca: Gastropoda: Cerithioidea: Pleuroceridae) from the Black Warrior River, Alabama, U.S.A. Proceedings of the Biological Society of Washington 116: 531-541.

Minton, R.L., S.P. Savarese, and D.C. Campbell. 2005. A new species of *Lithasia* (Mollusca: Caenogastropoda: Pleuroceridae) from the Harpeth River, Tennessee, U.S.A. Zootaxa 1054: 31-42.

Minton, R.L., A.P. Norwood, and D.M. Hayes. 2008. Quantifying phenotypic gradients in freshwater snails: A case study in *Lithasia* (Gastropoda: Pleuroceridae). Hydrobiologia 605: 173-182.

Miura, O., F. Köhler, T. Lee, J. Li, and D. Ó Foighil. 2013. Rare, divergent Korean *Semisulcospira* spp. mitochondrial haplotypes have Japanese sister lineages. Journal of Molluscan Studies 79: 86-89.

Neves, R.J., A.E. Bogan, J.D. Williams, S.A. Ahlstedt, and P.W. Hartfield. 1997. Status of aquatic mollusks in the southeastern United States: A downward spiral of diversity. Pp. 43-85 in: Aquatic Fauna in Peril: The Southeastern Perspective (G.W. Benz, and D.E. Collins, eds.). Special Publication 1, Southeast Aquatic Research Institute. Lenz Design and Communications, Decatur, Georgia.

Opinion 1195. 1981. *Pleurocera* Rafinesque, 1818 (Gastropoda): The type species is *Pleurocerus acutus* Rafinesque in Blainville, 1824. Bulletin of Zoological Nomenclature 38: 259-265.

Ponder, W.F. 1991. The anatomy of *Diala,* with an assessment of its taxonomic position (Mollusca: Cerithioidea). Pp. 499-519 in Proceedings of the Third International Marine Biological Workshop: The Marine Flora and Fauna of Albany, Western Australia, vol. 2 (F.E. Wells, D.I. Walker, H. Kirkman, and R. Lethbridge, eds.). Western Australian Museum, Perth.

Richardson, T.D., J.F. Scheiring, and K.M. Brown. 1988. Secondary production of two lotic snails (Pleuroceridae: *Elimia*). Journal of the North American Benthological Society 7: 234-241.

Sides, J.D. 2005. The systematics of freshwater snails of the genus *Pleurocera* (Gastropoda: Pleuroceridae) from the Mobile River Basin. PhD diss., Department of Biological Sciences, University of Alabama.

Strong, E.E. 2005. A morphological reanalysis of *Pleurocera acuta* Rafinesque, 1831, and *Elimia livescens* (Menke, 1930) (Gastropoda: Cerithioidea: Pleuroceridae). Nautilus 119: 119-132.

Strong, E.E., and T.J. Frest. 2007. On the anatomy and systematics of *Juga* from western North America (Gastropoda: Cerithioidea: Pleuroceridae). Nautilus 121: 43-65.

Strong, E.E., and F. Köhler. 2009. Morphological and molecular analysis of '*Melania*' *jacqueti* Dautzenberg and Fischer, 1906: From anonymous orphan to critical basal offshoot of the Semisulcospiridae (Gastropoda: Cerithioidea). Zoologica Scripta 38: 483-502.

Strong, E.E., O. Gargominy, W.F. Ponder, and P. Bouchet. 2008. Global diversity of gastropods (Gastropoda: Mollusca) in freshwater. Hydrobiologia 595: 149-166.

Strong, E.E., D.J. Colgan, J.M. Healy, C. Lydeard, W.F. Ponder, and M. Glaubrecht. 2011. Phylogeny of the gastropod superfamily Cerithioidea using morphology and molecules. Zoological Journal of the Linnean Society 162: 43-89.

Thompson, F.G. 2011. An annotated checklist and bibliography of the land and freshwater snails of Mexico and Central America. Bulletin of the Florida Museum of Natural History 50: 1-299.

Whelan, N.V., and E.E. Strong. 2014. Seasonal reproductive anatomy and sperm storage in pleurocerid gastropods (Cerithioidea: Pleuroceridae). Canadian Journal of Zoology 92: 989-995.

Whelan, N.V., and E.E. Strong. 2016. Morphology, molecules and taxonomy: Extreme incongruence in pleurocerids (Gastropoda, Cerithioidea, Pleuroceridae). Zoologica Scripta 45: 62-87.

Whelan, N.V., P.D. Johnson, and P.M. Harris. 2012a. Presence or absence of carinae in closely related populations of *Leptoxis ampla* (Anthony, 1855) (Gastropoda: Cerithioidea: Pleuroceridae) is not the result of ecophenotypic plasticity. Journal of Molluscan Studies 78: 231-233.

Whelan, N.V., P.D. Johnson, and P.M. Harris. 2012b. Rediscovery of *Leptoxis compacta* (Anthony, 1854) (Gastropoda: Cerithioidea: Pleuroceridae). PLoS ONE 7: 1-6.

Whelan, N.V., P.D. Johnson, and P.M. Harris. 2015. Life-history traits and shell morphology in the genus *Leptoxis* Rafinesque, 1819 (Gastropoda: Cerithioidea: Pleuroceridae). Journal of Molluscan Studies 81: 85-95.

Whelan, N.V., P.D. Johnson, J.T. Garner, and E.E. Strong. 2017. On the identity of *Leptoxis taeniata*—a misapplied name for the threatened Painted Rocksnail (Cerithioidea: Pleuroceridae). ZooKeys 697: 21-36.

11 Semisulcospiridae Morrison, 1952

DAVID C. CAMPBELL

The Semisulcospiridae includes the genus *Juga* H. Adams & A. Adams, 1854, from western North America (Burch and Tottenham, 1980) and several genera in eastern Asia, including *Hua* Chen, 1943, *Koreoleptoxis* Burch & Jung, 1988, and *Semisulcospira* Boettger, 1886 (Burch et al., 1988; Strong and Köhler, 2009). Some authors also recognize subgenera of *Juga* (*Calibasis* D. W. Taylor, 1966, *Oreobasis* D. W. Taylor, 1966, and the fossil *Idabasis* D. W. Taylor, 1966) or additional genera in Asia: *Biwamelania* Matsuoka, 1985 (although the name was used earlier, this is the first paper to validly describe it), *Koreanomelania* Burch & Jung, 1988, *Namrutua* Abbott, 1948, and *Senckenbergia* Yen, 1939. The name *Parajuga* Prozorova & Starobogatov, 2004, is in use in literature for species from southeastern Russia and nearby China and Korea, but it has not yet been officially described. Other genus names used historically but incorrectly for semisulcospirids include *Melania* Lamarck, 1799, *Goniobasis* Lea, 1862, *Elimia* H. Adams & A. Adams, 1854, and *Oxytrema* Rafinesque, 1819. *Juga* is the only freshwater cerithioidean native to Pacific drainages in the United

States, so records from that region can be confidently assigned to that genus. The situation is more complex in Asia, where the range of Semisulcospiridae overlaps with other families in the southern part of its range, any of which might be incorrectly called *Melania*.

The modern distribution ranges from central California north to Washington and east to Nevada and eastern Oregon in North America (Burch and Tottenham, 1980). In Asia, species range from the Amur basin in southeastern Russia to Vietnam and Taiwan, west to central China (Egorov, 2006; Strong and Köhler, 2009). However, fossils from Alaska and Siberia (Henderson, 1935 [as "*Goniobasis*"]; Zhadin, 1952) show that the family formerly lived in the gap between the modern Asian and North American ranges. As the Bering land bridge was present for most of the Cenozoic (Marincovich and Gladenkov, 1999), freshwater mollusks could have ranged across the region during warmer climates in the Miocene or Paleogene.

Semisulcospiridae differs from its closest relative,

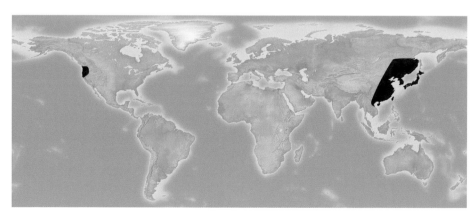

Distribution of Semisulcospiridae.

Pleuroceridae, by anatomical (notably in the gonoductal groove, seminal receptacle, prostate, and kidney) and molecular characters (Lydeard et al., 2002;
Strong and Frest, 2007; Strong and Köhler, 2009;
Strong et al., 2011). Freshwater cerithioideans from
many different families may have similar shell morphology, so anatomical or molecular data are necessary to definitively confirm the affinities of different
species. Most semisulcospirids have the typical high-
spired shell form common to most cerithioideans, but
the genus *Koreoleptoxis* was named for its stout shape
resembling the pleurocerid *Leptoxis* (Burch et al.,
1988). The similar shell forms cause particular confusion in southeastern Asia, where Semisulcospiridae,
Pachychilidae, and Thiaridae overlap in range and
many taxa remain known only from shells (Strong
and Köhler, 2009).

   Although some studies have examined intraspecific
variation in the family (e.g., Davis, 1972; Urabe, 2000),
many nominal species remain poorly known and the
total number of valid species in the family is unclear.
In particular, the type species of *Hua, Namrutua,* and
Senckenbergia require detailed study before relationships within the family become known. In Japan,
radiation in the ancient Lake Biwa has produced
several endemic species of *Semisulcospira* (Kamiya
et al., 2011), adding to the complexity of the family.
The current classification of *Juga* (e.g., Burch and Tottenham, 1980) largely follows Goodrich (1942, 1944).
However, Goodrich invoked no-longer-accepted
principles such as a supposed trend of "depauper-
ization" and treating polyphyletic groups as a single
species. Additionally, Goodrich provided little justifi-
cation for his decisions on synonymy. Thus, the species need to be reexamined in a modern phylogenetic
context, with new anatomical and molecular data.
There are no comprehensive revisions of all of the
Asian species. Starobogatov et al. (2004) provided a
key to the Russian species, but several species names
used in that publication still await formal description,
in addition to the genus name *Parajuga* and the family
name Jugidae (which, if it is officially named, would
be a synonym of Semisulcospiridae). Also, Starobogatov and collaborators generally recognized more species than most other authors, so the species deserve
further study to test for the degree of anatomical
and genetic differences. Vinarski and Kantor (2016)

Semisulcospira reiniana (Brot, 1876). Japan: Kyoto, Kyoto Prefecture. 32 mm. INHS 52321.

revise the taxonomy from Starobogatov et al. (2004),
using the genus *Juga* for the Russian species due to
the unavailability of *Parajuga.* However, the Russian
taxa studied so far with molecular data show affinities for other Asian taxa, not the North American
Juga. The Korean semisulcospirids were revised by
Köhler (2017). He synonymized *Koreanomelania* and
"*Parajuga*" with *Koreoleptoxis;* despite the difference
in shell shape, these taxa share oviparity and are
similar in molecular analyses. Davis (1969) reviewed
the Japanese species of semisulcospirid, though some
new taxa have been proposed since then (Watanabe
and Nishino, 1995). The *niponica* group of Davis (1969)
contains the species later separated as *Biwamelania.*
However, Köhler (2016) and Kamiya et al. (2011) found
that *Biwamelania* does not seem to be monophyletic
with respect to the rest of *Semisulcospira* and should
probably be treated as a synonym. Pace (1973) cat-
aloged the Taiwanese freshwater snails. No recent
study covers all of the Chinese species.

   The only molecular phylogenies published to date
use *cox1*, 16S, and 28S (Lee et al., 2007; Ó Foighil
et al., 2009; Strong and Köhler, 2009; Kim et al.,
2010; Strong et al., 2011; Miura et al., 2013; Hsu et al.,
2014; Chiu et al., 2016; Köhler, 2016; Zeng et al. 2016;
Köhler, 2017). However, these genes frequently have
anomalous alleles in this family (Lee et al., 2007;
Miura et al., 2013; Köhler, 2016; Köhler, 2017). Preliminary molecular data for *Juga* (Campbell et al., 2016)
suggest that Goodrich recognized too few species
in *Juga,* and that undescribed species are present.
Studies on Asian species are complicated by the
anomalous alleles (not always recognized as such in
older studies). Excluding the anomalous alleles, the

molecular patterns tend to show stronger matches with geography than with the species identifications (Lee et al., 2007), also suggesting a need for a more comprehensive phylogenetic study and revision. The recent publication of complete or nearly complete mitochondrial genomes (Zeng et al., 2015; An et al., 2015, 2017; Kim and Lee, 2018) should help with improved genetic studies. Strong and Köhler (2009), Strong et al. (2011), and Köhler (2016, 2017) provide the most comprehensive genus-level analyses. *Semisulcospira* and *Koreoleptoxis* (including *Koreanomelania* and *"Parajuga"*) constitute a clade separate from *Juga*. The data of Miura et al. (2013) and Köhler (2016) indicate that *Biwamelania* is also related to the other northern Asian forms and is a probable synonym of *Semisulcospira*. *Hua jacqueti* and a sequences from Du et al. (2016) on GenBank from southern China (based on the ranges of the listed species; the GenBank entries do not give location) place as a basal clade distinct from the other Asian taxa. 16S data places *Hua jacqueti* closer to *Juga* than to the other Asian taxa (Strong and Köhler, 2009), but with *cox1,* these species place as sister to the remaining Asian taxa (pers. obs.). The genus names *Hua, Namrutua,* and *Senckenbergia* are based on taxa from this region, but standard *Semisulcospira* overlaps with these taxa in the Yangtze drainage, so anatomical and genetic studies on the type species are needed to determine which, if any, is appropriate for this southern group.

Although our understanding of conservation needs will change as a more thorough phylogentic picture develops, at present many semisulcospirids appear to be at risk. Johnson et al. (2013) identified *Juga* as 91% imperiled, including one possibly extinct species (*Juga bulbosa,* listed as endangered in the species-by-species ranking but tallied as extinct in the summary statistics). In Japan, Nishino (2012) lists 15 *Semisulcospira* species endemic to Lake Biwa; of these, 14 are listed for Japan and/or for Shiga Prefecture as Near Threatened or Endangered. Suh et al. (2014) list one species as endangered and one as vulnerable in South Korea.

As with freshwater mollusks generally, multiple factors are involved, including excessive water use, limited natural ranges, dams, siltation, pollution, eutrophication, invasive species, and other habitat modification (Strong et al., 2008).

In some cases, however, deliberate control of semi-sulcospirid population is a goal due to their hosting of parasites of medical or veterinary importance. Several species of trematodes parasitize semisulcospirids, including *Paragonimus* spp. (Urabe, 2003). These lung flukes afflict humans and many other carnivorous or omnivorous mammals, primarily in eastern Asia. They use a crab or crayfish as a second intermediate host (Diaz, 2013). Another common semisulcospirid-hosted parasite in humans and other mammals is the small intestinal fluke *Metagonimus* (Shimazu, 2002). Besides potential problems from the flukes themselves, some flukes can carry rickettsid bacteria (*Neorickettsia* spp.). Important diseases in this group include Potomac horse fever and salmon poisoning (which is particularly deadly to canids) (Headley et al., 2011). Regrettably, many parasitological references are not up to date in molluscan systematics (by a few decades or more!), leading to potential confusion about the host species. In turn, inaccurate identification of the host species may lead to inappropriate attempts at control or other problems.

Semisulcopsirids and other freshwater cerithio-deans are important grazers and shredders in many freshwater ecosystems, especially rivers (Dillon, 2000). Antonio et al. (2010) found *Semisulcospira libertina* to be a generalist in its feeding habits. In turn, the snails may be eaten by various predators, including crabs (Kobayashi, 2012) and crayfish (US Fish and Wildlife Service, 1998). They can also be important competitors (Hawkins and Furnish, 1987). Thus, they play significant roles in shaping fluvial ecosystems.

LITERATURE CITED

An, H.S., D.H. Kim, H.K. Hwang, J.I. Myeong, and C.M. An. 2015. The complete mitochondrial genome of *Koreoleptoxis globus ovalis*. Unpublished manuscript. Data on GenBank. https://www.ncbi.nlm.nih.gov/nuccore/LC006055.1.

An, H.S., D.H. Kim, H.K. Hwang, J.I. Myeong, and C.M. An. 2017. Mitochondrial Genome sequence of *Koreanomelania nodifila*. Unpublished manuscript. Data on GenBank. https://www.ncbi.nlm.nih.gov/nuccore/KJ696780.1.

Antonio, E.S., A. Kasai, M. Ueno, Y. Kurikawa, K. Tsuchiya, H. Toyohara, Y. Ishihi, H. Yokoyama, and Y. Yamashita. 2010. Consumption of terrestrial organic matter by estuarine molluscs determined by analysis of their stable isotopes and cellulase activity. Estuarine, Coastal and Shelf Science 86: 401–407.

Burch, J.B., and J.L. Tottenham. 1980. North American freshwater snails: Species list, ranges and illustrations. Walkeriana 1: 81-215.

Burch, J.B., P.-R. Chung, and Y. Jung. 1988. A guide to the freshwater snails of Korea. Walkerana 2: 195-232.

Campbell, D.C., S.A. Clark, E.J. Johannes, C. Lydeard, and T.J. Frest. 2016. Molecular phylogenetics of the freshwater gastropod genus *Juga* (Cerithioidea: Semisulcospiridae). Biochemical Systematics and Ecology 65: 158-170.

Chiu, Y.-W., H. Bor, P.-H. Kuo, K.-C. Hsu, M.-S. Tan, W.-K. Wang, and H.-D. Lin. 2016. Origins of *Semisulcospira libertina* (Gastropoda: Semisulcospiridae) in Taiwan: Mitochondrial DNA Part A: DNA Mapping, Sequencing, and Analysis 28 (4): 518-525.

Davis, G.M. 1969. A taxonomic study of some species of *Semisulcospira* in Japan (Mesogastropoda: Pleuroceridae). Malacologia 7 (2-3): 211-294.

Davis, G.M. 1972. Geographic variation in *Semisulcospira libertina* (Mesogastropoda: Pleuroceridae). Proceedings of the Malacological Society, London, 40: 5-32.

Diaz, J.H. 2013. Paragonimiasis acquired in the United States: Native and nonnative species. Clinical Microbiology Reviews 26: 493-504.

Dillon, R.T., Jr. 2000. The Ecology of Freshwater Molluscs. Cambridge University Press, Cambridge, UK.

Du, Y.Y., X.P. Wu, and S. Ouyang. 2016. Phylogenetic relationships of Pleuroceridae Gastropoda (Mollusca) based on mitochondrial COI gene sequences. Unpublished manuscript. Sequences FJ437213-23 on GenBank.

Egorov, R. 2006. Illustrated Catalogue with Selected Identification Keys of the Recent Fresh- and Brackish-Water Pectinibranch Mollusks (Gastropoda: Pectinibranchia) of Russia and Adjacent Regions. Treasure of Russian Shells, supplement 4. Colus-Doverie, Moscow.

Goodrich, C. 1942. The Pleuroceridae of the Pacific coastal drainage, including the Western Interior Basin. Occasional Papers of the Museum of Zoology, University of Michigan, 469: 1-4.

Goodrich, C. 1944. Pleuroceridae of the Great Basin. Occasional Papers of the Museum of Zoology, University of Michigan, 485: 1-11.

Hawkins, C.P., and J.K. Furnish. 1987. Are snails important competitors in stream ecosystems? Oikos 49: 209-220.

Headley, S.A., D.G. Scorpio, O. Vidotto, and J.S. Dumler. 2011. *Neorickettsia helminthoeca* and salmon poisoning disease: A review. Veterinary Journal 187: 165-173.

Henderson, J. 1935. Fossil non-marine mollusca of North America. Geological Society of America Special Paper 3: vii, 1-313.

Hsu, K.C., H. Bor, H.D. Lin, P.H. Kuo, M.S. Tan, and Y.W.

Chiu. 2014. Mitochondrial DNA phylogeography of *Semisulcospira libertina* (Gastropoda: Cerithioidea: Pleuroceridae): Implications [of] the history of landform changes in Taiwan. Molecular Biology Reports 41: 3733-3743.

Johnson, P.D., A.E. Bogan, K.M. Brown, N.M. Burkhead, J.R. Cordeiro, J.T. Garner, P.D. Hartfield, et al. 2013. Conservation status of freshwater gastropods of Canada and the United States. Fisheries 38: 247-282.

Kamiya, S., M. Shimamoto, and T. Hashimoto. 2011. Allozyme analysis of Japanese *Semisulcospira* species (Gastropoda: Pleuroceridae) reveals that Lake Biwa endemic species are not monophyletic. American Malacological Bulletin 29 (1-2): 23-36.

Kim, W.-J., D.-H. Kim, J.S. Lee, I.-C. Bang, W.-O. Lee, and H. Jung. 2010. Systematic relationships of Korean freshwater snails of *Semisulcospira, Koreanomelania,* and *Koreoleptoxis* (Cerithioidea;[sic] Pleuroceridae) revealed by mitochondrial cytochrome oxidase I sequences. Korean Journal of Malacology 26: 275-283.

Kim, Y.K., and S. M. Lee. 2018. The complete mitochondrial genome of freshwater snail, *Semisulcospira coreana* (Pleuroceridae: Semisulcospiridae) [sic]. Mitochondrial DNA Part B: Resources 3 (1): 259-260.

Kobayashi, S. 2012. Dietary preference of the potamid crab *Geothelphusa dehaani* in a mountain stream in Fukuoka, northern Kyushu, Japan. Plankton & Benthos Research 7 (4): 159-166.

Köhler, F. 2016. Rampant taxonomic incongruence in a mitochondrial phylogeny of *Semisulcospira* freshwater snails from Japan (Cerithioidea: Semisulcospiridae). Journal of Molluscan Studies 82 (2): 268-281.

Köhler, F. 2017. Against the odds of unusual mtDNA inheritance, introgressive hybridisation and phenotypic plasticity: Systematic revision of Korean freshwater gastropods (Semisulcospiridae, Cerithioidea). Invertebrate Systematics 31: 249-268.

Lee, T., H.C. Hong, J.J. Kim, and D. Ó Foighil. 2007. Phylogenetic and taxonomic incongruence involving nuclear and mitochondrial markers in Korean populations of the freshwater snail genus *Semisulcospira* (Cerithioidea: Pleuroceridae). Molecular Phylogenetics and Evolution 43: 386-397.

Lydeard, C., W.E. Holznagel, M. Glaubrecht, and W.F. Ponder. 2002. Molecular phylogeny of a circum-global, diverse gastropod superfamily (Cerithioidea: Mollusca: Caenogastropoda): Pushing the deepest phylogenetic limits of mitochondrial LSU rDNA sequences. Molecular Phylogenetics and Evolution 22: 399-406.

Marincovich, L., and A. Yu. Gladenkov. 1999. Evidence for an early opening of the Bering Strait. Nature 397: 149-151.

Miura, O., F. Köhler, T. Lee, J. Li, and D. Ó Foighil. 2013. Rare, divergent Korean *Semisulcospira* spp. mitochondrial haplotypes have Japanese sister lineages. Journal of Molluscan Studies 79: 86-89.

Nishino, M. 2012. Ch2 introduction: Biodiversity of Lake Biwa. Pp. 31-35 in Lake Biwa: Interactions between Nature and People (H. Kawanabe, M. Nishino, and M. Maehata, eds.). Springer, Dordrecht, Netherlands.

Ó Foighil, D., T. Lee, D.C. Campbell, and S.A. Clark. 2009. All voucher specimens are not created equal: A cautionary tale involving the North American pleurocerids *Juga hemphilli* (Henderson 1935) and *Elimia dooleyensis* (Lea, 1862). Journal of Molluscan Studies 75: 305-306.

Pace, G.L. 1973. The freshwater snails of Taiwan (Formosa). Malacological Review Supplement 1: 1-118, plates 1-19, figures 1-17.

Shimazu, T. 2002. Life cycle and morphology of *Metagonimus miyatai* (Digenea: Heterophyidae) from Nagano, Japan. Parasitology International 51: 271-280.

Starobogatov, Y.I., L.A. Prozorova, V.V. Bogatov, and E.M. Sayenko. 2004. Molluscs. Pp. 9-491 in Key to Freshwater Invertebrates of Russia and Adjacent Lands, vol. 6, Molluscs, Polychaetes, Nemerteans [in Russian] (S.J. Tsalolikhin, ed.). Nauka, St. Petersburg.

Strong, E.E., and T.J. Frest. 2007. On the anatomy and systematics of *Juga* from western North America (Gastropoda: Cerithioidea: Pleuroceridae). Nautilus 121: 43-65.

Strong, E.E., and F. Köhler. 2009. Morphological and molecular analysis of '*Melania*' *jacqueti* Dautzenberg and Fischer, 1906: From anonymous orphan to critical basal offshoot of the Semisulcospiridae (Gastropoda: Cerithioidea). Zoologica Scripta 38: 483-502.

Strong, E.E., O. Gargominy, W.F. Ponder, and P. Bouchet. 2008. Global diversity of gastropods (Gastropoda: Mollusca) in freshwater. Hydrobiologia 595: 149-166.

Strong, E.E., D.J. Colgan, J.M. Healy, C. Lydeard, W.F. Ponder, and M. Glaubrecht. 2011. Phylogeny of the gastropod superfamily Cerithioidea using morphology and molecules. Zoological Journal of the Linnean Society 162: 43-89.

Suh, M.-H., B.-Y. Lee, S.T. Kim, C.-H. Park, H.-K. Oh, H.-Y. Kim, J.-H. Lee, and S.Y. Lee, eds. 2014. Korean Red List of Threatened Species. 2nd ed. National Institute of Biological Resources, Incheon, Korea.

Urabe, M. 2000. Phenotypic modulation by the substratum of shell sculpture in *Semisulcospira reiniana* (Prosobranchia: Pleuroceridae). Journal of Molluscan Studies 66: 53-59.

Urabe, M. 2003. Trematode fauna of prosobranch snails of the genus *Semisulcospira* in Lake Biwa and the connected drainage system. Parasitology International 52: 21-34.

US Fish and Wildlife Service. 1998. Recovery Plan for the Shasta Crayfish (*Pacifastacus fortis*). US Fish and Wildlife Service, Portland, Oregon.

Vinarski, M., and Y. Kantor. 2016. The Analytical Catalogue of Fresh and Brackish Water Molluscs of Russia and Adjacent Countries. A.N. Severtsov Institute of Ecology and Evolution of RAS, Moscow.

Watanabe, N., and M. Nishino. 1995. A study on taxonomy and distribution of the freshwater snail, genus *Semisulcospira* in Lake Biwa, with descriptions of eight new species. Lake Biwa Study Monographs 6: 1-33.

Zeng, T., J. Rong, W.-J. Wang, S.-B. Li, J.-K. Chen, and B.-S. Jin. 2016. Study on the genetic structure of *Semisulcospira libertina* populations from five rivers of Poyang Lake basin. Acta Hydrobiologica Sinica 40 (1): 211-216.

Zeng, T., W. Yin, R. Xia, C. Fu, and B. Jin. 2015. Complete mitochondrial genome of a freshwater snail, *Semisulcospira libertina* (Cerithioidea: Semisulcospiridae). Mitochondrial DNA 26: 897-898.

Zhadin, Z.D. 1952. Mollyuski presnykh i solonovatykh vod SSSR. [Mollusks of fresh and brackish waters of the USSR]. Academy of Sciences of the U.S.S.R., Moscow Translation by the Israel Program for Scientific Translation, Jerusalem, 1965.

12 Thiaridae Gill, 1871 (1823)

MATTHIAS GLAUBRECHT AND MARCO T. NEIBER

The Thiaridae Gill, 1871 (1823) (citation of author and date of the taxon following Recommendation 40A of the ICZN Code of Zoological Nomenclature), in earlier treatments subsumed under the name "Melaniidae," has been used for a long time as a "rubbish bin" to accommodate all freshwater Cerithioidea lineages. By splitting the "Melaniidae" in the sense of Thiele (1928, 1929) into three separate families, Morrison (1952, 1954) started a process of disassembling this large, from the present day perspective, polyphyletic assemblage of freshwater Cerithioidea. Taxonomic concepts have largely been in flux over the following decades (Starobogatov, 1970; Houbrick, 1988; Ponder & Warén, 1988; Brown, 1994). Only during the past two decades, after the removal of Melanopsidae, Pachychilidae, Paludomidae, Pleuroceridae, Semisulcospiridae, a better substantiated circumscription of "core" Thiaridae began to emerge on the basis of molecular and/or morphological evidence (e.g., Glaubrecht, 1993, 1996; Holznagel and Lydeard, 2000; Lydeard et al., 2002; Simone, 2001; Strong, 2011; Strong et al., 2011), although relationships with fossil forms as outlined in the classification of Bandel (2006) still remain problematic.

Thiaridae, as here conceived, form a monophyletic group (Lydeard et al., 2002; Strong et al., 2011; Gimnich, 2015) with its constituent species being probably autochthonous in Southeast and South Asia, Australia, and some Pacific Islands, as well as sub-Saharan Africa, both in lotic and lentic freshwater environments, with some species also tolerating brackish conditions in the lower courses and estuaries of rivers. Some species, such as *Melanoides tuberculata* (Müller, 1774), have an extraordinarily high invasive potential and today have an almost circum-global distribution in tropical and subtropical biomes (e.g., Brown, 1994; Glaubrecht, 1996). Morphologically, Thiaridae are characterized by a tubular nonglandular pallial oviduct, a spermatophore bursa, and a large and textured accessory pad in the midgut (Strong et al., 2011). Parthenogenetic reproduction, as was in particular described in *Melanoides tuberculata* or in *Mieniplotia scabra* (see Glaubrecht 1996), is also widely distributed among Australian thiarids (Glaubrecht

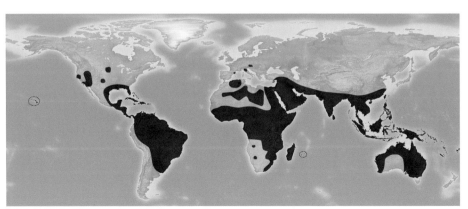

Distribution of Thiaridae.

Thiara cancellata Röding, 1798. Philippines, Davao del Norte, Talikud Island. 25.5 mm. Poppe 1025144.

et al., 2009). Thiaridae have realized two life history strategies that were characterized in Glaubrecht (1996, 1999, 2006) by the duration of ontogenetic stages to remain within the brood pouch of the female, which at the same time might also reflect a differential degree of parental investment, even including nourishment of the embryos and juveniles via nutritive tissue. While in some thiarids only very early ontogenetic stages, viz. eggs or embryos without shell, develop and are released as veligers (ovoviviparity), other thiarid species brood and even transform their subhemocoelic brood pouch into a "pseudoplacenta" (Glaubrecht, 1996) that, via matrotrophy, apparently helps to nourish the developing juveniles, as, e.g., in the Southeast Asian thiarid *Tarebia granifera* (Lamarck, 1822) (eu-viviparity, see Glaubrecht et al., 2009).

Diversity on the generic as well as the species level of Thiaridae is currently rather poorly understood, especially because the genus *Melanoides* Olivier, 1804, has been found to be polyphyletic with regard to some currently accepted genera (Facon et al., 2003; Genner et al., 2007; Van Bocxlaer et al., 2015). Being both intriguing and unexpected, these findings need further investigation using a broader taxon sampling and additional molecular markers that allow a better phylogenetic resolution. Glaubrecht et al. (2009) surveyed the thiarid diversity of Australia, recognizing about eight major groups subsumed under the generic names *Thiara* Röding, 1798, the type genus of the family, *Melanoides* Olivier, 1804, *Sermyla* Adams & Adams, 1854, *Melasma* Adams & Angas,

1864, *Plotiopsis* Brot, 1874, *Stenomelania* Fischer, 1885, *Ripalania* Iredale, 1943, and *Plotia* Röding, 1798 (now newly described as *Mieniplotia* Low & Tan, 2014, because *Plotia* is an objective synonym of *Pyramidella* Lamarck, 1799, *Tiaropsis* Brot, 1870, a homonym *Tiaropsis* Agassiz, 1850, a genus in Cnidaria, and *Pseudoplotia* Forcart, 1950, which has been used occasionally, is an unavailable name; see Low and Tan, 2014). In addition, the following taxa belong to the Thiaridae: *Tarebia* Adams & Adams, 1854, with the widespread and invasive species *T. granifera*, *Neoradina* Brandt, 1974, from Thailand (see Brandt, 1974; plus questionable taxa mentioned from the Nicobar and Andaman Islands; see Nevill, 1884; Preston, 1915), *Balanocochlis* Fischer, 1885, from Southeast Asia and some West Pacific islands (Starmühlner, 1976), and *Fijidoma* Morrison, 1952, endemic to the Fiji Islands.

In concert with uncertainties as to the delimitations of genera within the family, research on thiarids is further hampered by the incredibly large phenotypic plasticity of their shell that is, however, also known from other freshwater Cerithioidea. This conchological variability might lead to an overestimation of the number of species, as specifically shown for other lineages of limnic Cerithioidea, such as Melanopsidae (Glaubrecht 1993, 1996) or Pachychilidae (Köhler and Glaubrecht 2001, 2003, 2006; Glaubrecht and Köhler, 2004). In addition to the difficult assessment of large individual, ecological and/or geographical variability, an isolated treatment of only few or even singular specimens from widely separated localities always poses the danger of misinterpreting taxonomic diversity by underestimating conchological disparity, resulting in systematic-taxonomic uncertainties and unresolved phylogenetic affinities. These problems are exacerbated by the putatively widespread occurrence of parthenogenesis in different lineages and the associated problems of what is actually meant by the term "species." Taking the abovementioned considerations into account, accurate estimates of the species number in the family are impossible at the moment, and the number of about 135 validly described species given by Strong et al. (2008) ought to be interpreted only as a rough guess until detailed revisions will be available for the entire group. Nonetheless, taxonomic redundancy can be expected to be rather high.

A total of 59 species assigned to the Thiaridae are currently included in the IUCN Red List, 1 as Critically Endangered, 3 as Endangered, 4 as Vulnerable, 4 as Near Threatened, 22 as Least Concern, and 25 as Data Deficient (IUCN, 2017). About 20% of the species are included in one of the threat categories and 37% are assessed as Least Concern, whereas over 40% are assessed as Data Deficient, reflecting our current lack of knowledge with regard to distribution and taxonomic status of many of the constituent species of the family.

Thiarid species, especially *Melanoides tuberculata,* are known as important intermediate hosts of digenean trematodes. Based on a literature survey, Pinto and Melo (2011) list 37 fluke species belonging to 17 families and 25 genera from *Melanoides tuberculata;* among these, 11 are known to be human pathogens. Trematode diversity using *M. tuberculata* as intermediate host may even be much greater, as evidenced by 81 additional larval stages not assignable to adult trematode stages reported by Pinto and Melo (2011).

LITERATURE CITED

Bandel, K. 2006. Families of the Cerithioidea and related superfamilies (Palaeo-Caenogastropoda; Mollusca) from the Triassic to the Recent characterized by protoconch morphology—including the description of new taxa. Freiberger Forschungshefte C 511: 57-134.

Brandt, R.A.M. 1974. The non-marine aquatic Mollusca of Thailand. Archiv für Molluskenkunde 105: i-iv, 1-423.

Brown, D.S. 1994. Freshwater Snails of Africa and Their Medical Importance. 2nd ed. Taylor & Francis, London.

Facon, B., J.-P. Pointier, M. Glaubrecht, C. Poux, P. Jarne, and P. David. 2003. A molecular phylogeography approach to biological invasions of the New World by parthenogenetic thiarid snails. Molecular Ecology 12: 3027-3039.

Genner, M.J., J.A. Todd, E. Michel, D. Erpenbeck, A. Jimoh, D.A. Joyce, A. Piechocki, and J.-P. Pointier. 2007. Amassing diversity in an ancient lake: Evolution of a morphologically diverse parthenogenetic gastropod assemblage in Lake Malawi. Molecular Ecology 16: 517-530.

Gimnich, F. 2015. Molecular approaches to the assessment of biodiversity in limnic gastropods (Cerithioidea, Thiaridae) with perspectives on a Gondwanian origin. PhD diss., Humboldt Universität zu Berlin.

Glaubrecht, M. 1993. Mapping the diversity: Geographical distribution of the freshwater snail *Melanopsis* (Gastropoda: Cerithioidea: Melanopsidae) with focus on its

systematics in the Mediterranean Basin. Mitteilungen aus dem Hamburger Zoologischen Museum und Institut 90: 41-97.

Glaubrecht, M. 1996. Evolutionsökologie und Systematik am Beispiel von Süß- und Brackwasserschnecken (Mollusca: Caenogastropoda: Cerithioidea): Ontogenese-Strategien, paläontologische Befunde und Historische Zoogeographie. Backhuys, Leiden.

Glaubrecht, M. 1999. Systematics and the evolution of viviparity in tropical freshwater gastropods (Cerithioidea: Thiaridae sensu lato)—an overview. Courier Forschungsinstitut Senckenberg 125: 91-96.

Glaubrecht, M. 2006. Independent evolution of reproductive modes in viviparous freshwater Cerithioidea (Gastropoda, Sorbeoconcha)—a brief review. Basteria 69, Supplement 3: 28-32.

Glaubrecht, M., and F. Köhler. 2004. Radiating in a river: Systematics, molecular genetics and morphological differentiation of viviparous freshwater gastropods endemic to the Kaek River, central Thailand (Cerithioidea, Pachychilidae). Biological Journal of the Linnean Society 82: 275-311.

Glaubrecht, M., N. Brinkmann, and J. Pöppe. 2009. Diversity and disparity 'down under': Systematics, biogeography and reproductive modes of the 'marsupial' freshwater Thiaridae (Caenogastropoda, Cerithioidea) in Australia. Zoosystematics and Evolution 85: 199-275.

Holznagel, W.E., and C. Lydeard. 2000. A molecular phylogeny of North American Pleuroceridae (Gastropoda: Cerithioidea) based on mitochondrial 16S rDNA sequences. Journal of Molluscan Studies 66: 233-257.

Houbrick, R.S. 1988. Cerithiodean phylogeny. In Prosobranch Phylogeny: Proceedings of a Symposium Held at the 9th International Malacological Congress, Edinburgh, 1986 (W.F. Ponder, ed.). Malacological Review Supplement 4: 88-128.

IUCN Red List of Threatened Species. 2017. Version 2017-1. www.iucnredlist.org. Accessed 25 May 2017.

Köhler, F., and M. Glaubrecht. 2001. Toward a systematic revision of the Southeast Asian freshwater gastropod *Brotia* H. Adams, 1866 (Cerithioidea: Pachychilidae): An account of species from around the South China Sea. Journal of Molluscan Studies 67: 281-318.

Köhler, F., and M. Glaubrecht. 2003. Morphology, reproductive biology and molecular genetics of ovoviviparous freshwater gastropods (Cerithioidea: Pachychilidae) from the Philippines, with description of a new genus *Jagora.* Zoologica Scripta 32: 35-59.

Köhler, F., and M. Glaubrecht. 2006. A systematic revision of the Southeast Asian freshwater gastropod *Brotia* (Cerithioidea: Pachychilidae). Malacologia 48: 159-251.

Low, M.E.Y., and S.K. Tan. 2014. *Mieniplotia* gen. nov. for *Buccinum scabrum* O. F. Müller, 1774, with comments on the nomenclature of *Pseudoplotia* Forcart, 1950, and *Tiaropsis* Brot, 1870 (Gastropoda: Caenogastropoda: Cerithioidea: Thiaridae). Occasional Molluscan Papers 3: 15–17.

Lydeard, C., W.E. Holznagel, M. Glaubrecht, and W.F. Ponder. 2002. Molecular phylogeny of a circum-global, diverse gastropod superfamily (Cerithioidea: Mollusca: Caenogastropoda): Pushing the deepest phylogenetic limits of mitochondrial LSU rDNA sequences. Molecular Phylogenetics and Evolution 22: 399–406.

Morrison, J.P.E. 1952. World relations of the melanians (an abstract). American Malacological Union News Bulletin and Annual Report 1951: 6–9.

Morrison, J.P.E. 1954. The relationships of Old and New World melanians. Proceedings of the United States National Museum 103: 357–394.

Nevill, G. 1884. Hand List of Mollusca in the Indian Museum, Calcutta. Part II. Gastropoda. Prosobranchia-Neurobranchia (contd.). Trustees of the Indian Museum Calcutta, Calcutta.

Pinto, H.A., and A.L. de Melo. 2011. A checklist of trematodes (Platyhelminthes) transmitted by *Melanoides tuberculata* (Mollusca: Thiaridae). Zootaxa 2799: 15–28.

Ponder, W.F., and A. Warén. 1988. Classification of the Caenogastropoda and Heterostropha—a list of the family-group names and higher taxa. In Prosobranch phylogeny: Proceedings of a Symposium Held at the 9th International Malacological Congress, Edinburgh, 1986 (W.F. Ponder, ed.). Malacological Review Supplement 4: 288–328.

Preston, H.B. 1915. The Fauna of British India, including Ceylon and Burma: Mollusca (Freshwater Gastropoda & Pelecypoda.). Taylor & Francis, London.

Simone, L.R.L. 2001. Phylogenetic analyses of Cerithioidea (Mollusca, Caenogastropoda) based on comparative morphology. Arquivos de Zoologia 36: 147–263.

Starmühlner, F. 1976. Beiträge zur Kenntnis der Süßwasser-Gastropoden pazifischer Inseln. Ergebnisse der Österreichischen Indopazifik-Expedition des 1. Zoologischen Institutes der Universität Wien. Annalen des Naturhistorischen Museums in Wien 80: 473–656.

Starobogatov, Y.I. 1970. Fauna Molliuskov i Zoogeograficheskoe Raionirovanie Kontinental'nykh Vodoemov Zemnogo Shara. Nauka, Leningrad.

Strong, E.E. 2011. More than a gut feeling: Utility of midgut anatomy in phylogeny of the Cerithioidea (Mollusca: Caenogastropoda). Zoological Journal of the Linnean Society 162: 585–630.

Strong, E.E., O. Gargominy, W.F. Ponder, and P. Bouchet. 2008. Global diversity of gastropods (Gastropoda: Mollusca) in freshwater. Hydrobiologia 595: 149–166.

Strong, E.E., D.J. Colgan, J.M. Healy, C. Lydeard, W.F. Ponder, and M. Glaubrecht. 2011. Phylogeny of the gastropod superfamily Cerithioidea using morphology and molecules. Zoological Journal of the Linnean Society 162: 43–89.

Thiele, J. 1928. Revision des Systems der Hydrobiiden und Melaniiden. Zoologische Jahrbücher, Abteilung für Systematik, Ökologie und Geographie der Tiere 55: 351–402.

Thiele, J. 1929. Handbuch der systematischen Weichtierkunde. Teil 1 (Loricata; Gastropoda: Prosobranchia). Fischer, Jena, Germany.

Van Bocxlaer, B., C. Clewing, J.-P. Mongindo Etimosundja, A. Kankonda, O. Wembo Ndeo, and C. Albrecht. 2015. Recurrent camouflaged invasions and dispersal of an Asian freshwater gastropod in tropical Africa. BMC Evolutionary Biology 15: 33.

13 Amnicolidae Tryon, 1862

STEPHANIE A. CLARK

The Amnicolidae is currently divided into two subfamilies, Amnicolinae and Baicaliinae (Wilke et al., 2013). In the past the amnicolids have been treated either as a distinct family or as a subfamily of the Hydrobiidae (Burch and Tottenham, 1980; Hershler and Thompson, 1988; Kabat and Hershler, 1993; Bouchet and Rocroi, 2005). Recent comparative anatomical and molecular studies have shown that the family Hydrobiidae is polyphyletic and that many more families need to be recognized, such as the Amnicolidae (Wilke et al., 2001; Wilke et al., 2013, Wilke and Delicado, chap. 18, this volume). Wilke et al. (2013) also showed that the Bythinellinae and Emmericiinae, previously included in the Amnicolidae by Bouchet and Rocroi (2005), should be treated as separate families. However, the sister relationships among the various families previously considered to belong to the Hydrobiidae remain uncertain (Wilke et al., 2013).

Amnicolids have small conical to depressed conical dextral shells that vary in size from 2 to 5 mm in height. They have smooth, relatively featureless shells, and the operculum is paucispiral and pale yellow. Amnicolids have separate sexes and show some sexual dimorphism, with the females tending to be larger and broader than the males. Their tentacles are long and tapering and the animal is generally pale. The pallial roof is often pigmented with a small number of dark bands. The penis is bifurcate with an internal gland which sometimes extends into the haemocoel, while females have a spermathecal duct (Hershler and Thompson, 1988). They lay distinctive semilenticular, corneous egg capsules boarded with a thin lamina flange (Baker, 1928; Berry, 1943), which contain a single egg and are typically attached to various substrates including leaves, gravel, and pieces of woody debris (Clark, pers. obs.).

The family is widely distributed across the Palearctic, Nearctic, and northern Asia but to date has not been recorded from the southern hemisphere. However, the bulk of the generic and species diversity is found in the Palaearctic (Strong et al., 2008). The highest diversity is found in and around Lake Baikal (Kantor et al., 2010).

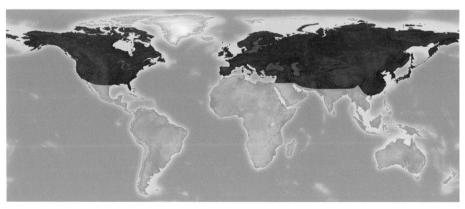

Distribution of Amnicolidae.

Amnicola limosus (Say, 1817). Canada: Lake Egmont, Halifax Regional Municipality, Nova Scotia. 3.9 mm. Poppe 966521.

There are about 120 described species, primarily from the Palearctic, with a smaller number from the Nearctic and northern Asia (note that Strong et al., 2008, give an estimate of about 200 species, but their estimate includes the Bythinellids and Emmericinids). The family is found in a wide range of aquatic habitats, from isolated springs and seepages, streams and rivers, and caves to very large lakes. Some species have wide ranges, such as *Amnicola limosa* (Burch and Tottenham, 1980), while others are known only from a few locations or a single location, such as *Erhaia daliensis* and *Amnicola cora* (Hubricht, 1979; Davis et al., 1985).

In North America the family is represented by at least 20 species in six genera (*Amnicola* Gould & Haldeman, 1840, *Antroselates* Hubricht, 1963, *Colligyrus* Hershler, 1999, *Dasyscias* F. G. Thompson & Hershler, 1991, *Lyogyrus* Gill, 1863, and *Taylorconcha* Hershler et al., 1994 (Walker, 1906; Hubricht, 1963; Thompson, 1968; Hubricht, 1971; Thompson and Hershler, 1991; Hershler, 1999; Liu et al., 2016). Most of the known diversity is found in the eastern half of the North America, but particularly in the southeastern states, with one genus, *Amnicola,* found from coast to coast and one genus, *Colligyrus,* known only from west of the Continental Divide (Johnson et al., 2013; Liu et al., 2015; Liu et al., 2016). Amnicolids are widely

distributed throughout North America; however they appear to be lacking from the southwestern states and further south into Mexico. They are found only in freshwater, from very small springs and streams to very large lakes (Burch and Tottenham, 1980; Clarke, 1973), with four species known only from caves in Arkansas, Missouri, Kentucky, Indiana, and Florida (*Amnicola cora, Amnicola stygius, Antroselates spiralis,* and *Dasyscias* franzi (Hubrich, 1963, 1971, 1979; Thompson and Hershler, 1991).

In Europe the family is represented by about 12 species in three genera: *Marstoniopsis* van Regteren Altena, 1936, *Parabythinella* Radoman, 1973, and *Terrestribythinella* Sitnikova et al., 1992 (Kantor et al., 2010; Welter-Schultes, 2012). They are found in springs, streams, and lakes (Radea et al., 2013).

In the Far East and in Lake Baikal, the family is represented by about 64 species and 11 genera: *Akiyoshia* Kuroda & Habe, 1954, *Baicalia* Martens, 1876, *Godlewskia* Crosse & Fischer, 1879, *Korotnewia* Kozhov, 1936, *Kolhymamnicola* Starobogatov & Budnikova, 1976, *Liobaicalia* Martens, 1876, *Maackia* Clessin, 1880, *Parabaikalia* Lindholm, 1909, *Psuedobaikalia* Lindholm, 1909, *Pyrgobaicalia* Starobogatov, 1972, and *Teratobaikalia* Lindholm, 1909 (Kantor et al., 2010).

In China and Japan the family is represented by at least 22 species in four genera: *Akiyoshia* Kuroda & Habe, 1954, *Chencuia* Davis, 1997, *Erahia* Davis & Kuo, 1985, and *Moria* Kuroda & Habe, 1958, from China and Japan (Davis et al., 1985; Davis et al., 1992; Davis and Rao, 1997, Kantor et al., 2010).

The North American amnicolid fauna is not particularly diverse, with most species having relatively large ranges; the exceptions are those adapted to caves. However, the taxonomy and systematics of the two most diverse genera, *Amnicola* and *Lyogyrus,* are in need of revision (Liu et al., 2015, 2016). Liu et al. (2016) found that two taxa previously thought to represent undescribed western US species belonged to eastern genera, with one synonymized with *Amnicola limosa,* but they were not able to fully resolve their taxonomic and systematics relationships.

The factors that affect the diversity and conservation of amnicolids include habitat destruction from dams, channel modification, siltation, water-quality degradation from point and nonpoint pollution, and the introduction of nonindigenous species. Lysne et al. (2008) explored challenges and opportunities of freshwater conservation and highlighted the need for involvement of affected interest groups from local communities to government agencies.

The IUCN Red List (2016) lists 22 species of Amnicolidae: *Amnicola* (2 species), *Baicalia* (14 species), *Emmericia* (4 species), and *Lyogyrus* (2 species), of which 4 are listed as Vulnerable, 13 as Data Deficient, and 5 as Least Concern. However, several other taxa now included in the Amnicolidae are currently listed as belonging to the Hydrobiidae. These include the following 16 taxa: *Akiyoshia* (4 species), *Colligyrus* (2 species), *Godlewskia* (1 species), *Kolhymamnicola* (1 species), *Marstoniopsis* (3 species), *Parabythinella* (3 species), *Taylorconcha* (1 species), and *Terrestribythinella* (1 species), of which 4 are listed as Critically Endangered, 4 as Vulnerable, 6 as Data Deficient, and 1 as Least Concern.

LITERATURE CITED

Baker, F.C. 1928. The fresh water Mollusca of Wisconsin. Part I. Gastropoda. Bulletin of the Wisconsin Geological and Natural History Survey 70: 1-507, plates 1-28.

Berry, E.G. 1943. The Amnicolidae of Michigan: Distribution, ecology, and taxonomy. Miscellaneous Publications of the Museum of Zoology, University of Michigan, 57: 1-68, plates 1-9.

Bouchet, P., and J.-P. Rocroi. 2005. Classification and nomenclator of gastropod families. Malacologia 47: 1-397.

Burch, J.B., and J.L. Tottenham. 1980. North American freshwater snails: Species list, ranges and illustrations. Walkerana 1: 81-215.

Clarke, A.H. 1973. The freshwater molluscs of the Canadian Interior Basin. Malacologia 13: 1-509.

Davis, G.M., and S. Rao. 1997. Discovery of *Erhaia* (Gastropoda, Pomatiopsidae) in northern India with description of a new genus of Erhaiini from China. Proceedings of the Academy of Natural Sciences of Philadelphia 148: 273-299.

Davis, G.M., Y.H. Kuo, K.E. Hoagland, P.L. Chen, H.M. Yang, and D.J. Chen. 1985. *Erhaia*, a new genus and new species of Pomatiopsidae from China (Gastropoda: Rissoacea). Proceedings of the Academy of Natural Sciences of Philadelphia 137: 48-78.

Davis, G.M., C.-E. Chen, C. Wu, T.-F. Kuang, X.-G. Xing, L. Li, W.-J. Liu, and Y.-L. Yan. 1992. The Pomatiopsidae of Hunan, China (Gastropoda: Rissoacea). Malacologia 34: 143-342.

Hershler, R. 1999. A systematic review of the hybrodiid

snails (Gastropoda: Rissooidea) of the Great Basin, western United States. Part II. Genera *Colligyrus, Eremopyrgus, Fluminicola, Pristinicola,* and *Tryonia*. Veliger 42: 306-337.

Hershler, R., and F.G. Thompson. 1988. Notes on morphology of *Amnicola limosa* (Say, 1817) (Gastropoda: Hydrobiidae) with comments on status of the subfamily Amnicolinae. Malacological Review 21: 81-92.

Hubricht, L. 1963. New species of Hydrobiidae. Nautilus 76: 138-140, plate 8.

Hubricht, L. 1971. New Hydrobiidae from Ozark caves. Nautilus 84: 93-96.

Hubricht, L. 1979. A new species of *Amnicola* from an Arkansas cave. Nautilus 93: 142.

IUCN Red List of Threatened Species. 2016. Version 2016-2. www.iucnredlist.org. Accessed 20 September 2016.

Johnson, P.D., A.E. Bogan, K.M. Brown, N.M. Burkhead, J.R. Cordeiro, J.T. Garner, P.D. Hartfield, et al. 2013. Conservation status of freshwater gastropods of Canada and the United States. Fisheries 38: 242-282.

Kabat, A.R., and R. Hershler. 1993. The prosobranch snail family Hydrobiidae (Gastropoda: Rissooidea): Review of classification and supraspecific taxa. Smithsonian Contributions to Zoology 547: 1-94.

Kantor, Y.I., M.V. Vinarski, A.A. Schileyko, and A.V. Sysoev. 2010. Catalogue of the continental mollusks of Russia and adjacent territories. Version 2.3.1. Published 2 March 2010 at www.ruthenica.com/categorie-8.html.

Liu, H.-P., R. Hershler, and C.S. Rossel. 2015. Taxonomic status of the Columbia Duskysnail (Truncatelloidea, Amnicolidae, *Colligyrus*). ZooKeys 514: 1-13.

Liu, H.-P., D. Marceau, and R. Hershler. 2016. Taxonomic identity of two amnicolid gastropods of conservation concern in lakes of the Pacific Northwest of the USA. Journal of Molluscan Studies 82: 464-471.

Lysne, S.J., K.E. Perez, K.M. Brown, R.L. Minton, and J.D. Sides. 2008. A review of freshwater gastropod conservation: Challenges and opportunities. Journal of the North American Benthological Society 27: 463-470.

Radea, C., A. Parmakelis, D. Papadogiannis, D. Charou, and K.A. Triantis. 2013. The hydrobioid freshwater gastropods (Caenogastropoda, Truncatelloidea) of Greece: New records, taxonomic re-assessments using DNA sequence data and an update of the IUCN Red List Categories. ZooKeys 350: 1-20.

Strong, E.E., O. Gargominy, W.F. Ponder, and P. Bouchet. 2008. Global diversity of gastropods (Gastropoda: Mollusca) in freshwater. Hydrobiologia 595: 149-166.

Thompson, F.G. 1968. The Aquatic Snails of the Family Hydrobiidae of Peninsular Florida. University of Florida Press, Gainesville.

Thompson, F.G., and R. Hershler. 1991. Two new hydrobiid snails (Amnicolinae) from Florida and Georgia, with a discussion of the biogeography of freshwater gastropods of south Georgia streams. Malacological Review 24: 55-72.

Walker, B. 1906. New and little known species of Amnicolidae. Nautilus 19: 114-117, plate 5.

Welter-Schultes, F.W. 2012. European non-marine molluscs, a guide for species identification. Planet Poster Editions, Göttingen, Germany.

Wilke, T., G.M. Davis, A. Falniowski, F. Giusti, M. Bodon, and M. Szarowska. 2001. Molecular systematics of Hydrobiidae (Mollusca: Gastropoda: Rissooidea): Testing monophyly and phylogenetic relationships. Proceedings of the Academy of Natural Sciences of Philadelphia 151: 1-21.

Wilke, T., M. Haase, R. Hershler, H.-P. Liu, B. Misof, and W.F. Ponder. 2013. Pushing short DNA fragments to the limit: Phylogenetic relationships of 'hydrobioid' gastropods (Caenogastropoda: Rissooidea). Molecular Phylogenetics Evolution 66: 715-736.

14 Assimineidae H. & A. Adams, 1856

HIROSHI FUKUDA

The Assimineidae are one of the most derived families among truncatelloideans along with Pomatiopsidae Stimpson, 1865, and Truncatellidae Gray, 1840 (Ponder, 1988). They share the steplike mode of locomotion and the omniphoric grooves that carry the fecal pellets and spawns from the inside of the pallial cavity to the outside by the right groove and, in turn, transmit clean water into the cavity by the left one. These unique characters are thought to be adaptations to an amphibious mode of life as discussed by Davis (1967) and Ponder (1988). These three families plus a marine family, Falsicingulidae Slavoshevskaya, 1975, of the northern Pacific are monophyletic according to the recent molecular phylogeny (Criscione and Ponder, 2013).

Assimineids are widespread in nearly all temperate and tropical regions. Many genera and species are known in coastal areas and islands in the Indo-Pacific region, while the occurrence records from the Atlantic (e.g., Marcus and Marcus, 1965; Abbott, 1974; van Aartsen, 2008) are rather few. The author does not know of reliable records from the Pacific coast of South America or the Atlantic coast of Africa.

Taxonomic revisions of assimineids have been published by several well-known malacologists, e.g., Martens (1866), Böttger (1887), Kobelt (1906), Thiele (1927), Habe (1942, 1943), and Abbott (1949, 1958). More than 240 species names were described in the genus *Assiminea* Fleming, 1828, before the 1950s (listed by Abbott, 1958) and about 100 in *Omphalotropis* Pfeiffer, 1851, in the nineteenth century (Kobelt, 1906). However, many of them are still poorly known, and both *Assiminea* and *Omphalotropis* are polyphyletic. Fukuda and Ponder (2003) listed 67 genus-group taxa (including three questionable ones) of the family, but 30 of them still lack anatomical information of soft parts, and their taxonomy is highly confused. In addition, there are likely a large number of undescribed species, judging from the Japanese fauna that has at least 70 undescribed species at present (e.g., Fukuda, 1996, 2012; Kameda et al., 2008; Government of Japan, 2014; Okinawa Prefectural Government, 2017).

Distribution of Assimineidae.

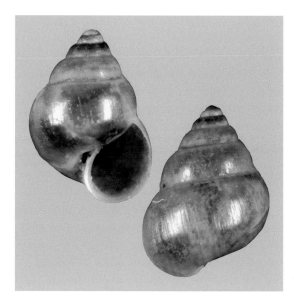

Eussoia cf. *aurifera* (Preston, 1912). Madagascar: Antalaha, Sava Region. 4.2 mm. Poppe 747411.

Assiminea grayana Fleming, 1828, the type species of *Assiminea,* is known from western Europe, but this species is identical with *A. violacea* Heude, 1882, of China in anatomy and DNA sequences (Fukuda and Kameda, in prep.). Therefore *A. grayana* is thought to have been introduced from China into Europe before the description of the species in the early nineteenth century. *A. parasitologica* Kuroda, 1958, a species endemic to Japan, is the sister species of *A. grayana* = *A. violacea* (Fukuda and Mitoki, 1997). The Japanese species has been recently introduced into Oregon through ballast water exchange and has replaced the native *Angustassiminea californica* (Tryon, 1865) (Laferriere et al., 2010), whereas *A. parasitologica* is regarded as threatened in Japan (Fukuda, 1996, 2012; Government of Japan, 2014).

Assiminea and *Angustassiminea* Habe, 1943, have a simple, conical, brownish, and nonumbilicate shell with a nearly smooth surface. Those shells are known as typical for the family, but actually, assimineid shells and opercula are very diversified, as in the Pomatiopsidae (Davis, 1979). They contrast strongly with truncatellids, which show relatively uniform cylindrical shells (Clench and Turner, 1948; Fukuda et al., 2017). In assimineids, the terrestrial spe-

cies of Guam, Palau, the Philippines, New Caledonia, Papua New Guinea, Micronesia, and Polynesia have remarkable diversity of their shells and opercula (e.g., Quadras and Moellendorff, 1894; Kobelt, 1906; Cooke and Clench, 1943; Kondo, 1944; Clench, 1946, 1948, 1955, 1958; Abbott, 1949; Frank, 1956; Zilch, 1967; Bauman, 1996; Bouchet and Abdou, 2003). Their shell shapes are discoidal, depressed, trochiform, globose, scalariform, elongate, or tall, with or without spiral/ longitudinal ribs. Some have long periostracul hairs on the surfaces. *Balambania* Crosse, 1891, from Cebu, the Philippines, is one of the most extreme examples. It shows an open spiral and the coils do not touch each other. Some species have horny, paucispiral, and pyriform opercula, but others have calcareous, multispiral, and round ones. Therefore, the definition of this family by the shells and opercula is impossible. In the current studies, the family is defined by the short rectangular, quadrangular, or trapezoid shape of the central teeth of the radula (Fukuda and Mitoki, 1995; Fukuda and Ponder, 2003).

Fukuda and Ponder (2003) tentatively classified Recent assimineids into three "Groups" based on the presence/absence of the basal cusps on the central tooth of the radula and the distinct cephalic tentacles. Bouchet and Rocroi (2005) assigned three subfamilies, Assimineinae, Ekadantinae Thiele, 1929, and Omphalotropidinae Thiele, 1927 (= Garrettiinae Kobelt, 1906), to Groups 1, 2, and 3, respectively, but the former two "subfamilies" are not distinguishable and thus synonymous according to some recent molecular analyses (Fukuda et al., 2015; Hallan and Fukuda, 2015; Hallan et al., 2015; Fukuda and Kameda, in prep.). The Omphalotropidinae are probably monophyletic, but the genus *Omphalotropis* is polyphyletic and should be separated into many different genera that are endemic to each island group in the Indo-Pacific. The "real" *Omphalotropis* consists of only some species of Mauritius (Griffiths and Florens, 2006), including *O. hieroglyphica* (Potiez & Michaud, 1838), the type species of the genus. Bouchet et al. (2017) listed Omphalotropidinae as the invalid synonym of Garrettiinae, but it is inadequate, because ICZN (1971) ruled that the name Omphalotropidinae is to be given precedence over Garrettiinae by any author who believes these names are synonymous.

The systematic positions of a few genera still remain uncertain: one of them is *Ditropisena* Iredale, 1933, from mangrove swamps of northern Australia, Japan, Palau, Mauritius, and the Comoros. The genus was wrongly assigned to the operculate land snail family Cyclophoridae Gray, 1847, because of the similarity of the depressed shell with wide umbilicus and round operculum. The soft-parts anatomy and molecular analysis clearly indicate that the genus is a member of assimineids (Fukuda, 1996, 2012; Fukuda and Ponder, 2003), but it is not included in the assimineine clade or in Omphalotropidine (Fukuda et al., in prep.). Another bizarre group consists of *"Assiminea" ovata* (Krauss, 1848), *"A." capensis* Bartsch, 1915, *"A." globulus* Connoly, 1939, and a few closely related species from tidal flats in South Africa. This species group was shown to belong to the Assimineidae based on molecular analyses by Criscione and Ponder (2013), but it has a unique combination of morphological characters: the central tooth of the radula is inverted and triangular, and the head-foot lacks the omniphoric grooves (Fukuda and Ponder, 2010).

Assimineids are gonochoristic, with males and females, but Australian *Rugapedia androgyna* Fukuda & Ponder, 2004, is known as a subsequent hermaphrodite because all juveniles (small in size) are males and all fully matured (large) individuals are females (Fukuda and Ponder, 2004). *Pseudogibbula duponti* Dautzenberg, 1891, of the Zaire River, Africa, was reported to be viviparous and parthenogenetic only with females (Brown, 1994).

The habitats of this family are remarkably wide as the result of an extraordinary radiation. Many species live in brackish waters on estuaries and fully terrestrial environments, but at least several species are known in upper subtidal to intertidal zones of seashores, freshwaters, and in caves. This is an extremely rare case among the entire kingdom Animalia. The only other known example of a species with marine, freshwater, and terrestrial habitats in a single family is the Chironomidae (Insecta: Diptera). Especially, *Angustassiminea* and *Cavernacmella* Habe, 1942, have marine, freshwater, and land species in a single genus. Most of *Angustassiminea* species are found in supratidal zones and estuaries, but two undescribed species of the Ryukyu Islands are restricted on cliffs at above 20 and 70 m from the sea, and one species of Minami-daito Island occurs only in a small freshwater pond (Fukuda, 2017; Kubo et al., 2017a, b). *Cavernacmella* is endemic to Japan, and more than 50 undescribed species have been recognized inside limestone caves (Kameda et al., 2008), but one congeneric species is found in the small freshwater stream in Yakushima Island and several species are known in the supratidal zone and on the forest floor (Wada and Chiba, 2011; Wada et al., 2013; Kameda and Fukuda, in prep.). Therefore, it is clear that the invasion from marine to freshwater and land occurred many times independently within the Assimineidae. The gills were lost in some terrestrial species, but still remain in others. The kidney in many species of the Omphalotropidinae of the central Pacific is extended anterior to the pallial end (Fukuda and Ponder, 2007; Tatara and Fukuda, 2011), although the kidney is restricted posterior to the end of the pallial cavity in most of caenogastropods except for cyclophoroideans and some littorinoideans. This is likely to be an adaptation to the habitat without water. Representative genera and species of the family in various habitats are as follows:

Marine: Some *Assiminea* in tidal flats of Japan (Fukuda, 2012), some *Angustassiminea* (Fukuda and Mitoki, 1996b), *Paludinella* Pfeiffer, 1841, from the Mediterranean and Indo-Pacific, including Oman, Thailand, Papua New Guinea, the Philippines, Guam, Palau, Taiwan, and Japan (Tatara and Fukuda, 2013), most of *Paludinellassiminea* Habe, 1995, from Japan (Fukuda and Mitoki, 1995), *Suterilla* Thiele, 1927, from New Zealand and Tasmania (Fukuda et al., 2006), and *"Assiminea" capensis* group from South Africa (Fukuda and Ponder, 2010).

Brackish water: Many genera and species are known in mangroves and reed fields on estuaries. *Assiminea, Angustassiminea, Austropilula* Thiele, 1927, *Conassiminea* Fukuda & Ponder, 2006, *Cryptassiminea* Fukuda & Ponder, 2005, *Ditropisena, Macrassiminea* Thiele, 1927, *Metassiminea* Thiele, 1927, *Optediceros* Leith, 1853, *Ovassiminea* Thiele, 1927, *Pseudomphala* Heude, 1882, *Rugapedia* Fukuda & Ponder, 2004, *Schuettiella* Brandt, 1974, *Sculptassiminea* Thiele, 1927, and some *Taiwanassiminea* Kuroda & Habe, 1950 (see Fukuda and Mitoki, 1996a, b, 1997; Fukuda and Ponder, 2004, 2005, 2006).

Freshwater: *Angustassiminea infima* (Berry, 1947) from Death Valley, California, and *"Assiminea" pecos* Taylor, 1987, from the Rio Grande region of the United States and Mexico (Hershler, 1987; Sada, 2001; Hershler et al., 2007; Hershler and Liu, 2008), *Aviassiminea palitans* Fukuda & Ponder, 2003, and *Austroassiminea letha* Solem et al., 1982, from Australia (Fukuda and Ponder, 2003), *Cyclotropis* Tapparone-Canefri, 1883, from New Guinea, Indonesia, Malaysia, Thailand, and Vietnam (Brandt, 1974; Fukuda et al., 2015), *Eussoia* Preston, 1912, from Kenya, Somalia, Mozambique, and Tanzania, and *Pseudogibbula* Dautzenberg, 1891, from Zaire (Brown, 1994), *Solenomphala* Heude, 1882, from China, Japan, and Southeast Asia (introduced into Italy; Benocci et al., 2014), *Suterilla fluviatilis* Fukuda et al., 2007, from Norfolk Island, Australia, *Tutuilana striata* Hubendick, 1952, of Samoa (Fukuda et al., 2006), and *Taiwanassiminea bedaliensis* (Rensch, 1934) of Indonesia (introduced into northern Australia, Thailand, Hawaii, and Japan; Brandt, 1974; Haynes et al., 2007; Fukuda et al., 2015).

Terrestrial: Most of the *"Omphalotropis"* group of the Pacific islands, including *Conacmella* Thiele, 1927, *Garrettia* Paetel, 1873, *Opinorelia* Iredale, 1944, *Pseudocyclotus* Thiele, 1894, *Quadrasiella* Moellendorff, 1894, and *Rupacilla* Thiele, 1927. *"O." coturnix* (Crosse, 1867) was recorded in mountainsides of New Caledonia at 1200 m from the sea. A part of *Cavernacmella* and *Taiwanassiminea* are also found in forests (Wada et al., 2013; Hallan and Fukuda, 2015).

Some marine and brackish-water species of *Assiminea* and *Angustassiminea* have a planktonic larval stage as a veliger (Sander, 1950; Sander and Sibrecht, 1967; Kurata and Kikuchi, 1999, 2000). *Angustassiminea* spp. inhabit the upper intertidal zone and are widely distributed around the world [e.g., *A. satumana* (Habe, 1942) of Japan, *A. lucida* (Pease, 1869) and *A. nitida* (Pease, 1865) of Polynesia, *A. vulgaris* (Webster, 1905) of New Zealand, *A. succinea* (Pfeiffer, 1840) of the Caribbean Sea to Brazil]. Their adult shells and animals are not easily distinguishable from each other (Fukuda and Mitoki, 1996b), and all or most of them might be conspecific with an extraordinarily wide distribution range as the result of a very long planktonic period. In contrast, all of the freshwater and terrestrial assimineids are thought to be direct development.

Some of the assimineids are known to be associated with transmission of *Paragonimus* Braun, 1899 (lung flukes). In Japan, *Assiminea parasitologica* and *"A." yoshidayukioi* Kuroda, 1959, are known as the primary intermediate hosts of *Paragonimus ohirai* Miyazaki, 1939, and *P. iloktsuenensis* Chen, 1940, but their human infection has not been reported (Yoshida and Kawashima, 1961; Yoshida, 1977). Abbott (1948, 1958) mentioned that the miracidium of *Schistosoma japonicum* (Katsurada, 1904), which is the cause of a serious human disease, schistosomiasis (bilharzia), and parasitic on *Oncomelania* spp. (Pomatiopsidae), was strongly attracted to assimineids of the Philippines, but all attempts to infect them have failed.

In recent years, nearly all habitats of assimineids have been threatened moderately to severely by human activities. Estuaries (mangrove swamps and reed fields) and seashores (tidal flats, rocky shores) are being destroyed by landfilling, reclamation, bank protection, pollution, and so on (Fukuda, 1996, 2012; Suzukida and Fukuda, 2003). Also, terrestrial and freshwater species with narrow distribution ranges (e.g., endemic to small islands and lakes) are jeopardized by deforestation, desertification, and invasion of alien species (Bauman, 1996; Bouchet and Abdou, 2003). A land flatworm, *Platydemus manokwari* de Beauchamp, 1963, has been rapidly introduced into many Pacific islands and predates terrestrial assimineids (Kubo et al., 2017b; Fukuda, pers. obs.) as well as many other groups of native land snails. Therefore, conservation of assimineids is strongly needed as an urgent issue. In the latest Red Data Book of Japan (Government of Japan, 2014), 30 assimineid species are listed under various categories: 1 species as Extinct, 5 as Endangered, 14 as Vulnerable, and 10 as Near Threatened.

LITERATURE CITED

Abbott, R.T. 1948. Handbook of medically important mollusks of the Orient and the western Pacific. Bulletin of the Museum of Comparative Zoology, Harvard, 100: 245-328, plates 1-5.

Abbott, R.T. 1949. New syncerid mollusks from the Marianas Islands (Gastropoda, Prosobranchiata, Synceridae). Occasional Papers of Bernice P. Bishop Museum 19: 261-274.

Abbott, R.T. 1958. The gastropod genus *Assiminea* in the

Philippines. Proceedings of the Academy of Natural Sciences of Philadelphia 110: 213-277, plates 15-25.

Abbott, R.T. 1974. American Seashells, 2nd ed. Van Nostrand Reinhold, New York.

Bauman, S. 1996. Diversity and decline of land snails on Rota, Mariana Islands. American Malacological Bulletin 12: 13-27.

Benocci, A., G. Manganelli, and F. Giusti. 2014. New records of non-indigenous molluscs in the Mediterranean Basin: Two enigmatic alien gastropods from the Tuscan Archipelago (Italy). Journal of Conchology 41: 617-626.

Böttger, O. 1887. Aufzählung der zur Gattung *Assiminea* Fleming gehörigen Arten. Jahrbücher der Deutschen Malakozoologischen Gesellschaft 14: 147-234, plate 6.

Bouchet, P., and A. Abdou. 2003. Endemic land snails from the Pacific islands and the museum record: Documenting and dating the extinction of the terrestrial Assimineidae of the Gambier Islands. Journal of Molluscan Studies 69: 165-170.

Bouchet, P., and J.-P. Rocroi. 2005. Classification and nomenclator of gastropod families. Malacologia 47: 1-397.

Bouchet, P., J.-P. Rocroi, B. Hausdorf, A. Kaim, Y. Kano, A. Nützel, P. Parkhaev, M. Schrödl, and E.E. Strong. 2017. Revised classification, nomenclator and typification of gastropod and monoplacophoran families. Malacologia 61: 1-526.

Brandt, R.A.M. 1974. The non-marine aquatic Mollusca of Thailand. Archiv für Molluskenkunde 105: 1-423.

Brown, D.S. 1994. Freshwater Snails of Africa and Their Medical Importance, 2nd ed. Taylor & Francis, London.

Clench, W.J. 1946. New genera and species of Synceridae from Ponape, Caroline Islands. Occasional Papers of Bernice P. Bishop Museum 18: 199-206.

Clench, W.J. 1948. Two new genera and a new species of Synceridae from the Caroline Islands. Occasional Papers of Bernice P. Bishop Museum 19: 191-194.

Clench, W.J. 1955. *Setaepoma*, a new genus in the Synceridae from the Solomon Islands. Nautilus 68: 134.

Clench, W.J. 1958. The land and freshwater molluscs of Rennell Island, Solomon Islands. Pp. 155-202, plates 16-19, in The Natural History of Rennell Island, British Solomon Islands (T. Wolff, ed.), vol. 2. Danish Science Press, Copenhagen.

Clench, W.J., and R.D. Turner. 1948. A catalogue of the family Truncatellidae with notes and descriptions of new species. Occasional Papers on Mollusks 1: 157-212.

Cooke, C.M., Jr., and W.J. Clench. 1943. Land shells (Synceridae) from the southern and western Pacific. Occasional Papers of Bernice P. Bishop Museum 17: 249-262.

Criscione, F., and W.F. Ponder. 2013. A phylogenetic analysis of rissooidean and cingulopsoidean families (Gastropoda: Caenogastropoda). Molecular Phylogenetics and Evolution 66: 1075-1082.

Davis, G.M. 1967. The systematic relationships of *Pomatiopsis lapidaria* and *Oncomelania hupensis formosana* (Prosobranchia: Hydrobiidae). Malacologia 6: 1-143.

Davis, G.M. 1979. The origin and evolution of the gastropod family Pomatiopsidae with emphasis on the Mekong River Triculinae. Academy of Natural Sciences of Philadelphia Monograph 20: 1-120.

Frank, A. 1956. Mollusques terrestres et fluviatiles de l'Archipel Néo-Calédonien. Mémoires de Muséum National d'Histoire Naturelle, Série A Zoologie 13: 1-200, plates 1-24.

Fukuda, H. 1996. Gastropoda. In Present status of estuarine locales and benthic invertebrates occurring in estuarine environment in Japan [in Japanese] (S. Hanawa and H. Sakuma, eds.). WWF Japan Science Report 3: 11-63.

Fukuda, H. 2012. Assimineidae. Pp. 46-54 in Threatened Animals of Japanese Tidal Flats: Red Data Book of Seashore Benthos [in Japanese] (Japanese Association of Benthology, ed.). Tokai University Press, Hadano, Japan.

Fukuda, H. 2017. *Angustassiminea* sp. A, B and C. Pp. 434-436 in Threatened Wildlife in Okinawa [in Japanese]. 3rd ed. (Animals), Red Data Okinawa (Nature Conservation Division, Department of Environmental Affairs, Okinawa Prefectural Government, ed.). Okinawa Prefectural Government, Naha.

Fukuda, H., and T. Mitoki. 1995. A revision of the family Assimineidae (Mollusca: Gastropoda: Neotaenioglossa) stored in the Yamaguchi Museum. Part 1. Subfamily Omphalotropidinae. Bulletin of the Yamaguchi Museum 21: 1-20.

Fukuda, H., and T. Mitoki. 1996a. A revision of the family Assimineidae (Mollusca: Gastropoda: Neotaenioglossa) stored in the Yamaguchi Museum. Part 2. Subfamily Assimineinae (1) Two species from Taiwan. Bulletin of the Yamaguchi Museum 22: 1-11.

Fukuda, H., and T. Mitoki. 1996b. A revision of the family Assimineidae (Mollusca: Gastropoda: Neotaenioglossa) stored in the Yamaguchi Museum. Part 3. Subfamily Assimineinae (2) *Angustassiminea* and *Pseudomphala*. Yuriyagai 4: 109-137.

Fukuda, H., and T. Mitoki. 1997. A revision of the family Assimineidae (Mollusca: Gastropoda: Neotaenioglossa) stored in the Yamaguchi Museum. Part 4. Subfamily Assimineinae (3) '*Assiminea*' *parasitologica*. Bulletin of the Yamaguchi Museum 23: 1-8.

Fukuda, H., and W.F. Ponder. 2003. Australian freshwater assimineids, with a synopsis of the Recent genus-group taxa of the Assimineidae (Mollusca: Caenogastropoda: Rissooidea). Journal of Natural History 37: 1977-2032.

Fukuda, H., and W.F. Ponder. 2004. A protandric assimineid gastropod: *Rugapedia androgyna* n. gen. and n. sp. (Mollusca: Caenogastropoda: Rissooidea) from Queensland, Australia. Molluscan Research 24: 75-88.

Fukuda, H., and W.F. Ponder. 2005. A revision of the Australian taxa previously attributed to *Assiminea buccinoides* (Quoy & Gaimard) and *Assiminea tasmanica* Tenison-Woods (Mollusca: Gastropoda: Caenogastropoda: Assimineidae). Invertebrate Systematics 19: 325-360.

Fukuda, H., and W.F. Ponder. 2006. *Conassiminea,* a new genus of the Assimineidae (Caenogastropoda: Rissooidea) from southeastern Australia. Journal of Molluscan Studies 72: 39-52.

Fukuda, H., and W.F. Ponder. 2007. A challenge to the relationships of '*Omphalotropis*' [in Japanese]. Venus 66: 106-107.

Fukuda, H., and W.F. Ponder. 2010. Bizarre unidentified rissooids occurring in tidal-flats of South Africa [in Japanese]. Venus 69: 86-87.

Fukuda, H., W.F. Ponder, and B.A. Marshall. 2006. Anatomy and relationships of *Suterilla* Thiele (Caenogastropoda: Assimineidae) with descriptions of four new species. Molluscan Research 26: 141-168.

Fukuda, H., W.F. Ponder, and A. Hallan. 2015. Anatomy, relationships and distribution of *Taiwanassiminea affinis* (Böttger) (Caenogastropoda: Assimineidae), with a reassessment of *Cyclotropis* Tapparone-Canefri. Molluscan Research 35: 24-36.

Fukuda, H., Y. Kameda, T. Hirano, H. Kubo, Y. Hayase, and T. Saito. 2017. Towards a taxonomic revision of the Truncatellidae (Caenogastropoda: Truncatelloidea) of Japan [in Japanese with English abstract]. Molluscan Diversity 5: 33-77.

Government of Japan (Ministry of Environment). 2014. Red Data Book 2014. Threatened Wildlife of Japan, vol. 6, Mollusks [in Japanese with English summary]. Wildlife Division, Nature Conservation Bureau, Ministry of Environment, Tokyo.

Griffiths, O.L., and V.F.B. Florens. 2006. A Field Guide to the Non-Marine Molluscs of the Mascarene Islands (Mauritius, Rodrigues and Réunion) and the Northern Dependencies of Mauritius. Bioculture Press, Mauritius.

Habe, T. 1942. Classification of Japanese Assimineidae [in Japanese]. Venus 12: 32-56, plates 1-4.

Habe, T. 1943. Supplemental note and corrections on the Japanese Assimineidae [in Japanese]. Venus 13: 96-106.

Hallan, A., and H. Fukuda. 2015. *Taiwanassiminea phantasma* n. sp.: A terrestrial assimineid (Gastropoda: Truncatelloidea) from Middle Osborn Island, Kimberley, Western Australia. Molluscan Research 35: 112-122.

Hallan, A., H. Fukuda, and Y. Kameda. 2015. Two new

species of *Ovassiminea* Thiele, 1927 (Truncatelloidea: Assimineidae) from tropical Australia. Molluscan Research 35: 262-274.

Haynes, K.A., C.T. Tran, and R.H. Cowie. 2007. New records of alien Mollusca in the Hawaiian Islands: Nonmarine snails and slugs (Gastropoda) associated with the horticultural trade. Bishop Museum Occasional Papers 96: 54-63.

Hershler, R. 1987. Redescription of *Assiminea infima* Berry, 1947, from Death Valley, California. Veliger 29: 274-288.

Hershler, R., and H.-P. Liu. 2008. Phylogenetic relationships of assimineid gastropods of the Death Valley–lower Colorado River region: Relicts of a late Neogene marine incursion? Journal of Biogeography 35: 1816-1825.

Hershler, R., H.-P. Liu, and B.K. Lang. 2007. Genetic and morphological variation of the Pecos Assiminea, an endangered mollusk of the Rio Grande region, United States and Mexico (Caenogastropoda: Rissooidea: Assimineidae). Hydrobiologia 579: 317-335.

ICZN (International Commission on Zoological Nomenclature). 1971. Opinion 973. *Realia* Baird, 1850 (Gastropoda): Suppressed under the plenary powers. Bulletin of Zoological Nomenclature 28: 149-150.

Kameda, Y., A. Kawakita, and M. Kato. 2008. One endemic in one cave: Extraordinary cryptic diversity of troglobiotic land snail *Cavernacmella kuzuuensis* [in Japanese]. Venus 67: 99.

Kobelt, W. 1906. Synopsis der Pneumonopomen-Familie Realiidae. Jahrbücher Nassauischer Verein für Naturkunde, Wiesbaden, 59: 49-144.

Kondo, Y. 1944. Dentition of six syncerid genera Gastropoda, Prosobranchiata, Synceridae (Assimineidae). Occasional Papers of Bernice P. Bishop Museum 17: 313-318.

Kubo, H., H. Fukuda, Y. Hayase, Y. Kameda, H. Ozawa, and R. Ueshima. 2017a. Investigations of the present status of threatened molluscan species in Okinawa Prefecture, southwestern Japan, for the second revision of *the Red Data Okinawa*—1. Yonaguni Island [in Japanese with English summary]. Molluscan Diversity 5: 1-14.

Kubo, H., H. Fukuda, Y. Hayase, Y. Kameda, T. Kurozumi, and R. Ueshima. 2017b. Investigations of the present status of threatened molluscan species in Okinawa Prefecture, southwestern Japan, for the second revision of *the Red Data Okinawa*—3. Kita-daito and Minami-daito islands [in Japanese with English summary]. Molluscan Diversity 5: 21-32.

Kurata, K., and E. Kikuchi. 1999. Life cycle and production of *Assiminea japonica* v. Martens and *Angustassiminea castanea* (Westerlund) at a red march in Gamo Lagoon, northern Japan (Gastropoda: Assimineidae). Ophelia 50: 191-214.

Kurata, K., and E. Kikuchi. 2000. Comparisons of life-history traits and sexual dimorphism between *Assiminea japonica* and *Angustassiminea castanea* (Gastropoda: Assimineidae). Journal of Molluscan Studies 66: 177–196.

Laferriere, A.M., H. Harris, and J. Schaefer. 2010. Early Detection of a New Invasive Mesogastropod, *Assiminea parasitologica* in Pacific Northwest Estuaries. South Slough National Estuarine Research Reserve with the Confederated Tribes of the Coos, Lower Umpqua and Siuslaw Indians, Coos Bay, Charleston, Oregon.

Marcus, E., and E. Marcus. 1965. On Brazilian supratidal and estuarine snails. Boletim de Faculdade de Filosofia, Ciéncias e Letras, Universidade de Sao Paulo, Zoologia 287: 29–82, plates 1–10.

Martens, E. von. 1866. Conchological gleanings. II. On some species of *Assiminea*. Annals and Magazine of Natural History, series 3, 17: 202–207.

Okinawa Prefectural Government (Nature Conservation Division, Department of Environmental Affairs). 2017. Threatened Wildlife in Okinawa, 3rd ed. (Animals)—Red Data Okinawa [in Japanese]. Okinawa Prefectural Government, Naha.

Ponder, W.F. 1988. The truncatelloidean (= rissoacean) radiation—a preliminary phylogeny. Malacological Review Supplement 4: 129–166.

Quadras, J.F., and O.F. von Moellendorff. 1894. Diagnoses specierum novarum a J. F. Quadras in insulis Mariannis collectarum scripserunt. Nachrichtsblatt der Deutschen Malakozoologischen Gesellschaft 26: 33–42.

Sada, D.W. 2001. Demography and habitat use of the badwater snail (*Assiminea infima*), with observation on its conservation status, Death Valley National Park, California, U.S.A. Hydrobiologia 466: 255–265.

Sander, K. 1950. Beobachtungen zur Fortpflanzung von *Assiminea grayana* Leach. Archiv für Molluskenkunde 79: 147–149.

Sander, K., and L. Sibrecht. 1967. Das Schlupfen der Veliger-

larve von *Assiminea grayana* leach (Gastropoda Prosobranchia). Zeitscrift für Morphologie und Ökologie der Tiere 60: 141–152.

Suzukida, K., and H. Fukuda. 2003. Systematics and conservation of the cryptic species comprising '*Assiminea japonica*' (Mollusca: Gastropoda: Rissooidea). Records of the South Australian Museum Monograph Series 7: 303–309.

Tatara, Y., and H. Fukuda. 2011. Diversity of habitats and evolution of Assimineidae [in Japanese]. Umiushi Tsuushin 70: 5–7.

Tatara, Y., and H. Fukuda. 2013. Revision of the species of *Paludinella* Pfeiffer, 1841 (Caenogastropoda: Assimineidae)—a preliminary report [in Japanese]. Venus 71: 134.

Thiele, J. 1927. Über die Schneckenfamilie Assimineidae. Zoologische Jahrbücher, Abteilung für Systematik, Ökologie und Geographie Tiere 53: 114–116, plate 1.

Van Aartsen, J.J. 2008. The Assimineidae of the Atlantic-Mediterranean seashores. Basteria 72: 165–181.

Wada, S., and S. Chiba. 2011. Seashore in the mountain: Limestone-associated land snail fauna on the oceanic Hahajima Island (Ogasawara Islands, western Pacific). Biological Journal of the Linnean Society 102: 686–693.

Wada, S., Y. Kameda, and S. Chiba. 2013. Long-term stasis and short-term divergence in the phenotypes of microsnails on oceanic islands. Molecular Ecology 22: 4801–4810.

Yoshida, Y. 1977. Illustrated Human Parasitology. Nanzando, Tokyo.

Yoshida, Y., and K. Kawashima. 1961. On the distribution of the snail hosts of *Paragonimus ohirai* Miyazaki, 1939 and *Paragonimus iloktsuenensis* Chen, 1940 in Japan. Japanese Journal of Parasitology 10: 152–160.

Zilch, A. 1967. Die Typen und Typoide des Natur-Museums Senckenberg. 36: Mollusca, Assimineidae. Archiv für Molluskenkunde 96: 67–100.

15 Bithyniidae Gray, 1857

WINSTON PONDER

Bithyniids are freshwater snails of small to medium size, with conical to ovate or globose shells. They differ from other truncatelloideans in having a calcareous operculum showing concentric growth rings, a distinctive bifurcate penis with a tubular internal gland that extends into the head, and a large gill used for filter feeding to supplement browsing. The tight-fitting operculum enables some species to survive out of water for weeks or even months (Lilly 1953).

Bithyniids live in a wide range of freshwater habitats in Europe, Asia, Africa, and mainland Australia, but many prefer swampy marshes and slow-flowing streams. They are absent from New Zealand and New Caledonia; although they are not found naturally in the Americas, one species (*Bithynia tentaculata*) has been introduced to North America, where it has become widespread in temperate areas. Bithyniids are known from the Neogene of Europe and Egypt (Wenz, 1938-1944).

Taylor (1966) suggested that bithyniids were related to the architaenioglossan Viviparidae. However, the cytology of the osphradium (Haszprunar, 1985) and sperm (Healy, 1988), as well as the general anatomy (Ponder, 1988) and molecular studies (Wilke et al., 2001; Wilke et al., 2013), show that this group is a member of the Truncatelloidea (sensu Criscione and Ponder, 2013).

This family was long known as Bulimidae, but the generic name *Bulimus* was suppressed by the International Commission on Zoological Nomenclature in favor of *Bithynia* and the family name Bithyniidae.

There has not been a modern review of bithyniid genera. There are many names, some based on fossils, and revision is needed. The genera include *Bithynia,* with species ranging through most of the western Palaearctic region and in the northwestern extremity of Africa (Brown, 1988). The genus-group taxa *Opisthorchophorus* Beriozkina & Starobogatov, 1994, *Paraelona* Beriozkina & Starobogatov, 1994, *Milletelona* Beriozkina & Starobogatov, 1994, *Digyrcidium* Locard, 1882, and *Codiella* Locard, 1894, recognized for eastern European *Bithynia*-like taxa (Beriozkina et al., 1995), are very similar to (and possibly synonyms of) *Bithynia.*

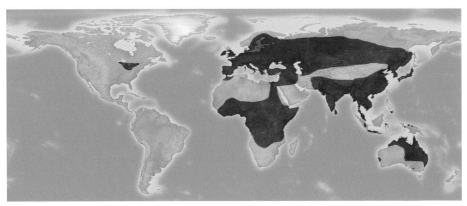

Distribution of Bithyniidae.

Gabbia is based on a temperate Australian species, and as currently recognized includes a wide range of shell morphologies (Ponder, 2003) and encompasses mainly tropical species from northern Australia, New Guinea, and Southeast Asia. Other Southeast Asian bithyniids are included in the genera *Bithynia, Wattebledia* Crosse, 1886, and *Hydrobioides* Nevill, 1885. The type species of *Digoniostoma* Annandale, 1920 (often treated as a subgenus of *Bithynia*), is from India, but species attributed to this "genus" occur throughout Southeast Asia (e.g., Jutting, 1956). In general features these species are very similar to some taxa attributed to *Bithynia* or *Gabbia*. Genera apparently endemic in India, but poorly known, are *Parabithynia* Pilsbry, 1928, *Mysorella* Godwin-Austen, 1919, and *Neosataria* Kulkarni & Khot, 2015 (see Kulkarni and Khot, 2015), while *Parafossarulus* Annandale, 1924, and *Pseudovivipara* Annandale, 1918, are found in China and two poorly known genera, *Emmericiopsis* Thiele, 1928, and *Petroglyphus* Moellendorff, 1894, are based on species from Indonesia (Sumatra) and the Philippines (Mindanao), respectively.

Pseudobithynia includes species from the Middle East and southern Europe that lack a penial appendage (Glöer and Pešić, 2006; Albrecht et al., 2008). Bithyniids are also found in tropical Africa, and these are placed in the genera *Congodoma* Mandahl-Barth, 1968, *Funduella* Mandahl-Barth, 1968, *Gabbiella* Mandahl-Barth, 1968 (and subgenera *Conogabbia* Mandahl-Barth, 1968, and *Omphalogabbia* Mandahl-Barth, 1968), *Incertihydrobia* Verdcourt, 1958, *Jubaia* Mandahl-Barth, 1968 (Brown, 1994), *Sierraia* Connolly, 1929, *Liminitesta* Mandahl-Barth, 1974, and *Soapilia* Binder, 1961. The latter three genera are from western Africa and are unusual in that they inhabit rapidly flowing rivers (Brown, 1988).

Surprisingly few molecular studies have been carried out on the group, and there are no published phylogenetic studies of bithyniids. The studies that have been carried out are focused on local issues with species that are intermediate hosts of significant trematodes. Studies using allozyme markers for some bithyniids from Thailand that are intermediate hosts of *Clonorchis sinensis* have been carried out and have identified cryptic taxa (Kiatsopit et al., 2011; Kiatsopit et al., 2013). An analysis by Kulsantiwong et al. (2013) showed that the clade formed by the Thai *Gabbia*

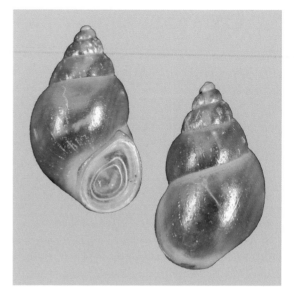

Bithynia tentaculata (Linnaeus, 1758). Romania: Danube Delta, Tulcea County. 7.8 mm. Poppe 917738.

species was sister to those Thai species included in *Bithynia* (the type species of these genera were not included). Other Southeast Asian genera (*Wattebledia* and *Hydrobioides*) formed separate clades. Species-specific primers have been designed from Random Amplified Polymorphic DNA (RAPD) products for one of these host species, *Bithynia funiculata* (Kulsantiwong et al., 2013). Polymorphic microsatellite loci have been characterized in *Bithynia tentaculata* in the United States by Henningsen et al. (2010), and the genetic structure of some northern US populations (Whalen 2011) has been analyzed.

Several species have narrow ranges, and some of those live in threatened habitats. Examples are some species restricted to spring habitats in Australia (Ponder, 2003) and springs and small lakes in southern Europe (e.g., Glöer and Maassen, 2009; Glöer et al., 2010).

In Asia, several bythniids are known to be intermediate hosts of significant human parasites, such as the Chinese liver fluke *Clonorchis sinensis* and other Asian liver flukes infecting humans (*Opisthorchis viverrini* and *Opisthorchis felineus*), as well as some flukes that only occasionally infect humans, for example, lecithodendriids (including intestinal flukes such as *Phaneropsulus*). The introduced *Bithynia tentaculata* is a host for the intestinal flukes of waterfowl,

Cyathocotyle bushiensis and *Sphaeridiotrema globulus,* in North America.

LITERATURE CITED

Albrecht, C., C. Wolff, P. Glöer, and T. Wilke. 2008. Concurrent evolution of ancient sister lakes and sister species: The freshwater gastropod genus *Radix* in lakes Ohrid and Prespa. Hydrobiologia 615: 157-167.

Beriozkina, G.V., O.V. Levina, and I. Starobogatov. 1995. Revision of Bithyniidae from European Russia and Ukraine. Ruthenica 5: 27-38.

Brown, D.S. 1988. *Sierraia:* Rheophilous West African river snails (Prosobranchia: Bithyniidae). Zoological Journal of the Linnean Society 93: 313-356.

Brown, D.S. 1994. Freshwater Snails of Africa and Their Medical Importance. 2nd ed. Taylor & Francis, London.

Criscione, F., and W.F. Ponder. 2013. A phylogenetic analysis of rissooidean and cingulopsoidean families (Gastropoda: Caenogastropoda). Molecular Phylogenetics and Evolution 66: 1075-1082.

Glöer, P., and W.J.M. Maassen. 2009. Three new species of the family Bithyniidae from Greece (Gastropoda: Bithyniidae). Mollusca 27: 41-48.

Glöer, P., and V. Pešić. 2006. On the identity of *Bithynia graeca* Westerlund, 1879 with the description of three new *Pseudobithynia* n. gen. species from Iran and Greece (Gastropoda: Bithyniidae). Malakologische Abhandlungen 24: 29-36.

Glöer, P., A. Falnioski, and V. Pešic. 2010. The Bithyniidae of Greece (Gastropoda: Bithyniidae). Journal of Conchology 40: 179.

Haszprunar, G. 1985. The fine morphology of the osphradial sense organs of the Mollusca. Part 1. Gastropoda—Prosobranchia. Philosophical Transactions of the Royal Society of London B: Biological Sciences 307: 457-496.

Healy, J.M. 1988. Sperm morphology and its systematic importance in the Gastropoda. Pp. 251-266 in Prosobranch Phylogeny (W.F. Ponder, D.J. Eernisse, and J.H. Waterhouse, eds.). Malacological Review Supplement, Ann Arbor, Michigan.

Henningsen, J.P., S.L. Lance, K.L. Jones, C. Hagen, J. Laurila, R.A. Cole, and K.E. Perez. 2010. Development and characterization of 17 polymorphic microsatellite loci in the faucet snail, *Bithynia tentaculata* (Gastropoda: Caenogastropoda: Bithyniidae). Conservation Genetics Resources 2: 247-250.

Kiatsopit, N., P. Sithithaworn, T. Boonmars, S. Tesana, A. Chanawong, W. Saijuntha, T.N. Petney, and R.H.

Andrews. 2011. Genetic markers for studies on the systematics and population genetics of snails, *Bithynia* spp., the first intermediate hosts of *Opisthorchis viverrini* in Thailand. Acta Tropica 118: 136-141.

Kiatsopit, N., P. Sithithaworn, W. Saijuntha, T.N. Petney, and R.H. Andrews. 2013. *Opisthorchis viverrini:* Implications of the systematics of first intermediate hosts, *Bithynia* snail species in Thailand and Lao PDR. Infection, Genetics and Evolution 14: 313-319.

Kulkarni, S., and R. Khot. 2015. *Neosataria,* replacement name for *Sataria* Annandale, 1920 (Mollusca: Gastropoda: Bithyniidae), preoccupied by *Sataria* Roewer, 1915 (Arachnida: Opiliones: Sclerosomatidae). Zootaxa 3974: 599-600.

Kulsantiwong, J., S. Prasopdee, S. Piratae, P. Khampoosa, A. Suwannatrai, W. Duangprompo, T. Boonmars, W. Ruangjirachuporn, J. Ruangsittichai, and V. Viyanant. 2013. Species-specific primers designed from RAPD products for *Bithynia funiculata,* the first intermediate host of liver fluke, *Opisthorchis viverrini,* in north Thailand. Journal of Parasitology 99: 433-437.

Lilly, M.M. 1953. The mode of life and the structure and functioning of the reproductive ducts of *Bithynia tentaculata* (L.). Proceedings of the Malacological Society of London 30: 87-110.

Ponder, W.F. 1988. The truncatelloidean (= rissoacean) radiation—a preliminary phylogeny. Malacological Review Supplement 4: 129-166.

Ponder, W.F. 2003. Monograph of the Australian Bithyniidae (Caenogastropoda: Rissooidea). Zootaxa 230: 1-126.

Taylor, D.W. 1966. A remarkable snail fauna from Coahuila, Mexico. Veliger 9: 152-228.

Wenz, W. 1938-1944. Gastropoda: Teil l, Band 6. Allgemeiner Teil und Prosobranchia. Pp. i-viii, 1-1639 in Handbuch der Paläozoologie, vol. 6, part 1 (O.H. Schindewolf, ed.). Gebrüder Borntraeger, Berlin.

Whalen, S.J. 2011. Analysis of the genetic structure of *Bithynia tentaculata* snail populations in Wisconsin and Minnesota. Master's thesis, Minnesota State University.

Wilke, T., G.M. Davis, A. Falniowski, F. Giusti, M. Bodon, and M. Szarowska. 2001. Molecular systematics of Hydrobiidae (Mollusca: Gastropoda: Rissooidea): Testing monophyly and phylogenetic relationships. Proceedings of the Academy of Natural Sciences of Philadelphia 151: 1-21.

Wilke, T., M. Haase, R. Hershler, H.-P. Liu, B. Misof, and W.F. Ponder. 2013. Pushing short DNA fragments to the limit: Phylogenetic relationships of 'hydrobioid' gastropods (Caenogastropoda: Rissooidea). Molecular Phylogenetics and Evolution 66: 715-736.

16 Cochliopidae Tryon, 1866

STEPHANIE A. CLARK

The Cochliopidae is currently divided into three subfamilies: Cochliopinae, Littoridininae, and Semisalsinae (Bouchet and Rocroi, 2005). In the past the cochliopids have been treated either as a distinct family or as a subfamily of the Hydrobiidae (Burch and Tottenham, 1980; Hershler and Thompson, 1992; Kabat and Hershler, 1993; Bouchet and Rocroi, 2005). Recent comparative anatomical and molecular studies have shown that the family Hydrobiidae is polyphyletic and that many more families need to be recognized, such as the Cochliopidae (Wilke et al., 2001, 2013; Wilke and Delicado, chap. 18, this volume). Wilke et al. (2013) found strong support with a large number of synapomorphies for Cochliopidae to be recognized at the family level. However, the sister relationships among the various families previously considered to belong to the Hydrobiidae remain uncertain (Wilke et al., 2013).

Cochliopids have small, planispiral to elongate conic and rarely uncoiled dextral shells that vary in size from 1 to 8 mm in height. They typically have smooth, relatively featureless shells, although a number of genera are sculptured with ribs, nodules, or frills, and the operculum can vary from paucispiral to multispiral, sometimes with a peg. Cochliopids have separate sexes and show some sexual dimorphism, with the females tending to be larger and broader than the males. Their tentacles are long and tapering and the animal is generally lightly pigmented. The female reproductive system has a spermathecal duct, the albumen gland is often highly reduced, and the capsule gland often forms a thin-walled brood pouch. The penis is often lobate and/or ornamented with complex glands. Cochliopids are morphologically extremely variable but can be distinguished from most other "hydroboid" families by the combination of the presence of a spermathecal duct and the absence of omniphoric grooves (Hershler and Thompson, 1992; Wilke et al., 2013). They reproduce either oviparously or ovoviviparously, with the young brooded in the pallial oviduct (Hershler and Thompson, 1992). Egg capsules are laid singly, in small chains, or in small masses linked by chalazae-like threads and can be attached to various substrates

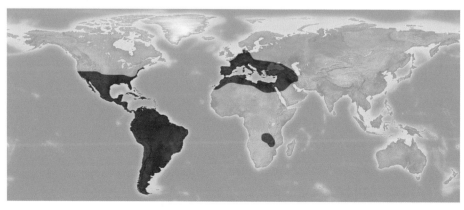

Distribution of Cochliopidae.

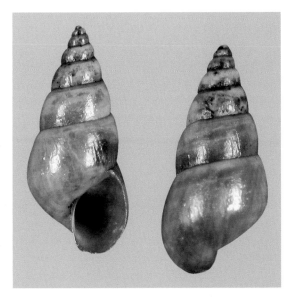

Heleobia australis (d'Orbigny, 1835). Uruguay: Laguna Redonda, near Franquia, Artigas Department. 5.7 mm. Poppe 840752.

including leaves, gravel, and woody debris (Hershler and Thompson, 1992).

The family is widely distributed throughout the Neotropical and Nearctic with smaller numbers from the Palearctic and Afrotropics (Hershler and Thompson, 1992; Strong et al., 2008). The highest diversity is found in central America (Hershler and Thompson, 1992), with a major radiation found in the Cuatro Ciénegas Spring complex in northern Mexico (Hershler, 1985).

There are about 260 described species, primarily found in the Neotropical realm (Strong et al., 2008). The family is found in a wide range of aquatic habitats, from isolated springs and seepages, saline springs, caves, streams, and rivers to brackish estuaries. Some species have wide ranges, such as *Littoridinops tenuipes* (Johnson et al., 2013), while most are known only from a few locations or a single location, such as *Cochliopina australis* (Thompson, 2011).

In North America the family is represented by at least 66 species in 19 genera: *Antrobia* Hubricht, 1971, *Aphaostracon* F. G. Thompson, 1968, *Balconorbis* Hershler & Longley, 1986, *Cochliopina* Morrison, 1946, *Eremopyrgus* Hershler, 1999, *Heleobops* F. G. Thompson, 1968, *Ipnobius* Hershler, 2001, *Juturnia* Hershler, Liu & Stockwell, 2002, *Littoridinops* Pilsbry, 1952, *Ono-*

bops F. G. Thompson, 1968, *Phreatoceras* Hershler & Longley, 1987, *Phreatodrobia* Hershler & Longley, 1986, *Pseudotryonia* Hershler, 2001, *Pyrgophorus* Ancey, 1888, *Spurwinkia* Davis & Mazurkiewicz, 1982, *Stygopyrgus* Hershler & Longley, 1986, *Texadina* Abbott & Ladd, 1951, *Texapyrgus* Thompson & Hershler, 1991, and *Tryonia* Stimpson, 1865 (Abbott and Ladd, 1951; Thompson, 1968; Hubricht, 1971; Davis et al., 1982; Hershler and Longley, 1986; Hershler and Sada, 1987; Hershler, 1989, 1999; Thompson and Hershler, 1991; Hershler, 2001; Hershler et al., 2002, 2011a, 2015). Most of the known diversity is found in the southwestern United States and Mexico, but the species are absent from the northern states, with the exception of the estuarine *Spurwinkia salsa*. They can be found in a range of aquatic habitats, from very small springs, caves, and streams to very large lakes, coastal marshes, and lagoons (Thompson, 1968; Burch and Tottenham, 1980; Hershler and Thompson, 1992). Most of the freshwater species are narrow range endemics, such as *Aphaostracon theiocrenetum* from Florida or *Balconorbis uvaldensis* from Texas, while the estuarine taxa tend to be widespread, such as *Littoridinops monroensis* (Thompson, 2000; Hershler and Longley, 1986). As more populations are examined both morphologically and genetically, additional diversity at both the generic and species levels is revealed (e.g., *Ipnobius* Hershler, 2001; *Juturnia* Hershler et al., 2002) or it is found that taxa should be placed in other families; e.g., Liu et al. (2016) showed that *Taylorconcha*, previously placed in the Cochliopidae, should be transferred to the Amnicolidae.

In Central America, the Caribbean, and South America, the family is represented by at least 180 species in 26 genera: *Aroapyrgus* H. B. Baker, 1931, *Coahuilix* D. W. Taylor, 1966, *Cochliopa* Stimpson, 1865, *Cochliopina* Morrison, 1946, *Durangonella* Morrison, 1945, *Emmericiella* Pilsbry, 1909, *Heleobia* Stimpson, 1865, *Heleobops* F. G. Thompson, 1968, *Juturnia* Hershler, Liu & Stockwell, 2002, *Lithococcus* Pilsbry, 1911, *Littoridina* Souleyet, 1852, *Littoridinops* Pilsbry, 1952, *Mesobia* F. G. Thompson & Hershler, 1991, *Mexipyrgus* D. W. Taylor, 1966, *Mexithauma* D. W. Taylor, 1966, *Minkleyella* Hershler, Liu & Landye 2011, *Nanivitrea* Thiele, 1928, *Onobops* F. G. Thompson, 1968, *Paludiscala* D. W. Taylor, 1966, *Phreatoceras* Hershler & Longley, 1987, *Pyrgophorus*

Ancey, 1888, *Subcochliopa* Morrison, 1946, *Tepalcatia* Thompson & Hershler, 2002, *Texadina* Abbott & Ladd, 1951, *Tryonia* Stimpson, 1865, and *Zetekina* Morrison, 1947 (Taylor, 1966; Hershler and Thompson, 1992; Thompson and Hershler, 1991, 2002; Hershler et al., 2011b; Thompson, 2011). They can be found in virtually all aquatic habitats, from very small springs, caves, and streams to very large lakes, coastal marshes, and lagoons (Hershler and Thompson, 1992; Thompson, 2011). Most of the freshwater species are narrow range endemics, such as *Cochliopa perforata* from Costa Rica or *Lithococcus multicarinatus* from Ecuador, while some are widespread, such as *Pyrgophorus coronatus,* from Mexico to Venezuela (Hershler and Thompson, 1992; Thompson, 2011). The most diverse genus is *Heleobia,* with more than 80 described species, with a large radiation "species flock" found in Lake Titicaca. The taxonomic and systematic relationships among the various genera and species are not fully resolved, as most species were described using shell morphology and/or anatomical differences (mostly the shape of the penis). However, recent anatomical and molecular studies are highlighting the need for further revisionary taxonomic studies (Kroll et al., 2012; Collado et al., 2013; Koch et al., 2015).

In Europe the family is represented by at least 17 species in the genus *Heleobia* Stimpson, 1865 (Hershler and Thompson, 1992). They are found from freshwater to estuarine conditions, such as *Heleobia dobrogica* from a cave in Romania (Falniowski et al., 2008), while *H. stagnorum* is widespread in estuarine Atlantic coastal zones of Europe and the Mediterranean region, extending to North Africa and east to Iran (Glöer, 2002). However, the taxonomic status and relationships between the described European species are yet to be resolved; for example, in a COI molecular analysis, Falniowski et al. (2008) analyzed COI, finding *H. dobrogica* to be very closely related to *H. dalmatica,* and suggested they are congeners and that the American species *Heleobops docimus* is very closely related (nested within their clade containing *H. dobrogica* and *H. dalmatica*).

In Africa the family is represented by at least three species in a single genus *Lobogenes* Pilsbry and Bequaert, 1927 (Pilsbry and Bequaert, 1927). They are currently known only from the Zaire and Zambezi River basins in the Democratic Republic of Congo (formerly Zaire and Zambia [Hershler and Thompson, 1992]).

A few species of cochliopids have been introduced to areas outside their native ranges and include *Tryonia porrecta, Littoridinops monroensis,* and *Spurwinkia salsa* in California, Arizona, and Hawaii (Hershler et al., 2007, 2015); *Pyrgophorus platyrachis* in Singapore (Ng et al., 2016), and *Pyrgophorus coronatus* in Jordan and Israel (Amr et al., 2014).

Among the factors that affect the diversity and conservation of cochliopids are habitat destruction from dams, channel modification, siltation, water-quality degradation from point and nonpoint pollution, and the introduction of nonindigenous species. Lysne et al. (2008) explore challenges and opportunities of freshwater conservation and highlight the need for involvement of many affected interest groups, from local communities to government agencies.

The majority of Cochliopids are narrow range endemics, and this combined with the vulnerability of habitats such as caves, springs, streams, and lakes places many species at risk of extinction. The IUCN Red List treats 49 species of Cochliopidae: *Aroapyrgus* (2 species), *Heleobia* (29), *Nurekia* (1) and *Tryonia* (17), of which three are listed as Critically Endangered, two as Endangered, five as Vulnerable, 32 as Data Deficient, 6 as Least Concern, and 1 as Extinct (IUCN, 2016). However, several other taxa now included in the Cochliopidae are currently listed as belonging to the Hydrobiidae. These include the following 37 taxa: Antrobia (1 species), *Aphaostracon* (6), *Coahuilix* (1), *Cochliopa* (1), *Cochliopina* (3), *Heleobops* (1), *Lithococcus* (1), *Littoridina* (10), *Littoridinops* (1), *Lobogenes* (3), *Mexipyrgus* (1), *Paludiscala* (1), *Phreatoceras* (1), *Phreatodrobia* (4), *Pseudotryonia* (1), and *Pyrgophorus* (1), of which 1 is listed as Critically Endangered, 1 as Endangered, 8 as Vulnerable, 1 as Near Threatened, 18 as Data Deficient, 7 as Least Concern, and 1 as Extinct.

LITERATURE CITED

Abbott, R.T., and H.S. Ladd. 1951. A new brackish-water gastropod from Texas (Amnicolidae: *Littoridina*). Journal of the Washington Academy of Sciences 41: 335–338.

Amr, Z., H. Nasarat, and E. Neubert. 2014. Notes on the current and past freshwater snail fauna of Jordan. Jordan Journal of Natural History 1: 83–115.

Bouchet, P., and J.-P. Rocroi. 2005. Classification and nomenclator of gastropod families. Malacologia 47: 1-397.

Burch, J.B., and J.L. Tottenham. 1980. North American freshwater snails: Species list, ranges and illustrations. Walkerana 1: 81-215.

Collado, G.A., M.A. Valladares, and M.A. Méndez. 2013. Hidden diversity in spring snails from the Andean Altiplano, the second highest plateau on Earth, and the Atacama Desert, the driest place in the world. Zoological Studies 52: 50-62.

Davis, G.M., M. Mazurkiewicz, and M. Mandracchia. 1982. *Spurwinkia:* Morphology, systematics, and ecology of a new genus of North American marshland Hydrobiidae (Mollusca: Gastropoda). Proceedings of the Academy of Natural Sciences of Philadelphia 134: 143-177.

Falniowski, A., M. Szarowska, I. Sirbu, A. Hillebrand, and M. Baciu. 2008. *Heleobia dobrogica* (Grossu & Negrea, 1989) (Gastropoda: Rissooidea: Cochliopidae) and the estimated time of its isolation in a continental analogue of hydrothermal vents. Molluscan Research 28: 165-170.

Glöer, P. 2002. Die Tierwelt Deutschlands, 73. Teil: Die Süßwassergastropoden Nord- und Mitteleuropas. Bestimmungsschlüssel, Lebensweise, Verbreitung. ConchBooks, Hackenheim, Germany.

Hershler, R. 1985. Systematic revision of the Hydrobiidae (Gastropoda: Rissoacea) of the Cuatro Ciénegas Basin, Mexico. Malacologia 26: 31-123.

Hershler, R. 1989. Springsnails (Gastropoda: Hydrobiidae) of Owens and Amargosa River (exclusive of Ash Meadows) drainages, Death Valley System, California-Nevada. Proceedings of the Biological Society of Washington 102: 176-248.

Hershler, R. 1999. A systematic review of the hydrobiid snails (Gastropoda: Rissooidea) of the Great Basin, western United States. Part II. Genera *Colligyrus, Eremopyrgus, Fluminicola, Pristinicola,* and *Tryonia.* Veliger 42: 306-337.

Hershler, R. 2001. Systematics of the North and Central American aquatic snail genus *Tryonia* (Rissooidea: Hydrobiidae). Smithsonian Contributions to Zoology 612: 1-53.

Hershler, R., and G. Longley. 1986. Phreatic hydrobiids (Gastropoda: Prosobranchia) from the Edwards (Balcones Fault Zone) Aquifer Region, south-central Texas. Malacologia 27: 127-172.

Hershler, R., and D.W. Sada. 1987. Springsnails (Gastropoda: Hydrobiidae) of Ash Meadows, Amargosa Basin, California-Nevada. Proceedings of the Biological Society of Washington 100: 776-843.

Hershler, R., and F.G. Thompson. 1992. A review of the aquatic gastropod subfamily Cochliopinae (Prosobran-chia: Hydrobiidae). Malacological Review Supplement 5: 1-140.

Hershler, R., H.-P. Liu, and C.L. Stockwell. 2002. A new genus and species of aquatic gastropods (Rissooidea: Hydrobiidae) from the North American Southwest: Phylogenetic relationships and biogeography. Proceedings of the Biological Society of Washington 115: 171-188.

Hershler, R., C.L. Davis, C.L. Kitting, and H.-P. Liu. 2007. Discovery of introduced and cryptogenic gastropods in the San Francisco Estuary, California. Journal of Molluscan Studies 73: 323-332.

Hershler, R., H.-P. Liu, and J.J. Landye. 2011a. New species and records of springsnails (Caenogastropoda: Cochliopidae: *Tryonia*) from the Chihuahuan Desert (Mexico and United States), an imperiled biodiversity hotspot. Zootaxa 3001: 1-32.

Hershler, R., H.-P. Liu, and J.J. Landye. 2011b. Two new genera and four new species of freshwater cochliopid gastropods (Rissoidea) from northeastern Mexico. Journal of Molluscan Studies 77: 8-23.

Hershler, R., H.-P. Liu, and J.S. Simpson. 2015. Assembly of a micro-hotspot of caenogastropod endemism in the southern Nevada desert, with a description of a new species of *Tryonia* (Truncatelloidea, Cochliopidae). ZooKeys 492: 107-122.

Hubricht, L. 1971. New Hydrobiidae from Ozark caves. Nautilus 84: 93-96.

IUCN Red List of Threatened Species. 2016. Version 2016-2. www.iucnredlist.org. Accessed 20 September 2016.

Johnson, P.D., A.E. Bogan, K.M. Brown, N.M. Burkhead, J.R. Cordeiro, J.T. Garner, P.D. Hartfield, et al. 2013. Conservation status of freshwater gastropods of Canada and the United States. Fisheries 38: 247-282.

Kabat, A.R., and R. Hershler. 1993. The prosobranch snail family Hydrobiidae (Gastropoda: Rissooidea): Review of classification and supraspecific taxa. Smithsonian Contributions to Zoology 547: 1-94.

Koch, E., S.M. Martin, and N.F. Ciocco. 2015. A molecular contribution to the controversial taxonomical status of some freshwater snails (Caenogastropoda: Rissooidea, Cochliopidae) from the Central Andes desert to Patagonia. Iheringia 105: 69-75.

Kroll, O., R. Hershler, C. Albrecht, E.M. Terrazas, R. Apaza, C. Fuentealba, C. Wolff, and T. Wilke. 2012. The endemic gastropod fauna of Lake Titicaca: Correlation between molecular evolution and hydrographic history. Ecology and Evolution 2: 1517-1530.

Liu, H.P., D. Marceau, and R. Hershler. 2016. Taxonomic identity of two amnicolid gastropods of conservation concern in lakes of the Pacific Northwest of the USA. Journal of Molluscan Studies 82: 464-471.

Lysne, S.J., K.E. Perez, K.M. Brown, R.L. Minton, and J.D. Sides. 2008. A review of freshwater gastropod conservation: Challenges and opportunities. Journal of the North American Benthological Society 27: 463-470.

Ng, T.H., J.H. Liew, J.Z.E. Song, and D.C.J. Yeo. 2016. First record of the cryptic invader *Pyrgophorus platyrachis* Thompson, 1968 (Gastropoda: Truncatelloidea: Cochliopidae) outside the Americas. BioInvasions Records 5: 75-80.

Pilsbry, H.A., and J. Bequaert. 1927. The aquatic mollusks of the Belgian Congo: With a geographical and ecological account of Congo malacology. Bulletin of the American Museum of Natural History 53: 69-659, plates 10-67.

Strong, E.E., O. Gargominy, W.F. Ponder, and P. Bouchet. 2008. Global diversity of gastropods (Gastropoda: Mollusca) in freshwater. Hydrobiologia 595: 149-166.

Taylor, D.W. 1966. A remarkable snail fauna from Coahuila, Mexico. Veliger: 152-228, plates 8-19.

Thompson, F.G. 1968. The Aquatic Snails of the Family Hydrobiidae of Peninsular Florida. University of Florida Press, Gainesville, Florida.

Thompson, F.G. 2000. An identification manual for the freshwater snails of Florida. Walkerana 10: 1-96.

Thompson, F.G. 2011. An annotated checklist and bibliography of the land and freshwater snails of Mexico and Central America. Bulletin of the Florida Museum of Natural History 50: 1-299.

Thompson, F.G., and R. Hershler. 1991. New hydrobiid snails from North America. Proceedings of the Biological Society of Washington 104: 669-683.

Thompson, F.G., and R. Hershler. 2002. *Tepalcatia,* a new genus of hydrobiid snails (Prosobranchia: Rissooidea) from the Rio Balsas basin, central Mexico. Proceedings of the Biological Society of Washington 115: 189-204.

Wilke, T., G.M. Davis, A. Falniowski, F. Giusti, M. Bodon, and M. Szarowska. 2001. Molecular systematics of Hydrobiidae (Mollusca: Gastropoda: Rissooidea): Testing monophyly and phylogenetic relationships. Proceedings of the Academy of Natural Sciences of Philadelphia 151: 1-21.

Wilke, T., M. Haase, R. Hershler, H.-P. Liu, B. Misof, and W.F. Ponder. 2013. Pushing short DNA fragments to the limit: Phylogenetic relationships of 'hydrobioid' gastropods (Caenogastropoda: Rissooidea). Molecular Phylogenetics Evolution 66: 715-736.

17 Helicostoidae Pruvot-Fol, 1937

THOMAS WILKE

The Helicostoidae comprises a single species, *Helicostoa sinensis*. It was originally described by Lamy (1926) from specimens cemented to a limestone block deposited at the mineral collection of the Muséum National d'Histoire Naturelle in Paris. The type locality was poorly described as "Kouei-Tchéou, ville située sur le Yang Tsé Kiang, à plus de 1200 kilom. de Chang-Hai" (Guizhou [Province], a town on the Yangtze River, more than 1200 km away from Shanghai). The material was likely collected in the 1910s in the Three Gorges area of the Yangtze River during a period of exceptionally low waters when the rapids were exposed (Philippe Bouchet, pers. comm.). As the site was likely flooded during the construction of the Three Gorges Reservoir, it remains questionable whether the species is still extant.

Helicostoa sinensis is a small (10–12 mm in diameter) vermetid-like snail that lives attached to hard substrate. It might therefore be the only sessile freshwater gastropod species. Based on operculum, soft body, and particularly radula remains, Pruvot-Fol (1937) considered the new family Helicostoidae to belong to the "taeneoglossate prosobranchs." However, due to the strong shell dimorphism observed—some specimens were flat and discoidal; others were more trochiform—an unambiguous family-level assignment remained impossible. Later, Heppell (1995) tentatively placed *Helicostoa* into the marine and brackish caenogastropod family Vermetidae. This assignment was rejected by Bouchet and Rocroi (2005), who considered the family Helicostoidae to belong to the caenogastropod superfamily Rissooidea (today Truncatelloidea). The latter hypothesis was also supported by Bieler and Petit (2011).

Given the peculiar life history of *H. sinensis,* the lack of detailed anatomical information, and the absence of well-preserved materials, phylogenetic studies of freshly collected material might be necessary in order to assess the systematic status of this family (Heppell, 1995). If no fresh specimens can be obtained, ancient DNA analyses of the type specimens, deposited at the Muséum National d'Histoire Naturelle in Paris, might be promising.

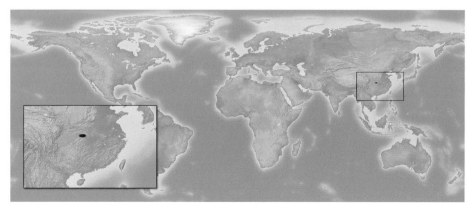

Distribution of Helicostoidae.

LITERATURE CITED

Bieler, R., and R. Petit. 2011. Catalogue of Recent and fossil 'worm-snail' taxa of the families Vermetidae, Siliquariidae, and Turritellidae (Mollusca: Caenogastropoda). Zootaxa 2948: 1-103.

Bouchet, P., and J.-P. Rocroi. 2005. Classification and nomenclator of gastropod families. Malacologia 47: 1-397.

Heppell, D. 1995. *Helicostoa:* A forgotten Chinese gastropod enigma. Pp. 29-30 in Abstracts, Twelfth International Malacological Congress (A. Guerra, E. Rolan, and F. Rocha, eds.). Unitas Malacologica, Vigo, Spain.

Lamy, E. 1926. Sur un coquille énigmatique. Journal de Conchyliologie 70: 51-56.

Pruvot-Fol, A. 1937. Etude d'un prosobranche d'eau douce: *Helicostoa sinensis* Lamy. Bulletin de la Société Zoologique de France 62: 250-257.

18 Hydrobiidae Stimpson, 1865

THOMAS WILKE AND DIANA DELICADO

With at least 900 valid species (Miller et al., 2018), the Hydrobiidae is one of the largest gastropod families. Moreover, it is suggested that the current number of named species may represent only 25% of their actual diversity (Strong et al., 2008). Prior to the introduction of molecular tools, the Hydrobiidae was used in an even broader sense. However, as its monophyly had long been questioned, Davis (1979) coined the term "hydrobioid" for hydrobiid-like taxa that share some anatomical and morphological characteristics (see also Hershler and Ponder, 1998). Since then, several commonly accepted subfamilies were excluded from the Hydrobiidae and raised to family level (see Bouchet et al., 2017). Examples include the Amnicolidae, the Cochliopidae, the Lithoglyphidae, and the Moitessieriidae (Wilke et al., 2001), as well as the Tateidae (Criscione and Ponder, 2013; Wilke et al., 2013). The latest phylogeny-based definition of the Hydrobiidae (Wilke et al., 2013) also suggested that the Australasian subfamily Clenchiellinae, the North American subfamily Fontigentinae, and the European genus *Bythiospeum* do not belong to the Hydrobiidae. The morphologically and anatomically variable Hydrobiidae was strongly supported in these genetic analyses. However, although the family is characterized by a closed ventral wall of the female capsule gland and sometimes by a pigmented coiled oviduct, there are apparently no uniquely shared morphological characters defining this group. The analysis also suggested that the Hydrobiidae is restricted to North America, Europe, northern and southern Africa, and western and central Asia.

The mostly small snails (shell size 0.5–15.0 mm) are adapted to various habitats. The majority occur in freshwater ecosystems such as springs (Bodon et al., 2001), streams and creeks (Hershler, 1994), and lakes (Radoman, 1983). Approximately 50 species inhabit brackish and very few even marine environments (Barnes, 1999). Interestingly, some brackish-water taxa may tolerate hypersaline conditions (Drake and Arias, 1995), suggesting that high salinity alone does not prevent them from colonizing open marine

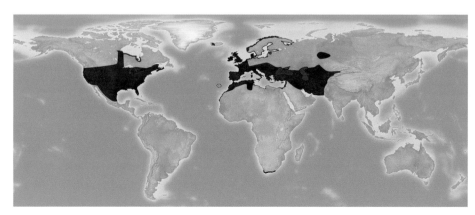

Distribution of Hydrobiidae.

settings but that other factors such as wave action, currents, and possible food and substrate availability do (Wilke and Davis, 2000).

The biodiversity of hydrobiids is distributed unevenly across the Nearctic and western Palearctic (e.g., Kabat and Hershler, 1993). The highest species and genus richness can be found in spring systems on the Apennine, Iberian, and Balkan peninsulas (Radoman, 1983; Bodon et al., 2001; Arconada and Ramos, 2003; Strong et al., 2008), in the endorheic watersheds of the North American Great Basin (Hershler, 1998), in ancient lakes such as lakes Ohrid and Prespa on the Balkan Peninsula (Radoman, 1983; Föller et al., 2015), and in brackish-water systems of the Pontocaspian Region (Logvinenko and Staroboga-tov, 1968; note that the Caspian Sea is also an ancient lake).

In North America, the family is represented by at least 186 species in 12 genera: *Birgella* Baker, 1926, *Cincinnatia* Pilsbry, 1891, *Ecrobia* Stimpson, 1865, *Floridobia* Thompson & Hershler, 2002, *Marstonia* Baker, 1926, *Notogillia* Pilsbry, 1953, *Nymphophilus* Taylor, 1966, *Probythinella* Thiele, 1928, *Pyrgulopsis* Call & Pilsbry, 1886, *Rhapinema* Thompson, 1970, *Spilochlamys* Thompson, 1968, and *Stiobia* Thompson & McCaleb, 1978 (Thompson, 1968; Davis et al., 1989; Hershler, 1996; Thompson and Hershler, 2002; Hershler et al., 2003, 2011, 2014; Hershler and Liu, 2009). By far the most species-rich genus of freshwater mollusks on the continent is *Pyrgulopsis,* with 146 species described (Hershler et al., 2014, 2016a, b, 2017). However, the phylogenetic relationships within this large genus are not fully resolved, and the taxon is potentially rendered paraphyletic both by the Mexican genus *Nymphophilus* and by the eastern North American genus *Floridobia* (Hershler et al., 2003; Liu and Hershler, 2005).

The vast majority of North American hydrobiids live in freshwaters. Their current distribution ranges from the coasts of California in the west (*Pyrgulopsis;* Hershler, 1994) to the coasts of Maine in the east (*Cincinnatia, Floridobia, Marstonia, Probythinella;* Hershler, 1996; Thompson and Hershler, 2002; Hershler et al., 2011), and from the Hudson Bay in Canada in the north (*Probythinella;* Hershler, 1996) to northern Mexico in the south (*Cincinnatia, Nymphophilus;* Thompson, 1979; Hershler et al., 2011). The brackish-

Macedopyrgula wagneri (Polinski, 1929). Macedonia: Lake Ohrid. 3.8 mm. UGSB 4953. Photo by T. Wilke.

water *Probythinella protera* occurs along the northern and eastern coasts of the Gulf of Mexico (Hershler, 1996), and the estuarine *Ecrobia truncata* along the coasts of the northwestern Atlantic from the Gulf of Saint Lawrence (Layton et al., 2014) to the Chesapeake Bay in Maryland (Davis et al., 1989). According to Layton et al. (2014), it is also found in the southern and western parts of the Hudson Bay (Arctic Ocean).

Africa is home to at least 13 nominal hydrobiid genera: *Belgrandia* Bourguignat, 1870, *Belgrandiella* Wagner, 1928, *Bullaregia* Khalloufi, Bejaóui & Delicado, 2017, *Ecrobia* Stimpson, 1865, *Hadziella* Kuščer, 1932, *Hauffenia* Pollonera, 1898, *Heideella* Backhuys & Boeters, 1974, *Horatia* Bourguignat, 1887, *Hydrobia* Hartmann, 1821, *Mercuria* Boeters, 1971, *Peringia* Paladilhe, 1874, *Pseudamnicola* Paulucci, 1878, and *Pseudoislamia* Radoman, 1979 (Boeters, 1976; Brown, 1994; Ghamizi et al., 1997; Bodon et al., 1999; Yacoubi-Khebiza et al., 2001; Wilke et al., 2002; Glöer et al., 2010, 2015; Khalloufi et al., 2017). The total number of species remains largely unknown (Brown, 1994), but might be in the lower double-digit range. Many species are known only from empty shells. Due to the lack of anatomical data, they were mostly attributed to European genera (discussed in Brown, 1994; Bodon et al., 1999). This concerns, for example, "*Hydrobia*" *aponensis* from the Mediterranean coast (Van Damme,

1984) as well as several "*Hydrobia*" spp. from eastern Africa (reviewed in Brown, 1994). However, *H. aponensis* was assigned to the genus *Semisalsa* (family Cochliopidae; Kroll et al., 2012) and most remaining African "*Hydrobia*" may also belong to families other than the Hydrobiidae. A notable exception is the estuarine "*Hydrobia*" *knysnaensis* from the South African coast (Barnes, 2004), which is a true hydrobiid but might have to be assigned to a new genus (Wilke et al., 2013).

Most African freshwater hydrobiids are restricted to springs, subterranean systems, and rivers in northwestern Africa (Algeria, Morocco, and Tunisia), particularly across the Atlas Mountains, underlining their Palearctic affiliation. This also applies to most estuarine species. *Peringia ulvae* occurs along the African Atlantic coast as far south as northern Mauritania (T. Wilke, unpublished data), and the genera *Ecrobia* and *Hydrobia* can be found along the Atlantic coast of northwestern Africa as well as along the Mediterranean coast of northern Africa (e.g., Wilke et al., 2000). As mentioned above, the only confirmed sub-Saharan hydrobiid taxon is the estuarine "*Hydrobia*" *knysnaensis* from South Africa.

The worldwide highest hydrobiid biodiversity can be found in the mountainous regions of Europe (e.g., Strong et al., 2008). Though both total numbers of genera and species are not known, the spring and groundwater systems in the Pyrenees (e.g., Arconada and Ramos, 2003), in the southern Alps (e.g., Giusti and Pezzoli, 1981), and in the Balkans (e.g., Radoman, 1983; Georgiev et al., 2017; Osikowski et al., 2017) alone harbor several hundred species in more than 60 genera. In addition, Lake Ohrid on the Balkan Peninsula is home to ca. 40 endemic hydrobiid species in ca. 21 genera (Radoman, 1983; Föller et al., 2015). Highly biodiverse regions also include the tributaries of river systems draining into the Mediterranean (e.g., Beran, 2011) and into the Black Sea (reviewed in Vinarski and Kantor, 2016).

Freshwater hydrobiids are common across southern, southwestern, and southeastern Europe. They also occur on many Mediterranean islands (e.g., Boeters, 1988; Glöer et al., 2015; Szarowska et al., 2016) and on the Atlantic Island of Madeira (Glöer et al., 2015). Their extant northern range limits are defined by occurrences of the estuarine *Mercuria anatina* in southern Ireland, southeastern England, and the midwestern parts of the Netherlands (e.g., Kerney, 1999). Brackish hydrobiids (genera *Adriohydrobia* Radoman, 1977, *Ecrobia* Stimpson, 1865, *Hydrobia* Hartmann, 1821, and *Peringia* Paladilhe, 1874) can be found along most European coasts. Their confirmed northwestern and northeastern distribution boundaries are in western Iceland and in the Russian White Sea, respectively (Wilke and Davis, 2000).

Overall, the systematic relationships among European hydrobiids are poorly understood, making even rough estimates of the total biodiversity speculative. Moreover, genetic information is available only for a portion of taxa, and large-scale comparative studies are lacking. Nonetheless, the high number of genus-level taxa for European hydrobiids, in general, and for the Balkans and the Pontocaspian area, in particular, is remarkable. However, preliminary molecular studies indicate that several genera are paraphyletic and/or that the genetic distances among genus-level taxa may be very small (e.g., Wilke et al., 2007; Föller et al., 2015; Osikowski et al., 2017). Thus, the taxonomy of European hydrobiids is clearly in need of revision in a comparative context using genetic, anatomical, ecological, and biogeographical information. Very likely, such integrated studies will further increase the number of described species. However, the total number of genus-level taxa may not necessarily grow, due to the necessity for synonymizations.

For Asia, more than 40 hydrobiid genera have been reported (Radoman, 1983; Yıldırım et al., 2006; Glöer and Georgiev, 2012; Glöer and Pešić, 2012; Şahin et al., 2012; Vinarski et al., 2014; Delicado et al., 2016; Glöer et al., 2016; Vinarski and Kantor, 2016). So far, the exact number of Asian species is not known. However, Vinarski and Kantor (2016) list approximately 120 hydrobiid taxa for the Asian part of Russia and adjacent territories. Supplementing these data with information from Turkey (e.g., Yıldırım et al., 2006), Israel (e.g., Schütt, 1991), and Iran (e.g., Glöer and Pešić, 2012), as well as with newer data from central Asia (e.g., Glöer et al., 2014) brings the total number of hydrobiid species in Asia to roughly 200.

Many of them are members of the subfamily Pyrgulinae, a group that is particularly biodiverse in the Pontocaspian area (e.g., Logvinenko and Star-

obogatov, 1968). Others resemble representatives of
the subfamily Pseudamicolinae in shell shape (e.g.,
Anatoludamnicola Şahin, Koca & Yıldırım, 2012, *Bucharamnicola* Izzatullaev et al., 1985, *Martensamnicola*
Izzatullaev et al., 1985, *Sarkhia* Glöer & Pešić, 2012,
Sogdamnicola Izzatullaev et al., 1985, *Tadzhikamni-
cola* Izzatullaev, 2004, *Turkmenamnicola* Izzatullaev
et al., 1985, *Valvatamnicola* Izzatullaev et al., 1985).
However, using genetic and anatomical data, some
Pseudamnicola-like species were recently assigned to
new pyrgulinid genera (Delicado et al., 2016). Moreover, preliminary phylogenetic studies of several
populations of the species-rich genus *Caspiohydrobia*
Starobogatov, 1970, indicate that they all belong to
the Caspian *Ecrobia grimmi* (T. Wilke, unpublished
data; see also Vinarski and Kantor, 2016). These
findings not only emphasize the need for a careful
revision of the central Asian hydrobiids; they also
underline the importance of supplementary genetic
data for a better understanding of the range limits of
those taxa.

The Hydrobiidae are widely distributed in the
midwestern parts of Asia. Numerous species occur
in the Caucasus as well as in the Caspian and eastern
Mediterranean basins. However, the northern, southern, and eastern distribution boundaries are not well
understood. The northern range limit of Asian hydrobiids is currently defined by occurrences of *Ecrobia
grimmi* in the Russian Lake Salamtka (see above), east
of the Ural Mountains. The southern limits are characterized by occurrences of *Pseudamnicola solitaria*
in the Dead Sea area of Israel and Jordan (Schütt,
1991), as well as of *Ecrobia grimmi* in Lake Sawa, Iraq
(Haase et al., 2010), and in the Hormozgan Province
of southern Iran (Glöer and Pešić, 2012). The known
eastern limits of hydrobiids in Asia are determined
by findings of *Chirgisia alaarchaensis* in north-central
Kyrgyzstan (Glöer et al., 2014) and by *Ecrobia* sp.
(= *Caspiohydrobia*) in the Lake Issyk-Kul area (Maxim
Vinarski, pers. comm.). Though there are several
reports of "hydrobioid" snails in eastern and southeastern Asia, most if not all of them might belong to
different families such as the Tateidae, Amnicolidae,
and Stenothyridae (see also Criscione and Ponder,
2013; Wilke et al., 2013).

Despite an increasing knowledge of hydrobiid
diversity and distribution, the infra-family systemat-

ics of the Hydrobiidae is not well understood and
many nominal subfamilies are in need of revision.
Based on molecular data, Wilke et al. (2013) found
the following subfamilies to be highly supported: the
Hydrobiinae (see also Wilke, 2003), the Pseudamni-
colinae (see also Delicado et al., 2015), the Pyrgulinae
(see also Wilke et al., 2007; Delicado et al., 2016),
the Nymphophilinae (see also Thompson, 1979;
Hershler et al., 2003; Liu and Hershler, 2005), and the
Islamiinae (Radoman, 1973, 1983). The validity of the
Belgrandiinae, Belgrandiellinae, and Horatiinae could
not be confirmed beyond a reasonable doubt. Moreover, *Mercuria* did not cluster in any of the above
taxa, and Boeters and Falkner (2017) recently assigned
it to its own subfamily, the Mercuriinae. The Caspi-
inae and Shadiniinae were not part of the analysis of
Wilke et al. (2013) but might represent valid hydrobiid
subfamilies according to Anistratenko (2013) and
Anistratenko et al. (2017), respectively.

Systematic and biogeographical studies indicate
that numerous hydrobiid species are narrow range
endemics. This high degree of endemism, together
with the vulnerability of habitats such as subterranean waters, springs, and brackish-water lagoons,
makes many species prone to extermination. From
the 1107 species of "Hydrobiidae" (including some
Amnicolidae, Cochliopidae, Moitessieriidae, Tateidae,
and related "hydrobioids") reported in the IUCN
Red List (IUCN, 2017), 31 are listed as Extinct, 152
as Critically Endangered, 115 as Endangered, 259 as
Vulnerable, 84 as Near Threatened, and 296 as Data
Deficient. Only 170 (= 15%) are of Least Concern.
Major threads include natural system modifications
(479 species), mainly due to dams and water management, and pollution (446 species) due to domestic
and urban waste waters as well as agricultural and
forestry effluents.

LITERATURE CITED

Anistratenko, V.V. 2013. On the taxonomic status of the
    highly endangered Ponto-Caspian gastropod genus *Caspia*
    (Gastropoda: Hydrobiidae: Caspiinae). Journal of Natural
    History 47: 51–64.
Anistratenko, V.V., T. Peretolchina, T. Sitnikova, D. Palatov,
    and D. Sherbakov. 2017. A taxonomic position of Arme-
    nian endemic freshwater snails of the genus *Shadinia*
    Akramowski, 1976 (Caenogastropoda: Hydrobiidae):

Combining morphological and molecular evidence. Molluscan Research 37: 212-221.

Arconada, B., and M.Λ. Ramos. 2003. The Ibero-Balearic region: One of the areas of highest Hydrobiidae (Gastropoda, Prosobranchia, Rissooidea) diversity in Europe. Graellsia 59: 91-104.

Barnes, R.S.K. 1999. What determines the distribution of coastal hydrobiid mudsnails within North-Western Europe? Marine Ecology 20: 97-110.

Barnes, R.S.K. 2004. The distribution and habitat in the Knysna Estuary of the endemic South African mudsnail *Hydrobia knysnaensis* and the influence of intraspecific competition and ambient salinity on its abundance. African Journal of Aquatic Science 29: 205-211.

Beran, L. 2011. Non-marine molluscs (Mollusca: Gastropoda, Bivalvia) of the Zrmanja River and its tributaries (Croatia). Natura Croatica 20: 397-409.

Bodon, M., M. Ghamizi, and F. Giusti. 1999. The Moroccan stygobiont genus *Heideella* (Gastropoda, Prosobranchia: Hydrobiidae). Basteria 63: 89-105.

Bodon, M., G. Manganelli, and F. Giusti. 2001. A survey of the European valvatiform hydrobiid genera with special reference to *Hauffenia* Pollonera, 1898 (Gastropoda: Hydrobiidae). Malacologia 43: 103-215.

Boeters, H.D. 1976. Hydrobiidae Tunesiens. Archiv für Molluskenkunde 107: 89-105.

Boeters, H.D. 1988. Moitessieriidae und Hydrobiidae in Spanien und Portugal. Archiv für Molluskenkunde 118: 181-261.

Boeters, H.D., and G. Falkner. 2017. The genus *Mercuria* Boeters, 1971 in France (Gastropoda: Caenogastropoda: Hydrobiidae). West-European Hydrobiidae, Part 13. Zoosystema 39: 227-261.

Bouchet, P., J. Rocroi, B. Hausdorf, A. Kaim, Y. Kano, A. Nützel, P. Parkhaev, M. Schrödl, and E. Strong. 2017. Revised classification, nomenclator and typification of gastropod and monoplacophoran families. Malacologia 61: 1-526.

Brown, D. 1994. Freshwater Snails of Africa and Their Medical Importance. 2nd ed. Taylor & Francis, London.

Criscione, F., and W.F. Ponder. 2013. A phylogenetic analysis of rissooidean and cingulopsidean families (Gastropoda: Caenogastropoda). Molecular Phylogenetics and Evolution 66: 1075-1082.

Davis, G.M. 1979. The origin and evolution of the gastropod family Pomatiopsidae, with emphasis on the Mekong River Triculinae. Academy of Natural Sciences of Philadelphia Monograph 20: 1-120.

Davis, G.M., M. McKee, and G. Lopez. 1989. The identity of *Hydrobia truncata* (Gastropoda, Hydrobiinae)—Comparative anatomy, molecular-genetics, ecology. Proceedings of the Academy of Natural Sciences of Philadelphia 141: 333-359.

Delicado, D., A. Machordom, and M.A. Ramos. 2015. Effects of habitat transition on the evolutionary patterns of the microgastropod genus *Pseudamnicola* (Mollusca, Hydrobiidae). Zoologica Scripta 44: 403-417.

Delicado, D., V. Pešić, and P. Glöer. 2016. Unraveling a new lineage of Hydrobiidae genera (Caenogastropoda: Truncatelloidea) from the Ponto-Caspian region. European Journal of Taxonomy 208: 1-29.

Drake, P., and A.M. Arias. 1995. Distribution and production of three *Hydrobia* species (Gastropoda: Hydrobiidae) in a shallow coastal lagoon in the Bay of Cadiz, Spain. Journal of Molluscan Studies 61: 185-196.

Föller, K., B. Stelbrink, T. Hauffe, C. Albrecht, and T. Wilke. 2015. Constant diversification rates of endemic gastropods in ancient Lake Ohrid: Ecosystem resilience likely buffers environmental fluctuations. Biogeosciences 12: 7209-7222.

Georgiev, D., A. Osikowski, S. Hofman, A. Rysiewska, and A. Falniowski. 2017. Contribution to the morphology of the Bulgarian stygobiont Truncatelloidea (Caenogastropoda). Folia Malacologia 25: 15-25.

Ghamizi, M., J.-C. Vala, and H. Bouka. 1997. Le genre *Pseudamnicola* au Maroc avec description de *Pseudamnicola pallaryi* n. sp. (Gastropoda: Hydrobiidae). Haliotis 26: 33-49.

Giusti, F., and E. Pezzoli. 1981. Notulae Malacologicae, XXV: Hydrobioidea nuove o poco conosciute dell'Italia appenninica (Gastropoda: Prosobranchia). Archiv für Molluskenkunde 111: 207-222.

Glöer, P., and D. Georgiev. 2012. Three new gastropod species from Greece and Turkey (Mollusca: Gastropoda: Rissooidea) with notes on the anatomy of *Bythinella charpentieri cabirius* Reischütz 1988. North-Western Journal of Zoology 8: 278-282.

Glöer, P., and V. Pešić. 2012. The freshwater snails (Gastropoda) of Iran, with descriptions of two new genera and eight new species. ZooKeys 219: 11-61.

Glöer, P., S. Bouzid, and H.D. Boeters. 2010. Revision of the genera *Pseudamnicola* Paulucci 1878 and *Mercuria* Boeters 1971 from Algeria with particular emphasis on museum collections (Gastropoda: Prosobranchia: Hydrobiidae). Archiv für Molluskenkunde 139: 1-22.

Glöer, P., H.D. Boeters, and V. Pešić. 2014. Freshwater molluscs of Kyrgyzstan with description of one new genus and species (Mollusca: Gastropoda). Folia Malacologica 24: 73-81.

Glöer, P., H.D. Boeters, and F. Walther. 2015. Species of the genus *Mercuria* Boeters, 1971 (Caenogastropoda: Truncatelloidea: Hydrobiidae) from the European Mediterranean region, Morocco and Madeira, with descriptions of new species. Folia Malacologica 23: 279-291.

Glöer, P., U. Bößneck, F. Walther, and M.T. Neiber. 2016. New taxa of freshwater molluscs from Armenia (Caenogastropoda: Truncatelloidea: Hydrobiidae). Folia Malacologica 24: 3-8.

Haase, M., M.D. Naser, and T. Wilke. 2010. *Ecrobia grimmi* in brackish Lake Sawa, Iraq: Indirect evidence for long-distance dispersal of hydrobiid gastropods (Caenogastropoda: Rissooidea) by birds. Journal of Molluscan Studies 76: 101-105.

Hershler, R. 1994. A Review of the North American freshwater snail genus *Pyrgulopsis* (Hydrobiidae). Smithsonian Contributions to Zoology 554: 1-115.

Hershler, R. 1996. Review of the North American aquatic snail genus *Probythinella* (Rissooidea: Hydrobiidae). Invertebrate Biology 115: 120-144.

Hershler, R. 1998. A systematic review of the hydrobiid snails (Gastropoda: Rissooidea) of the Great Basin, western United States. Part I. Genus *Pyrgulopsis*. Veliger 41: 1-132.

Hershler, R., and H.-P. Liu. 2009. New species and records of *Pyrgulopsis* (Gastropoda: Hydrobiidae) from the Snake River basin, southeastern Oregon: Further delineation of a highly imperiled fauna. Zootaxa 2006: 1-22.

Hershler, R., and W.F. Ponder. 1998. A review of morphological characters of hydrobioid snails. Smithsonian Contributions to Zoology 600: 1-55.

Hershler, R., H.-P. Liu, and F.G. Thompson. 2003. Phylogenetic relationships of North American nymphophiline gastropods based on mitochondrial DNA sequences. Zoologica Scripta 32: 357-366.

Hershler, R., F.G. Thompson, and H.-P. Liu. 2011. A large range extension and molecular phylogenetic analysis of the monotypic North American aquatic gastropod genus *Cincinnatia* (Hydrobiidae). Journal of Molluscan Studies 77: 232-240.

Hershler, R., V. Ratcliffe, H.-P. Liu, B. Lang, and C. Hay. 2014. Taxonomic revision of the *Pyrgulopsis gilae* (Caenogastropoda, Hydrobiidae) species complex, with descriptions of two new species from the Gila River basin, New Mexico. ZooKeys 429: 69-85.

Hershler, R., H.-P. Liu, C. Babbitt, M.G. Kellogg, and J.K. Howard. 2016a. Three new species of western California springsnails previously confused with *Pyrgulopsis stearnsiana* (Caenogastropoda, Hydrobiidae). ZooKeys 601: 1-19.

Hershler, R., H.-P. Liu, and L.E. Stevens. 2016b. A new springsnail (Hydrobiidae: *Pyrgulopsis*) from the lower Colorado River Basin, northwestern Arizona. Western North American Naturalist 76: 72-81.

Hershler, R., H.-P. Liu, C. Forsythe, P. Hovingh, and K. Wheeler. 2017. Partial revision of the *Pyrgulopsis kolobensis* complex (Caenogastropoda: Hydrobiidae), with resurrection of *P. pinetorum* and description of three new species from the Virgin River drainage, Utah. Journal of Molluscan Studies 83: 161-171.

IUCN Red List of Threatened Species. 2017. Version 2017-3. www.iucnredlist.org. Accessed 9 June 2018.

Kabat, A.R., and R. Hershler. 1993. The prosobranch snail family Hydrobiidae (Gastropoda: Rissooidea): Review of classification and supraspecific taxa. Smithsonian Contributions to Zoology 547: 1-94.

Kerney, M. 1999. Atlas of the Land and Freshwater Molluscs of Britain and Ireland. Harley Books, Colchester, Essex.

Khalloufi, N., M. Bejaóui, and D. Delicado. 2017. A new genus and species of uncertain phylogenetic position within the family Hydrobiidae (Caenogastropoda, Truncatelloidea) discovered in Tunisian springs. European Journal of Taxonomy 328: 1-15.

Kroll, O., R. Hershler, C. Albrecht, E.M. Terrazas, T.R. Apaza, C. Fuentealba, C. Wolff, and T. Wilke. 2012. The endemic gastropod fauna of Lake Titicaca: Correlation between molecular evolution and hydrographic history. Ecology and Evolution 2: 1517-1530.

Layton, K.K., A.L. Martel, and P.D. Hebert. 2014. Patterns of DNA barcode variation in Canadian marine molluscs. PLoS ONE 9:e95003.

Liu, H.-P., and R. Hershler. 2005. Molecular systematics and radiation of western North American nymphophiline gastropods. Molecular Phylogenetics and Evolution 34: 284-298.

Logvinenko, N.V., and Y.I. Starobogatov. 1968. Phylum: Mollusca. Pp. 308-413 in Atlas of Invertebrates of the Caspian Sea [in Russian] (Y.A. Birshtein, ed.). Pizhevaya promyshlennostje, Moscow.

Miller, J.P., M.A. Ramos, T. Hauffe, and D. Delicado. 2018. Global species richness of hydrobiid snails determined by climate and evolutionary history. Freshwater Biology 63: 1225-1239.

Osikowski, A., S. Hofman, D. Georgiev, A. Rysiewska, and A. Falniowski. 2017. Unique, ancient stygobiont clade of Hydrobiidae (Truncatelloidea) in Bulgaria: The origin of cave fauna. Folia Biologica 65: 79-93.

Radoman, P. 1973. New classification of fresh and brackish water Prosobranchia from the Balkans and Asia Minor. PosebnaIzdanja, Prirodnjacki Musej u Beogradu 32: 1-30.

Radoman, P. 1983. Hydrobioidea, a Superfamily of Prosobranchia (Gastropoda). I. Systematics. Serbian Academy of Sciences and Arts, Belgrade.

Şahin, S.K., S.B. Koca, and M.Z. Yıldırım. 2012. New genera

Anatolidamnicola and *Sivasi* (Gastropoda: Hydrobiidae) from Sivas and Malatya (Turkey). Acta Zoologica Bulgarica 64: 341-346.

Schütt, H. 1991. A contribution to the knowledge of some inland water hydrobiid snails in Israel (Gastropoda, Prosobranchia). Basteria 55: 129-137.

Strong, E.E., O. Gargominy, W.F. Ponder, and P. Bouchet. 2008. Global diversity of gastropods (Gastropoda: Mollusca) in freshwater. Hydrobiologia 595: 149-166.

Szarowska, M., A. Osikowski, S. Hofman, and A. Falniowski. 2016. *Pseudamnicola* Paulucci, 1878 (Caenogastropoda: Truncatelloidea) from the Aegean Islands: A long or short story? Organisms, Diversity & Evolution 16: 121-139.

Thompson, F.G. 1968. The Aquatic Snails of the Family Hydrobiidae of Peninsular Florida. University of Florida Press, Gainesville.

Thompson, F.G. 1979. The systematic relationships of the hydrobioid snail genus *Nymphophilus* Taylor 1966 and the status of the subfamily Nymphophilinae. Malacological Review 12: 41-49.

Thompson, F.G., and R. Hershler. 2002. Two genera of North American freshwater snails: *Marstonia* Baker, 1926, resurrected to generic status, and *Floridobia,* new genus (Prosobranchia: Hydrobiidae: Nymphophilinae). Veliger 45: 269-271.

Van Damme, D. 1984. The freshwater Mollusca of northern Africa. Distribution, biogeography and palaeoecology. Pp. 1-164 in Developments in Hydrobiology, vol. 25 (H.J. Dumont, ed.). W. Junk, Dordrecht.

Vinarski, M.V., and Y.I. Kantor. 2016. Analytical Catalogue of Fresh and Brackish Water Molluscs of Russia and Adjacent Countries. KMK Scientific Press, Moscow.

Vinarski, M.V., D.M. Palatov, and P. Glöer. 2014. Revision of 'Horatia' snails (Mollusca: Gastropoda: Hydrobiidae sensu lato) from South Caucasus with description of two new genera. Journal of Natural History 48: 2237-2253.

Wilke, T. 2003. *Salenthydrobia* gen. nov. (Rissooidea: Hydrobiidae): A potential relict of the Messinian salinity crisis. Zoological Journal of the Linnean Society 137: 319-336.

Wilke, T., and G.M. Davis. 2000. Infraspecific mitochondrial sequence diversity in *Hydrobia ulvae* and *Hydrobia ventrosa* (Hydrobiidae: Rissooidea: Gastropoda): Do their different life histories affect biogeographic patterns and gene flow? Biological Journal of the Linnean Society 70: 89-105.

Wilke, T., E. Rolán, and G.M. Davis. 2000. The mudsnail genus *Hydrobia* s.s. in the northern Atlantic and western Mediterranean: A phylogenetic hypothesis. Marine Biology 137: 827-833.

Wilke, T., G.M. Davis, A. Falniowski, F. Giusti, M. Bodon, and M. Szarowska. 2001. Molecular systematics of Hydrobiidae (Mollusca: Gastropoda: Rissooidea): Testing monophyly and phylogenetic relationships. Proceedings of the Academy of Natural Sciences of Philadelphia 151: 1-21.

Wilke, T., M. Pfenninger, and G.M. Davis. 2002. Anatomical variation in cryptic mudsnail species: Statistical discrimination and evolutionary significance. Proceedings of the Academy of Natural Sciences of Philadelphia 152: 45-66.

Wilke, T., C. Albrecht, V.V. Anistratenko, S.K. Şahin, and M.Z. Yıldırım. 2007. Testing biogeographical hypotheses in space and time: Faunal relationships of the putative ancient Lake Egirdir in Asia Minor. Journal of Biogeography 34: 1807-1821.

Wilke, T., M. Haase, R. Hershler, H.-P. Liu, B. Misof, and W.F. Ponder. 2013. Pushing short DNA fragments to the limit: Phylogenetic relationships of 'hydrobioid' gastropods (Caenogastropoda: Rissooidea). Molecular Phylogenetics and Evolution 66: 715-736.

Yacoubi-Khebiza, M., M. Messouli, N. Coineau, and A. Fakher el Abiari. 2001. Contribution to the knowledge of the groundwater communities from northern Morocco. 13th International Congress of Speleology, Brasilia, 407-410.

Yıldırım, M.Z., S.B. Koca, and Ü. Kebapçi. 2006. Supplement to the Prosobranchia (Mollusca: Gastropoda) fauna of fresh and brackish waters of Turkey. Turkish Journal of Zoology 30: 197-204.

19 Lithoglyphidae Tryon, 1866

STEPHANIE A. CLARK

The Lithoglyphidae is currently divided into two subfamilies, Lithoglyphinae and Benedictiinae (Bouchet and Rocroi, 2005). In the past the lithoglyphids have been treated either as a distinct family or as a subfamily of the Hydrobiidae (Burch and Tottenham, 1980; Thompson, 1984; Kabat and Hershler, 1993; Bouchet and Rocroi, 2005). Recent comparative anatomical and molecular studies have shown that the family Hydrobiidae is polyphyletic and that many more families need to be recognized, such as the Lithoglyphidae (Wilke et al., 2001; Wilke et al., 2013, Wilke and Delicado, chap. 18, this volume). However, the sister relationships among the various families previously considered to belong the Hydrobiidae remain uncertain (Wilke et al., 2013). In addition Wilke et al. (2013) found the two currently recognized subfamilies, Lithoglyphinae and Benedictiinae, to be paraphyletic when the North American genus *Fluminicola* was included in the analyses. Thus the subfamilial relationships need further clarification.

Lithoglyphids have small to medium-sized, conical to depressed conical, globose or almost planispiral dextral shells that vary in size from 2 to 15 mm in height. Their shells are smooth, relatively featureless, and often thickened; the operculum is paucispiral and pale yellow. Lithoglyphids have separate sexes and show sexual dimorphism, with the females tending to be larger and broader than the males. Their tentacles are long and tapering, and the animal is generally darkly pigmented. They have the ventral wall of the female capsule gland closed, and the males have a bladelike penis lacking large appendages and specialized glands. They lay relatively large, round corneous egg capsules either singly or in clusters; each capsule contains a single egg, and the capsules are typically attached to various substrates, including leaves, gravel, woody debris, rocks, and shells both living and empty (Thompson, 1984; Hershler and Frest, 1996; Wilke et al., 2001, 2013).

The family is distributed across the Nearctic and Palearctic (Strong et al., 2008). However, most of the diversity is found in the United States, especially in the western and southeastern states, with a few species in British Columbia, Canada, and San Luís

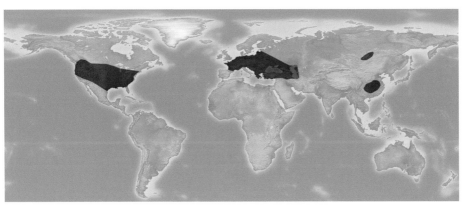

Distribution of Lithoglyphidae.

Lithoglyphus naticoides (Pfeiffer, 1828). Belgium (Nete River Basin): Lier, Antwerp Province. 5.1 mm. Poppe 410673.

Potosí, Mexico (Hershler et al., 2007; Thompson, 2011; Johnson et al., 2013).

There are about 200 described species, primarily from the Nearctic, with a smaller number from the Palearctic (Strong et al., 2008; Kantor et al., 2010; Welter-Schultes, 2012). The family is found in a wide range of freshwater habitats: isolated springs, seepages, caves, streams, rivers, and lakes. Some species have wide ranges, such as *Lithoglyphus naticoides* (Welter-Schultes, 2012), while others are known only from a few locations or a single location, such as *Antrorbis breweri* (Hershler & F. G. Thompson, 1990) from a single cave in Alabama (Hershler and Thompson, 1990).

In North America the family is represented by at least 70 species in seven genera: *Antrorbis* Hershler & F. G. Thompson, 1990, *Clappia* Walker, 1909, *Fluminicola* Stimpson, 1865, *Gillia* Stimpson, 1865, *Lepyrium* Dall, 1896, *Pterides* Pilsbry, 1909, and *Somatogyrus* Gill, 1863 (Thompson, 1969, 1984; Gordon, 1986; Hershler and Thompson, 1990; Hershler and Frest, 1996; Hershler et al., 2007; Thompson, 2011; Hershler and Liu, 2012; Liu et al., 2013). The genus *Somatogyrus* as currently recognized contains 37 species and is widespread throughout the eastern portion of the United States; it is found in springs, streams, and rivers, and sometimes two or three species occur sympatrically

(Burch and Tottenham, 1980). The genus is morphologically diverse but is poorly known anatomically and genetically, and the ranges of most species are not fully resolved. The bulk of the species were named in the period 1904-1909 from the Coosa River drainage in Georgia and Alabama and the Alabama portion of the Tennessee River, before both of these rivers were extensively dammed over the past century. They were all described from the shell alone. Many of the historical locations are submerged by the reservoirs formed by the dams, and it is possible some species might be extinct. However, the genus is still found extensively in the flowing portions of the main stem of the rivers and their tributaries. The genus needs to be fully revised taxonomically, especially using anatomical and genetic data. For example, preliminary anatomical and genetic data show that *Somatogyrus strengi* Pilsbry & Walker, 1906, is a member of the Amnicolidae and that populations from the Cahaba and Coosa rivers are more genetically related within drainages than between drainages (Clark, pers. obs.).

The genus *Fluminicola* as currently recognized contains 25 species and is widespread in the western United States and southern British Columbia, west of the Rocky Mountains; it is found in small to large springs, streams, and rivers, often two or more species occur sympatrically, and most are known from a single location or a few geographically proximate springs (Hershler and Frest, 1996; Frest and Johannes, 1998, 2005; Hershler et al., 2007). The genus is morphologically diverse but anatomically conservative, but as taxa are analyzed genetically, more diversity has been revealed (Hershler et al., 2007; Hershler and Liu, 2012).

The genus *Pterides* contains three species and is known from a small area of San Luís Potosí, Mexico. They were described from shells only, are morphologically atypical for the North American lithoglyphids, and are the only members of the family known from Mexico (Thompson, 2011).

The genus *Potamolithus* is known from Brazil, Chile, Uruguay, Paraguay, and Argentina and was previously classified in the Lithoglyphidae (Davis and da Silva, 1984). However, recent molecular studies have shown that *Potamolithus* is a member of the Tateidae (Wilke et al., 2013; Koch et al., 2015; Ponder, chap. 23, this volume).

In Europe the family is represented by approximately 10 species in three genera: *Dabriana* Radoman, 1974, *Lithoglyphus* Pfeiffer, 1828, and *Tanousia* Servain, 1881 (Kantor et al., 2010; Welter-Schultes, 2012).

In the Far East, Lake Baikal, and China, the family is represented by at least 24 species in five genera: *Benedictia* Lindholm, 1927, *Kobeltocochlea* Lindholm, 1909, *Lithoglyphopsis* Thiele, 1928, *Pseudobenedictia* Sitnikova, 1987, and *Yaroslawiella* Sitnikova, 2001 (Yen, 1939; Kantor et al., 2010).

At least one species (*Lithoglyphus naticoides*) has been introduced to areas outside of its native range, the Ponto-Caspian region, and is spreading across Europe (Mastitsky and Samoilenko, 2006; Mouthon, 2007; Tyutin and Slynko, 2007).

Among the factors that affect the diversity and conservation of lithoglyphids are habitat destruction from dams, channel modification, siltation, water-quality degradation from point and nonpoint pollution, and the introduction of nonindigenous species. Lysne et al. (2008) explore challenges and opportunities of freshwater conservation and highlight the need for involvement of many affected interest groups, from local communities to government agencies.

The majority of lithoglyphids are narrow range endemics, and this combined with the vulnerability of habitats such as springs, caves, streams, and lakes places many species at risk of extinction. The IUCN Red List (2016) lists 17 species of Lithoglyphidae: *Benedictia* (3 species), *Fluminicola* (6), *Kobeltocochlea* (4), *Lithoglyphopsis* (3) *Pseudobenedictia* (1), of which 2 are listed as Vulnerable, 13 as Data Deficient, and 2 as Least Concern. However, several other taxa now included in the Lithoglyphidae are currently listed as belonging to the Hydrobiidae. These include the following 36 taxa: *Antrobia* (1 species), *Clappia* (2), *Lepyrium* (1), *Lithoglyphus* (4) and *Somatogyrus* (28) of which 10 are listed as Critically Endangered, 2 as Vulnerable, 1 as Near Threatened, 14 as Data Deficient, 3 as Least Concern, and 6 as Extinct.

LITERATURE CITED

Bouchet, P., and J.-P. Rocroi. 2005. Classification and nomenclator of gastropod families. Malacologia 47: 1-397.

Burch, J.B., and J.L. Tottenham. 1980. North American freshwater snails: Species list, ranges and illustrations. Walkerana 1: 81-215.

Davis, G.M., and M.C.P. da Silva. 1984. *Potamolithus*: Morphology, convergence, and relationships among hydrobioid snails. Malacologia 25: 73-108.

Frest, T.J., and E.J. Johannes. 1998. Freshwater Mollusks of the Upper Klamath Drainage, Oregon. Yearly Report 1998. Unpublished report prepared for Oregon Natural Heritage Program, Portland, Oregon, and U. S. Department of the Interior, Bureau of Reclamation, Klamath Project, Klamath Falls, Oregon. Deixis Consultants, Seattle, Washington. pp. i-vii, 1-91, appendices.

Frest, T.J., and E.J. Johannes. 2005. Grazing effects on springsnails, Cascade-Siskiyou National Monument, Oregon 2004. Unpublished report prepared for the World Wildlife Fund Cascade-Siskiyou Ecoregion Office, Ashland, Oregon. Deixis Consultants, Seattle, Washington. pp. i-iv, 1-182.

Gordon, M.E. 1986. A new *Somatogyrus* from the southwestern Ozarks with a brief review of the Hydrobiidae from the interior highlands (Gastropoda: Prosobranchia). Nautilus 100: 71-77.

Hershler, R., and T.J. Frest. 1996. A review of the North American freshwater snail genus *Fluminicola* (Hydrobiidae). Smithsonian Contributions to Zoology 583: 1-41.

Hershler, R., and H.-P. Liu. 2012. Molecular phylogeny of the western North American pebblesnails, genus *Fluminicola* (Rissooidea: Lithoglyphidae), with description of a new species. Journal of Molluscan Studies 78: 321-329.

Hershler, R., and F.G. Thompson. 1990. *Antrorbis breweri* a new genus and species of hydrobiid cavesnail (Gastropoda) from Coosa River Basin, northeastern Alabama. Proceedings of the Biological Society of Washington 103: 197-204.

Hershler, R., H.-P. Liu, T.J. Frest, and E.J. Johannes. 2007. Extensive diversification of pebblesnails (Lithoglyphidae: *Fluminicola*) in the upper Sacramento River basin, northwestern USA. Zoological Journal of the Linnean Society 149: 371-422.

IUCN Red List of Threatened Species. 2016. Version 2016-2. www.iucnredlist.org. Accessed 20 September 2016.

Johnson, P.D., A.E. Bogan, K.M. Brown, N.M. Burkhead, J.R. Cordeiro, J.T. Garner, P.D. Hartfield, et al. 2013. Conservation status of freshwater gastropods of Canada and the United States. Fisheries 38: 247-282.

Kabat, A.R., and R. Hershler. 1993. The prosobranch snail family Hydrobiidae (Gastropoda: Rissooidea): Review of classification and supraspecific taxa. Smithsonian Contributions to Zoology 547: 1-94.

Kantor, Y.I., M.V. Vinarski, A.A. Schileyko, and A.V. Sysoev. 2010. Catalogue of the continental mollusks of Russia and adjacent territories. Version 2.3.1. published online 2 March 2010 at www.ruthenica.com/categorie-8.html.

Koch, E., S.M. Martin, and N.F. Ciocco. 2015. A molecular contribution to the controversial taxonomical status of some freshwater snails (Cacnogastropoda: Rissooidea, Cochliopidae) from the Central Andes desert to Patagonia. Iheringia 105: 69-75.

Liu, H.-P., J. Walsh, and R. Hershler. 2013. Taxonomic clarification and phylogeography of *Fluminicola coloradensis* Morrison, a widely ranging western North American pebblesnail. Monographs of the Western North American Naturalist 6: 87-100.

Lysne, S.J., K.E. Perez, K.M. Brown, R.L. Minton, and J.D. Sides. 2008. A review of freshwater gastropod conservation: Challenges and opportunities. Journal of the North American Benthological Society 27: 463-470.

Mastitsky, S.E., and V.M. Samoilenko. 2006. The gravel snail, *Lithoglyphus naticoides* (Gastropoda: Hydrobiidae), a new Ponto-Caspian species in Lake Lukomskoe (Belarus). Aquatic Invasions 1: 161-170.

Mouthon, J. 2007. *Lithoglyphus naticoides* (Pfeiffer) (Gastropoda: Prosobranchia): Distribution in France, population dynamics and life cycle in the Saône river at Lyon (France). Annales de Limnologie-International Journal of Limnology 43: 53-59.

Strong, E.E., O. Gargominy, W.F. Ponder, and P. Bouchet. 2008. Global diversity of gastropods (Gastropoda: Mollusca) in freshwater. Hydrobiologia 595: 149-166.

Thompson, F.G. 1969. Some hydrobiid snails from Georgia and Florida. Quarterly Journal of the Florida Academy of Sciences 32: 241-265.

Thompson, F.G. 1984. North American freshwater snail genera of the hydrobiid subfamily Lithoglyphinae. Malacologia 25: 109-141.

Thompson, F.G. 2011. An annotated checklist and bibliography of the land and freshwater snails of Mexico and Central America. Bulletin of the Florida Museum of Natural History 50: 1-299.

Tyutin, A.V., and Yu. V. Slynko. 2007. The first finding of the Black Sea Snail *Lithoglyphus naticoides* (Gastropoda) and its associated species-specific Trematoda in the Upper Volga Basin. Russian Journal of Biological Invasions 1: 45-49.

Welter-Schultes, F.W. 2012. European non-marine molluscs, a guide for species identification. Planet Poster Editions, Göttingen, Germany.

Wilke, T., G.M. Davis, A. Falniowski, F. Giusti, M. Bodon, and M. Szarowska. 2001. Molecular systematics of Hydrobiidae (Mollusca: Gastropoda: Rissooidea): Testing monophyly and phylogenetic relationships. Proceedings of the Academy of Natural Sciences of Philadelphia 151: 1-21.

Wilke, T., M. Haase, R. Hershler, H.-P. Liu, B. Misof, and W.F. Ponder. 2013. Pushing short DNA fragments to the limit: Phylogenetic relationships of 'hydrobioid' gastropods (Caenogastropoda: Rissooidea). Molecular Phylogenetics Evolution 66: 715-736.

Yen, T.C. 1939. Die chinesischen land-und süsswasser: Gastropoden des Naturmuseums Senckenberg. Abhandlungen der Senckenbergischen Naturforschenden Gesellschaft 444: 1-234.

20 Moitessieriidae Bourguignat, 1863

THOMAS WILKE

The Moitessieriidae sensu stricto comprise 12 genera from Europe: *Baldufa* Alba et al., 2010, *Bosnidilhia* Boeters, Glöer & Pešić, 2013, *Clameia* Boeters & Gittenberger, 1990, *Corseria* Boeters & Falkner, 2009, *Henrigirardia* Boeters & Falkner, 2003, *Moitessieria* Bourguignat, 1863, *Palacanthilhiopsis* Bernasconi, 1988, *Paladilhia* Bourguignat, 1865, *Palaospeum* Boeters, 1999, *Sardopaladilhia* Manganelli et al., 1998, *Sorholia* Boeters & Falkner, 2009, and *Spiralix* Boeters, 1972 (Boeters et al., 2013). Due to a subterranean life style with occasional occurrences in springs and surface runoffs, moitessierids typically inhabit mountainous areas and often have a patchy distribution. Most of the approximately 55 assumed species (Strong et al., 2008) can be found in circum-Alpine areas: northern Italy, southern France, possibly southern Germany, Austria, northern Slovenia, and western Hungary. Several species have also been described from the French and Spanish Pyrenees. The Moitessieriidae sensu stricto reach their northern distribution limits in central France. The southern limits are defined by occurrences of *Sardopaladilhia* and *Spiralix* in south-eastern Spain, *Corseria* on the Mediterranean Island of Corse, and *Sardopaladilhia* on the neighboring island of Sardinia (Boeters and Falkner, 2009; Bodon and Giusti, 1991), as well as *Clameia* on the Greek island Euboea (Boeters and Gittenberger, 1990). Few species have been reported from the Balkans, including *Paladilhia elongata* from Croatia (Oberösterreichisches Landesmuseum, Linz, coll. #6605494), *Bosnidilhia vreloana* from Bosnia and Herzegovina (Boeters et al., 2013), and five species of "*Paladilhiopsis*" from northern Albania (Grego et al., 2017). Moreover, Ghamizi et al. (1999) described the new genus *Atebbania* Ghamizi et al., 1999, from groundwaters in southern Morocco and noted a close relationship to the genus *Moitessieria*. Though *Atebbania* awaits further studies, it may thus be possible that the Moitessieriidae also occurs in northwestern Africa.

Many moitessierids have a restricted range, and some are known only from few or even single locations. A notable exception is *Moitessieria rolandiana,* which is widely distributed in southwestern France. However, it is possible that the restricted occurrence

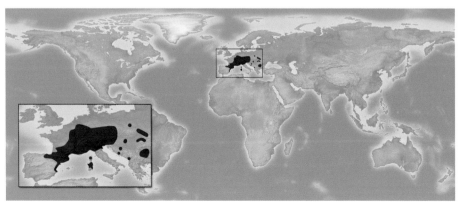

Distribution of Moitessieriidae.

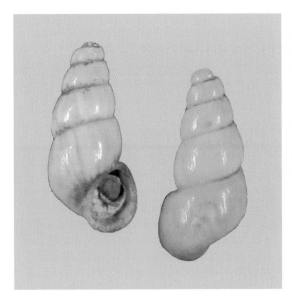

Bythiospeum suevicum (Geyer, 1905). Germany: Neckar River, near Horb, Baden-Württemberg. 3.2 mm. Poppe 724630.

of some species, often associated with low abundances, is due to deficient data, as sampling these tiny subterranean snails remains challenging. Moreover, the taxonomy of the surprisingly species-rich Moitessieriidae is largely based on subtle variations in shell shape and sculpture. It often remains unclear whether these differences are genetically fixed, reflecting species-level differentiations, or whether they simply represent eco-phenotypical variations driven by, for example, differences in water chemistry. The assumed restricted occurrence of some species may therefore be the result of a taxonomic bias.

Taxonomic challenges appear not only at the species but also at the genus level. The differentiation of the 12 European genera is primarily based on protoconch structure, the presence of a pallial tentacle, the presence and position of a distal seminal receptacle, and the form of the intestinal loop (Boeters et al., 2013). However, some of the respective character states are unknown for several nominal genera. Moreover, complementary genetic information for the Moitessieriidae s.s. remains sparse. In fact, as of July 2017, DNA sequences for only two species are available in the public nucleotide database GenBank. Therefore, both species- and genus-level relationships within the Moitessieriidae require further clarifications.

Interestingly, preliminary phylogenetic analyses indicate a potential sister-group relationship between the Moitessieriidae s.s. and the genus *Bythiospeum* Bourguignat, 1882 (Wilke et al., 2001, 2013). The latter was long considered to belong to the family Hydrobiidae (but see Falniowski and Šteffek, 1989), an assumption that is clearly rejected by multilocus genetic data. In fact, in some of these phylogenetic analyses, *Bythiospeum* clusters together with *Moitessieria* and *Sardopaladilhia;* in others the relationship with the Moitessieriidae s.s. remains unresolved (Wilke et al., 2013). Pending further phylogenetic investigations, the genus *Bythiospeum* is here considered to be a member of the Moitessieriidae sensu lato.

Bythiospeum spp. share not only genetic synapomorphies but also many ecological and biogeographical character states with other moitessierids. They mainly occur in subterranean waters, springs, and wells, and have a patchy distribution. Most of the approximately 62 *Bythiospeum* species listed in the IUCN Red List (IUCN 2017) can be found in circum-Alpine areas: southwestern and western France, Switzerland, southern Germany, Austria, Slovenia, western Hungary, and northern Italy. Compared to the Moitessieriidae s.s., the distribution area of *Bythiospeum* is shifted to the north and west. Accordingly, only very few species occur southwest (e.g., *B. articense*), south (*B. vallei*), or southeast (e.g., *B. grobbeni* and *B. hungaricum*) of the Alps. The genus reaches its northwestern distribution limit in northwestern Germany (*B. husmanni*). The northeastern limit is defined by occurrences of *B. neglectissimum* in southern Poland and western Ukraine, and the eastern limit by occurrences of *B. carpathica* and B. *leruthi* in Romania. Interestingly, more intensified research on cave snails in northern, western, and central Bulgaria has led to the recent description of several new species of *Bythiospeum* and related genera (e.g., Georgiev and Glöer, 2011, 2015a, b; Georgiev, 2012a), further extending the southeastern distribution limits of the group.

Similar to the Moitessieriidae s.s., the assumed high biodiversity of *Bythiospeum* and the strongly restricted range of many species are remarkable. However, due to a lack of detailed morphological and anatomical data for most species, it remains questionable whether all taxa described are valid (e.g., Glöer, 2002; Glöer et al., 2015). In fact, preliminary genetic analyses indicate that some nominal species differ by only one or two substitutions in the barcoding gene COI, potentially suggesting conspecificity (Richling et al., 2017). Moreover, Hirsch et al. (2012) found partly overlapping geographical clusters of genetically similar taxa, supporting the assumption that the biodiversity of *Bythiospeum* may be lower than commonly assumed.

The same might apply to the genus-level diversity. Whereas some older publications treat *Paladilhiopsis* Pavlović, 1913, as a distinct genus, more recent studies suggest that it might be a junior synonym of *Bythiospeum* (Haase, 1995; Boeters et al., 2013; Glöer et al., 2015). However, as stated above, genetic information that could help better resolve these genus-level relationships is lacking. This is also the case for the monotypic genus *Falniowskia* Bernasconi, 1991, as its type species, *F. neglectissima,* was originally described as *Bythiospeum neglectissimum;* for the genus *Devetakia* Georgiev & Glöer, 2011, which differs from *Bythiospeum* in some shell characters (Georgiev and Glöer, 2011); and for the genus *Balkanospeum* Georgiev, 2012, which can be distinguished from *Bythiospeum* by a penis outgrowth (Georgiev, 2012b).

Due to the restricted range of many moitessierids and the vulnerability of subterranean waters, springs, and wells to anthropogenic disturbances, many species are imperiled. Of the 118 species (55 species of Moitessieriidae s.s. and 63 species of *Bythiospeum* and related genera) recorded in the IUCN Red List (IUCN, 2017), 12 are listed as Critically Endangered, 12 as Endangered, 29 as Vulnerable, and 11 as Near Threatened. A relatively high number (37 species) are Data Deficient, and some of them are considered to be potentially Extinct. Main threads contributing to the demise of the Moitessieriidae s.l. include habitat modification (70 species) due to, for example, abstraction of groundwater, and pollution (83 species) due to domestic and urban waste waters as well as agricultural and forestry effluents.

LITERATURE CITED

Bodon, M., and F. Giusti. 1991. The genus *Moitessieria* in the island of Sardinia and in Italy: New data on the systematics of *Moitessieria* and *Paladilhia* (Prosobranchia: Hydrobiidae). Malacologia 33: 1-30.

Boeters, H.D., and G. Falkner 2009. Unbekannte westeuropäische Hydrobiidae, 15. Neue und alte Quell- und Grundwasserschnecken aus Frankreich (Gastropoda: Moitessieriidae und Hydrobiidae). Heldia 5: 149-162.

Boeters, H.D., and E. Gittenberger. 1990. Once more on the Moitessieriidae (Gastropoda Prosobranchia), with the description of *Clameia brooki* gen. et spec. nov. Basteria 54: 123-129.

Boeters, H.D., P. Glöer, and V. Pešić. 2013. Some new freshwater gastropods from southern Europe (Mollusca: Gastropoda: Truncatelloidea). Folia Malacologia 21: 225-235.

Falniowski, A., and J. Šteffek. 1989. A new species of *Bythiospeum* (Prosobranchia: Hydrobioidea: Moitessieriidae) from south Poland. Folia Malacologia 3: 95-101.

Georgiev, D. 2012a. Two new species of stygobiotic snails from the genus *Bythiospeum* (Gastropoda: Hydrobiidae) from Bulgaria. Acta Zoologica Bulgarica, Supplement 4: 17-20.

Georgiev, D. 2012b. New taxa of Hydrobiidae (Gastropoda: Risooidea) from Bulgarian cave and spring waters. Acta Zoologica Bulgarica 64: 113-121.

Georgiev, D., and P. Glöer. 2011. Two new species of a new genus *Devetakia* gen. nov. (Gastropoda: Hydrobiidae) from the caves of Devetashko Plateau, North Bulgaria. Acta Zoologica Bulgarica 63: 11-15.

Georgiev, D., and P. Glöer. 2015a. A new stygobiotic snail species from North Bulgaria. Historia Naturalis Bulgarica 21: 9-11.

Georgiev, D., and P. Glöer. 2015b. Two new stygobiont snail species (Gastropoda, Hydrobiidae) from a spring in West Bulgaria. Ecologica Montenegrina 2: 93-97.

Ghamizi, M., M. Bodon, M. Boulal, and F. Giusti. 1999. *Atebbania bernasconii,* a new genus and species from subterranean waters of the Tiznit Plan, southern Morocco (Gastropoda: Hydrobiidae). Journal of Molluscan Studies 65: 89-98.

Glöer, P. 2002. Die Süßwassergastropoden Nord- und Mitteleuropas. Bestimmungsschlüssel, Lebensweise, Verbreitung. Die Tierwelt Deutschlands, 73. Conchbooks, Hackenheim, Germany.

Glöer, P., J. Grego, Z.P. Erőss, and Z. Fehér. 2015. New records of subterranean and spring molluscs (Gastropoda: Hydrobiidae) from Montenegro and Albania with the description of five new species. Ecologica Montenegrina 4: 70-82.

Grego, J., P. Glöer, Z.P. Erőss, and Z. Fehér. 2017. Six new subterranean freshwater gastropod species from northern Albania and some new records from Albania and Kosovo (Mollusca, Gastropoda, Moitessieriidae and Hydrobiidae), Subterranean Biology 23: 85-107.

Haase, M. 1995. The stygobiont genus *Bythiospeum* in Austria: A basic revision and anatomical description of *B.*

cf. *geyeri* from Vienna (Caenogastropoda: Hydrobiidae). American Malacological Bulletin 11: 123-137.

Hirsch, J., F. Brümmer, R.O. Schill, M. Pfannkuchen, H.-J. Niederhöfer, G. Falkner, M. Schopper, and M. Blum. 2012. Erste Erkenntnisse zur Phylogenie der Brunnenschnecken in Südwestdeutschland. Mitteilungen der Deutschen Malakozoologischen Gesellschaft 87: 39.

IUCN Red List of Threatened Species. 2017. Version 2017-3. www.iucnredlist.org. Accessed 9 June 2018.

Richling, I., Y. Malkowsky, J. Kuhn, H.-J. Niederhöfer, and H.D. Boeters. 2017. A vanishing hotspot—the impact of molecular insights on the diversity of central European *Bythiospeum* Bourguignat, 1882 (Mollusca: Gastropoda: Truncatelloidea). Organisms, Diversity & Evolution 17: 67-85.

Strong, E.E., O. Gargominy, W.F. Ponder, and P. Bouchet. 2008. Global diversity of gastropods (Gastropoda: Mollusca) in freshwater. Hydrobiologia 595: 149-166.

Wilke, T., G.M. Davis, A. Falniowski, F. Giusti, M. Bodon, and M. Szarowska. 2001. Molecular systematics of Hydrobiidae (Mollusca: Gastropoda: Rissooidea): Testing monophyly and phylogenetic relationships. Proceedings of the Academy of Natural Sciences of Philadelphia 151: 1-21.

Wilke, T., M. Haase, R. Hershler, H.-P. Liu, B. Misof, and W.F. Ponder. 2013. Pushing short DNA fragments to the limit: Phylogenetic relationships of 'hydrobioid' gastropods (Caenogastropoda: Rissooidea). Molecular Phylogenetics and Evolution 66: 715-736.

21 Pomatiopsidae Stimpson, 1865

THOMAS WILKE

The Pomatiopsidae sensu lato comprises at least 29 genera distributed in Asia (*Blanfordia* A. Adams, 1863, *Cecina* Adams, 1861, *Delavaya* Heude, 1889, *Fenouilia* Heude, 1889, *Fukuia* Abbott & Hunter, 1949, *Gammatricula* Davis & Liu 1990, *Halewisia* Davis, 1979, *Hemibia* Heude, 1890, *Hubendickia* Brandt, 1968, *Hydrorissoia* Bavay, 1895, *Jinhongia* Davis, 1990, *Jullienia* Crosse & Fischer, 1876, *Karelainia* Davis, 1979, *Lacunopsis* Deshayes, 1876, *Manningiella* Brandt, 1970, *Neotricula* Davis, 1986, *Oncomelania* Gredler, 1881, *Pachydrobia* Crosse & Fischer, 1876, *Pachydrobiella* Thiele, 1928, *Pseudoiglica* Grego, 2018, *Robertsiella* Davis & Greer, 1980, *Saduniella* Brandt, 1970, *Thamkhondonia* Grego, 2018, *Tricula* Benson, 1843, *Wuconchona* Kang, 1983); South America (*Idiopyrgus* Pilsbry, 1911); North America (*Pomatiopsis* Tryon, 1862); Africa (*Tomichia* Benson, 1851); and Australia (*Coxiella* E. A. Smith, 1894) (e.g., Davis, 1979; Kabat and Hershler, 1993; Liu et al., 2014; Grego, 2018).

Most pomatiopsid species occur in aquatic systems such as rivers, streams, lakes, wetlands, and springs. However, some or all representatives of the genera *Fukuia, Oncolemania, Pomatiopsis,* and *Tomichia* are amphibious, and the genus *Blanfordia* is even terrestrial (Kameda and Kato, 2011). Few taxa occur in brackish or saline ecosystems, such as *Cecina* spp., *Coxiella* spp., and some *Tomichia* spp.

With approximately 170 species (Strong et al., 2008), the Pomatiopsidae is among the most species-rich freshwater gastropod families. The highest biodiversity can be found in Southeast Asia and the Japanese archipelago (>140 species), followed by sub-Saharan Africa with approximately 10-11 species, southern Australia with ca. 9 species, the northwestern Palearctic with 1-8 species, North America with 5-6 species, and South America with ca. 2 species.

Though the Pomatiopsidae are distributed in aquatic systems across the globe (with the exception of Antarctica and Europe), almost all genus-level taxa are restricted to single continents and many are even endemic. The family is represented in South America by the genus *Idiopyrgus* (= *Aquidaunia*; Davis, 1979, 1993; Simone, 2006). Species of this genus exclusively occur in Brazil, from the state Mato Grosso do Sul

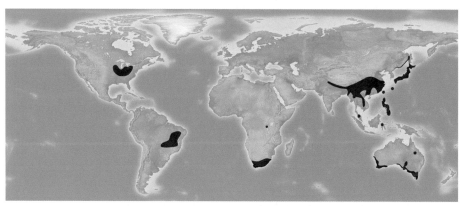

Distribution of Pomatiopsidae.

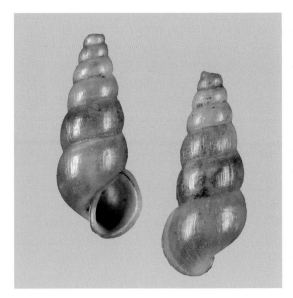

Oncomelania nosophora (Robson, 1915). Japan: Yamanashi Prefecture. 7.2 mm. Poppe 411916.

in the southwest to Minas Gerais in the southeast and Pernambuco in the northeast (e.g., Davis, 1979; Malek, 1983).

North America is inhabited by a singly indigenous genus, *Pomatiopsis*. It has a disjunct distribution; two or three species occur along the coastal mountain ranges from central Oregon to central California; two other species in eastern North America from Iowa and Missouri in the west to Virginia in the east, and from New York state and Ontario in the north to Alabama in the south. Records east and south of the latter range are either historic or await confirmation. The Salish Sea, the network of coastal waterways in the Pacific Northwest (Washington State, British Columbia), is inhabited by a representative of another pomatiopsid genus, the amphibious *Cecina manchurica*. However, this species is not native to North America and was first recorded in the area in 1961 (Morrison, 1963).

Africa, too, is home to a single genus, *Tomichia*, with 10 or 11 species (Davis, 1981; Brown, 1994). The genus was originally described from coastal areas in South Africa, and several species occur in fresh and saline inland waters from the Namibian border in the west to the central KwaZulu-Natal Province in the east. The other congeners live, highly disjunct from the South African taxa, in the African Rift, from the southwestern shores of Lake Tanganyika (West et al.,

2003) to the highlands west of Lake Kivu (reviewed in Brown, 1994).

The brackish and saline water representatives of the sole Australian genus *Coxiella* occur in permanent and temporary salt lakes or pools throughout most of the continent's southern parts. The genus ranges from the Shark Bay area in the northwest to Croajingolong National Park in the east and southern Tasmania in the south. It also has been reported from the inland lake systems Eyre and Callabonna as well as from Lake Buchanan in northeastern Australia (reviewed in Williams and Mellor, 1991).

In contrast to other continents, numerous pomatiopsid genera occur in Asia. The highest biodiversity, with several endemic genera, can be found in the lower Mekong River area of Laos, Cambodia, Thailand, and Vietnam (e.g., *Jullienia, Karelainia, Lacunopsis, Pachydrobia, Pseudoiglica, Thamkhondonia*), the upper Yangtze River area in China (e.g., *Delavaya, Fenouilia*), and the Japanese archipelago (e.g., *Blanfordia, Fukuia*). The western limit of the distribution area of the Asian species is defined by occurrences of *Tricula montana* in the Indian part of the southwestern Himalayas (Davis et al., 1986), the southern limit by occurrences of the intermediate schistosomiasis host *Oncomelania lindoensis* in the Lake Lindo area of Sulawesi (Davis and Carney, 1973), and the northwestern limit by occurrences of *Cecina* spp. on the Kuril Islands and in the northern Strait of Tartary (both Russia; Kussakin, 1975; Gulbin, 2010). There are other highly disjunct occurrences of pomatiopsid snails in Asia, such as *Robertsiella* spp. in central Malaysia (Ambu et al., 1984; Attwood et al., 2003).

The peculiar biogeographical patterns of pomatiopsids across the planet, i.e., the occurrence of endemic genera on the southern continents and a high biodiversity in Southeast Asian river systems, have raised questions about a Gondwanaland origin of the Pomatiopsidae. Davis (1979) proposed that pomatiopsids flourished more than 120 mya on the former supercontinent of Gondwana. They were introduced to Asia with the Indian Plate, and this collision and the subsequent orogeny of the Himalaya Mountains initiated the river systems that transported pomatiopsid snails into Southeast Asia from the west to the east. North America was then colonized from Asia via Kamchatka and Alaska.

This hypothesis has been questioned by Attwood (2009), who proposed an Australian origin of the Pomatiopsidae. Accordingly, "Proto-Pomatiopsidae" evolved in eastern Gondwana (Australia) and entered marine habitats. From there, they colonized Africa and South America during the Cretaceous and subsequently reached North America. During a much later period of time (i.e., 40–20 mya), Australian taxa independently colonized southeast Asia via island chains, subsequently reached Japan, and spread throughout China in an east-to-west direction (see also Liu et al., 2014).

Both hypotheses require that the Pomatiopsidae are monophyletic, that Asian taxa are closely related to African or Australian taxa, and that the North American taxa derived either from Asian or South American species. Interestingly, none of these assumptions is supported by recent multilocus phylogenetic analyses (Kameda and Kato, 2011; Wilke et al., 2013). Notably, Australian *Coxiella* and African *Tomichia* form a clade highly distinct from all Asian and North American pomatiopsids, and North American *Pomatiopsis* is an ancient sister to the Asian taxa. These findings call into question not only the above hypotheses about the early evolution of the Pomatiopsidae but also the traditional taxonomy of the family.

Previously, Davis (1979) suggested two nominal subfamilies: the Pomatiopsinae (= Cecininae, = Coxiellidae, = Hemibiinae, = Oncomelaniidae, = Tomichiinae) and the Triculinae. However, neither of these subfamilies appears to be monophyletic (see Kameda and Kato, 2011; Wilke et al., 2013, Liu et al., 2014). Moreover, as mentioned above, *Coxiella* and *Tomichia* form a monophyletic group distinct from the Pomatiopsidae s.s. According to Wilke et al. (2013), the Tomichiinae (= Coxiellidae) would be an available name for this group. However, as suggested by the latter authors, additional studies are necessary prior to the recognition of the *Coxiella-Tomichia* clade as a distinct family.

Not only higher level classifications within the Pomatiopsidae are poorly understood but also genus- and species-level relationships. This particularly concerns the Southeast Asian taxa. On the one hand, genetic analyses indicate that some genera, such as *Jullienia* (Liu et al., 2014), are highly distinct, raising questions whether the Tribe Jullieniini should be elevated to subfamily level (also see Davis, 1979; Bouchet et al., 2017). On the other hand, several genera appear to be paraphyletic, including *Tricula, Pachydrobia, Neotricula,* and *Hubendickia* (Liu et al., 2014) as well as *Blanfordia* and *Fukuia* (Kameda and Kato, 2011). Therefore, further analyses involving genetic, anatomical, and ecological information would be necessary to better resolve the relationships of Asian pomatiopsids. This is particularly true for the >20 nominal genera of the paraphyletic subfamily Triculinae (including poorly understood genera such as *Kunmingia, Lithoglyphopsis, Neoprososthenia, Parapyrgula, Saduniella,* and *Taihua*) as well as for the dubious genus *Rehderiella* from Thailand, which was assigned by Brandt (1974) to the nominal subfamily Rehderiellinae.

Genus- and species-level clarifications are also required for South African and South American pomatiopsids. Most authors agree that South America is inhabited by a single genus of Pomatiopsidae, *Idiopyrgus* (= *Aquidauania*). Recently, Simone (2012) described the pomatiopsid genus *Spiripockia* from the Lapa dos Peixes Cave in Bahia State, Brazil. The author noted a close relationship to another Brazilian troglobiont taxon, *Potamolithus troglobius*. However, based on anatomical data, *Potamolithus* was first placed into the family Lithoglyphidae by Davis and Pons da Silva (1984) and later assigned to the family Tateidae based on molecular data (Wilke et al., 2013). Pending further investigations, *Spiripockia* is therefore not considered to belong to the Pomatiopsidae.

As for South African *Tomichia,* preliminary molecular analyses indicate that the genus is rendered paraphyletic by Australian *Coxiella* (Wembo et al., 2014). Accordingly, the coastal South African species are highly distinct from species in the South Kivu Province (African Rift), calling for the designation of a new genus. These findings for the African taxa may also have important implications for an understanding of pomatiopsid evolution in general.

Despite numerous taxonomic problems within the Pomatiopsidae, the family has received considerable attention by evolutionary biologists and epidemiologists. Part of this high interest comes from the role of pomatiopsid snails as intermediate hosts for the human blood flukes *Schistosoma* spp. At least

10 pomatiopsid genera transmit schistosomiasis in Southeast Asia: *Neotricula, Pachydrobia, Gammatricula, Manningiella, Hubendickia, Lacunopsis, Lithoglyphopsis, Jinhongia, Robertsiella,* and *Oncomelania* (Davis, 1993; Davis et al., 1999; Attwood et al., 2005). Of particular interest is the latter taxon. It is the most widespread pomatiopsid genus and causes the most cases of human schistosomiasis in Asia (Ross et al., 2001). Though species-level relationships within *Oncomelania* are not fully resolved (e.g., Wilke et al., 2006; Zhao et al., 2010), there are probably four species of *Oncomelania* that today spread the disease: *O. hupensis* ssp. in eastern China, *O. robertsoni* in western China, *O. quadrasi* in the Philippines, and *O. lindoensis* on Sulawesi.

Another parasitic disease transmitted by pomatiopsid snails is paragonimiasis, a zoonosis that is caused by infections with lung flukes (*Paragonimus* spp.). The disease is prevalent in Southeast Asia, with species of *Tricula, Neotricula, Gammatricula,* and *Oncolemania* serving as intermediate hosts (e.g., Hata et al., 1988; Davis et al., 1999). In North America, paragonimiasis is transmitted by *Pomatiopsis lapidaria,* intermediate host for the lung fluke *Paragonimus kellicotti* (Diaz, 2013). However, the number of human infections in the United States and southern Canada remains low. There is also a report of a possible paragonimiasis transmission in South Africa by *Tomichia natalensis* (Appleton, 2014).

Whereas some pomatiopsid species such as *Oncomelania* spp. are controlled because of public health concerns, other species are imperiled mainly due to habitat modifications and pollution. Of the 117 species recorded in the IUCN Red List (IUCN, 2017), 3 are listed as Critically Endangered, 5 as Endangered, 16 as Vulnerable, 18 as Near Threatened, and 35 as Least Concern. Ca. 30% of the listed pomatiopsid species (40) are Data Deficient, calling for intensified research to better understand conservation status and conservation needs of these tiny snails with often restricted distribution areas.

LITERATURE CITED

Ambu, S., G.J. Greer, and K.C. Shekhar. 1984. Natural and experimental infection of mammals with a *Schistosoma japonicum*-like schistosome from Peninsular Malaysia. Tropical Biomedicine 1: 103-107.

Appleton, C.C. 2014. Paragonimiasis in KwaZulu-Natal province, South Africa. Journal of Helminthology 88: 123-128.

Attwood, S.W. 2009. Mekong schistosomiasis: Where did it come from and where is it going? Pp. 276-297 in The Mekong: Biophysical Environment of an International River Basin (I.C. Campbell, ed.). Academic Press, New York.

Attwood, S.W., S. Ambu, X.-H. Meng, E.S. Upatham, F.-S. Xu, and V.R. Southgate. 2003. The phylogenetics of triculine snails (Rissooidea: Pomatiopsidae) from South-East Asia and southern China: Historical biogeography and the transmission of human schistosomiasis. Journal of Molluscan Studies 69: 263-271.

Attwood, S.W., H.S. Lokman, and K.Y. Ong. 2005. *Robertsiella silvicola,* a new species of triculine snail (Caenogastropoda: Pomatiopsidae) from Peninsular Malaysia, intermediate host of *Schistosoma malayensis* (Trematoda: Digenea). Journal of Molluscan Studies 71: 379-391.

Bouchet, P., J. Rocroi, B. Hausdorf, A. Kaim, Y. Kano, A. Nützel, P. Parkhaev, M. Schrödl, and E. Strong. 2017. Revised classification, nomenclator and typification of gastropod and monoplacophoran families. Malacologia 61: 1-526.

Brandt, R.A.M. 1974. The non-marine Mollusca of Thailand. Archiv für Molluskenkunde 105: i-iv, 1-423.

Brown, D. 1994. Freshwater Snails of Africa and Their Medical Importance. 2nd ed. Taylor & Francis, London.

Davis, G.M. 1979. The origin and evolution of the gastropod family Pomatiopsidae, with emphasis on the Mekong River Triculinae. Academy of Natural Sciences of Philadelphia Monograph 20: 1-120.

Davis, G.M. 1981. Different modes of evolution and adaptive radiation in the Pomatiopsidae (Prosobranchia: Mesogastropoda). Malacologia 21: 209-262.

Davis, G.M. 1993. Evolution of prosobranch snails transmitting Asian *Schistosoma;* coevolution with *Schistosoma:* A review. Pp. 145-204 in Progress in Clinical Parasitology, vol. 3 (T. Sun, ed.). Springer, New York.

Davis, G.M., and M.C.P. Pons da Silva. 1984. *Potamolithus:* Morphology, convergence, and relationships among hydrobioid snails. Malacologia 25: 73-108.

Davis, G.M., and W.P. Carney. 1973. Description of *Oncomelania hupensis lindoensis,* first intermediate host of *Schistosoma japonicum* in Sulawesi (Celebes). Proceedings of the Academy of Natural Sciences of Philadelphia 125: 1-34.

Davis, G.M., N.V. Subba Rao, and K.E. Hoagland. 1986. In search of *Tricula* (Gastropoda: Prosobranchia): *Tricula* defined, and a new genus described. Proceedings of the Academy of Natural Sciences of Philadelphia 138: 426-442.

Davis, G.M., T. Wilke, Y. Zhang, X.J. Xu, C.P. Qiu,

C. Spolsky, D.C. Qiu, M.Y. Xia, and Z. Feng. 1999. Snail-*Schistosoma, Paragonimus* interaction in China: Population ecology, genetic diversity, coevolution and emerging diseases. Malacologia 41: 355-377.

Diaz, J.H. 2013. Paragonimiasis acquired in the United States: Native and nonnative species. Clinical Microbiology Reviews 26: 493-504.

Gulbin, V.V. 2010. Russian waters of the East Sea. II. Caenogastropoda: Sorbeoconcha, Hypsogastropoda. Korean Journal of Malacology 26: 127-143.

Grego, J. 2018. First record of subterranean rissoidean gastropod assemblages in Southeast Asia (Mollusca, Gastropoda, Pomatiopsidae). Subterranean Biology 25: 9-34.

Hata, H., Y. Orido, M. Yokogawa, and S. Kojima. 1988. *Schistosoma japonicum* and *Paragonimus ohirai:* Antagonism between *S. japonicum* and *P. ohirai* in *Oncomelania nosophora.* Experimental Parasitology 65: 125-130.

IUCN Red List of Threatened Species. 2017. Version 2017-3. www.iucnredlist.org. Accessed 9 June 2018.

Kabat, A.R., and R. Hershler. 1993. The prosobranch snail family Hydrobiidae (Gastropoda: Rissooidea): Review of classification and supraspecific taxa. Smithsonian Contributions to Zoology 547: 1-94.

Kameda, Y., and M. Kato. 2011. Terrestrial invasion of pomatiopsid gastropods in the heavy-snow region of the Japanese Archipelago. BMC Evolutionary Biology 11: 118.

Kussakin, O.G. 1975. A list of macrofauna in the intertidal zone of the Kuril Islands, with remarks on zoogeographical structure of the region. Publications of the Seto Marine Biological Laboratory 22: 47-74.

Liu, L., G.-N. Huo, H.-B. He, B. Zhou, and S.W. Attwood. 2014. A phylogeny for the Pomatiopsidae (Gastropoda: Rissooidea): A resource for taxonomic, parasitological and biodiversity studies. BMC Evolutionary Biology 14: 29.

Malek, E.A. 1983. The South American hydrobioid genus *Idiopyrgus* Pilsbry, 1911. Nautilus 97: 16-20.

Morrison, J.P.E. 1963. *Cecina* from the State of Washington. Nautilus 76: 150-151.

Ross, A.G.P., A.C. Sleigh, Y. Li, G.M. Davis, G.M. Williams, Z. Jiang, Z. Feng, and D.P. McManus. 2001. Schistosomiasis in the People's Republic of China: Prospects and challenges for the 21st century. Clinical Microbiology Reviews 14: 70-295.

Simone, L.R.L. 2006. Land and Freshwater Molluscs of Brazil: An Illustrated Inventory of the Brazilian Malacofauna, Including Neighboring Regions of South America, Respect to the Terrestrial and Freshwater Ecosystems. EGB, Sao Paulo.

Simone, L.R.L. 2012. A new genus and species of cavernicolous Pomatiopsidae (Mollusca, Caenogastropoda) in Bahia, Brazil. Papéis Avulsos de Zoologia 52: 515-524.

Strong, E.E., O. Gargominy, W.F. Ponder, and P. Bouchet. 2008. Global diversity of gastropods (Gastropoda: Mollusca) in freshwater. Hydrobiologia 595: 149-166.

Wembo, O., C. Clewing, D. Delicado, B. Baluku, and C. Albrecht. 2014. Phylogeny and biogeography of enigmatic freshwater microsnails endemic to Eastern Congo (South Kivu). Abstracts of the 1st International Conference on Biodiversity in the Congo Basin 2014, Kisangani, p. 119.

West, K., E. Michel, J.A. Todd, D.S. Brown, and J. Clabaugh. 2003. The Gastropods of Lake Tanganyika: Diagnostic Key and Taxonomic Classification with Notes on the Fauna. SIL Occasional Publications 2, Chapel Hill, North Carolina.

Wilke, T., G.M. Davis, D. Qiu, and R.C. Spear. 2006. Extreme mitochondrial sequence diversity in the intermediate schistosomiasis host *Oncomelania hupensis robertsoni:* Another case of ancestral polymorphism? Malacologia 48: 143-157.

Wilke, T., M. Haase, R. Hershler, H.-P. Liu, B. Misof, and W.F. Ponder. 2013. Pushing short DNA fragments to the limit: Phylogenetic relationships of 'hydrobioid' gastropods (Caenogastropoda: Rissooidea). Molecular Phylogenetics and Evolution 66: 715-736.

Williams, W.D., and M.W. Mellor. 1991. Ecology of *Coxiella* (Mollusca, Gastropoda, Prosobranchia), a snail endemic to Australian salt lakes. Palaeogeography, Palaeoclimatology, Palaeoecology 84: 339-355.

Zhao, Q.P., M.S. Jiang, D.T.J. Littlewood, and P. Nie. 2010. Distinct genetic diversity of *Oncomelania hupensis,* intermediate host of *Schistosoma japonicum* in mainland China as revealed by ITS sequences. PLoS Neglected Tropical Diseases 4: e611.

22 Stenothyridae Tryon, 1866

STEPHANIE A. CLARK

Stenothyrids have in the past been included in the Viviparidae, Littorinidae, Rissoidae, and Hydrobiidae (Ponder & de Keyzer, 1998). More recently they have been recognized as a distinct family (Davis et al., 1986, 1988; Ponder and de Keyzer, 1998). The Stenothyridae has been placed in the superfamily Rissooidea (Ponder & de Keyzer, 1998; Bouchet & Rocroi, 2005). However, recent molecular data (16S and 28S) indicate that the current concept of the Rissoidea (sensu Bouchet and Rocroi, 2005) is not monophyletic but comprises two major clades, Rissooidea and Truncatelloidea (Criscione & Ponder, 2013). Criscione & Ponder (2013) placed Stenothyridae in the Truncatelloidea, and more detailed anatomical and molecular data suggests that the Stenothyridae is sister to the Truncatellidae (Criscione & Ponder, 2013; Wilke et al., 2013; Golding, 2014). However, the relationships among the various families currently included in the Rissooidea and Truncatelloidea are still not fully resolved (Criscione & Ponder, 2013; Wilke et al., 2013; Golding, 2014).

Stenothyrids have small, conical to ovate-conic or pupiform, dextral shells that vary in size from about 1.4 to 5.0 mm in height. Their thin, opaque to translucent, relatively featureless shells often have spiral rows of small pits. The aperture is circular and the operculum bears two flangelike projections on the inner surface. Stenothyrids have separate sexes and show some sexual dimorphism, with the females tending to be larger and broader than the males. Their tentacles are long and tapering, the female has a spermathecal groove, the penis lacks glands but sometimes has a stylet, the foot has a metapodial tentacle, and the animal is generally mottled grey to black with cream to yellow patches (Davis et al., 1986, 1988; Ponder & de Keyzer, 1998; Wilke et al., 2013; Golding, 2014). They lay few relatively large, sessile eggs (Annandale and Prashad, 1921).

The family is distributed primarily in the Oriental realm, with smaller numbers of taxa from the Palearctic and Australasian zoogeographical regions (Strong et al., 2008). The highest diversity is found in the Mekong River, with 19 species (Brandt, 1974; Strong et al., 2008).

There are about 65 described species in three

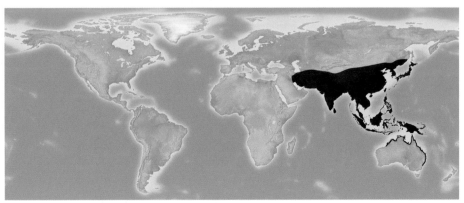

Distribution of Stenothyridae.

genera: *Farsithyra* Glöer & Pešić, 2009, from Iran; *Gangetia* Ancey, 1890, from India to Thailand; and *Stenothyra* Benson, 1856, from India, Thailand, Laos, Vietnam, Philippines, China, Japan, Papua New Guinea, and Australia (Brandt, 1974; Hoagland & Davis, 1979; Glöer & Pešić, 2009; Golding, 2014). The family is found in a wide range of aquatic habitats, from isolated springs to large rivers and mangrove habitats. Some species have wide ranges, such as *Stenothyra australis* (Golding, 2014), while others are known only from a few locations, such as *Farsithyra farsensis* (Glöer & Pešić, 2009).

On the Arabian Peninsula the family is represented by two species in two genera: *Gangetia* and *Stenothyra* (Neubert, 1998). Both species are found in estuarine habitats, with *Gangetia miliacea* considered to be introduced, while *Stenothyra arabica,* which was not described until 1998, is considered endemic to Oman (Neubert, 1998).

In Iran the family is represented by at least one species in the genus *Farsithyra* (Glöer and Pešić, 2009). The only described species is found in a small spring in the Fars region of Iran, but Glöer and Pešić (2009) believe additional taxa of *Farsithyra* occur in Iran.

From India to Thailand the family is represented by at least 30 species in two genera: *Gangetia* and *Stenothyra* (Brandt, 1974; Hoagland & Davis, 1979; Ramakrishna and Dey, 2007). They are found in a wide range of aquatic habitats, from streams to rivers and mangrove habitats. The bulk of the diversity is from the Mekong River drainage of Thailand, Laos, and Cambodia (Brandt, 1974; Hoagland & Davis, 1979).

In China and Japan the family is represented by at least 20 species in the genus *Stenothyra* (Kuroda, 1962; Davis et al., 1986, 1988; Okutani, 2000). They are found in both freshwater and estuarine habitats, with one species, *Stenothyra thermaecola,* found in hot springs in Japan (Kuroda, 1962). Davis et al. (1988) believe that the total number of species of *Stenothyra* is much higher than currently recognized.

In Australia, East Timor, and New Guinea, the family is represented by about eight species and subspecies in the genus *Stenothyra* (Golding, 2014). They are all found in estuarine habitats, especially in mangrove forests; to date no taxa have been found in freshwater habitats.

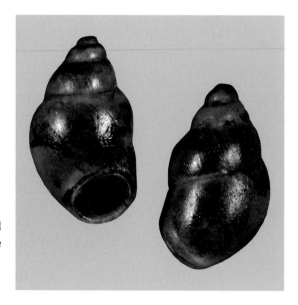

Stenothyra ventricosa Quoy & Gaimard, 1834. Indonesia: Kapuk, Jakarta, Java. 3 mm. Poppe 844908.

Among the factors that affect the diversity and conservation of stenothyrids are habitat destruction from dams, channel modification, siltation, water-quality degradation from point and nonpoint pollution, removal of mangrove habitats, and the introduction of nonindigenous species. Lysne et al. (2008) explore challenges and opportunities of freshwater conservation and highlight the need for involvement of many affected interest groups from local communities to government agencies.

The IUCN Red List (2016) lists 51 species of Stenothyridae: *Gangetia* (2 species) and *Stenothyra* (49), of which 1 is listed as Endangered, 2 as Vulnerable, 2 as Near Threatened, 18 as Data Deficient, and 28 as Least Concern.

LITERATURE CITED

Annandale, N., and B. Prashad. 1921. The Indian molluscs of the estuarine subfamily Stenothyrinae. Records of the Indian Museum 22: 121-136, plate 16.

Bouchet, P., and J.-P. Rocroi. 2005. Classification and nomenclator of gastropod families. Malacologia 47: 1-397.

Brandt, R.A.M. 1974. The non-marine aquatic Mollusca of Thailand. Archiv für Molluskenkunde 105: i-iv, 1-423.

Criscione, F., and W.F. Ponder. 2013. A phylogenetic analysis of rissooidean and cingulopsoidean families (Gastropoda: Caenogastropoda). Molecular Phylogenetics and Evolution 66: 1075-1082.

Davis, G.M., Y.H. Guo, K.E. Hoagland, L.C. Zheng, H.M. Yang, and Y.F. Zhou. 1986. Anatomy of *Stenothyra divalis* from the People's Republic of China and description of a new species of *Stenothyra* (Prosobranchia: Rissoacea: Stenothyridae). Proceedings of the Academy of Natural Sciences of Philadelphia 138: 318-349.

Davis, G.M., C.E. Chen, X.G. Xing, and C. Wu. 1988. The Stenothyridae of China. No. 2: *Stenothyra hunanensis*. Proceedings of the Academy of Natural Sciences of Philadelphia 140: 247-266.

Glöer, P., and V. Pešić. 2009. New freshwater gastropod species of the Iran (Gastropoda: Stenothyridae, Bithyniidae, Hydrobiidae). Museum für Tierkunde, Dresden. Mollusca 27: 33-89.

Golding, R.E. 2014. Molecular phylogeny and systematics of Australian and East Timorese Stenothyridae (Caenogastropoda: Truncatelloidea). Molluscan Research 34: 102-126.

Hoagland, K.E., and G.M. Davis. 1979. The stenothyrid radiation of the Mekong River. 1. The *Stenothyra mcmulleni* complex (Gastropoda: Prosobranchia). Proceedings of the Academy of Natural Sciences of Philadelphia 131: 191-230.

IUCN Red List of Threatened Species. 2016. Version 2016-2. www.iucnredlist.org. Accessed 11 October 2016.

Kuroda, T. 1962. Notes on the Stenothyridae (aquatic Gastro-

poda) from Japan and adjacent regions. Venus 22: 59-69, plate 4.

Lysne, S.J., K.E. Perez, K.M. Brown, R.L. Minton, and J.D. Sides. 2008. A review of freshwater gastropod conservation: Challenges and opportunities. Journal of the North American Benthological Society 27: 463-470.

Neubert, E. 1998. Annotated checklist of the terrestrial and freshwater molluscs of the Arabian Peninsula with descriptions of new species. Fauna of Arabia 17: 333-461.

Okutani, T. 2000. Marine Mollusks in Japan. Tokai University Press, Tokyo. i-xlviii, 1-1173.

Ponder, W.F., and R.G. de Keyzer. 1998. Superfamily Rissooidea. In: Mollusca: The southern synthesis. Vol. 5, Part B (P.L Beesley, G.J.B. Ross, and A. Wells, eds.). Fauna of Australia. CSIRO Publishing, Melbourne, Australia, pp. 745-766.

Ramakrishna and Dey, A. 2007. Handbook on Indian freshwater molluscs. Zoological Survey of India 23: i-xxiii, 1-399.

Strong, E.E., O. Gargominy, W.F. Ponder, and P. Bouchet. 2008. Global diversity of gastropods (Gastropoda: Mollusca) in freshwater. Hydrobiologia 595: 149-166.

Wilke, T., M. Haase, R. Hershler, H.-P. Liu, B. Misof, and W.F. Ponder. 2013. Pushing short DNA fragments to the limit: Phylogenetic relationships of 'hydrobioid' gastropods (Caenogastropoda: Rissooidea). Molecular Phylogenetics Evolution 66: 715-736.

23 Tateidae Thiele, 1925

WINSTON PONDER

Tateids are one of the more diverse freshwater mollusk families globally and by far the most speciose family of freshwater mollusks in Australia, New Zealand, and New Caledonia. They also occur on various Pacific Islands, including Fiji and Vanuatu, New Guinea, and Sulawesi (Indonesia), and in temperate South America.

All species are small (<10 mm), and most have been discovered and described in the past 30 years. Many have very restricted distributions, and some of those are of conservation concern. A few tateids live in brackish water, but the great majority live in freshwater.

Although now known as tateids, this group was until recently referred to a related family, the Hydrobiidae; but over most of their known distribution, typical hydrobiids do not occur together with tateids. These two families are morphologically very similar and were distinguished mainly on the basis of molecular data (Ponder et al., 2008; Wilke et al., 2013). Thus, using morphology alone, tateids cannot reliably be distinguished from hydrobiids. At the species, or even generic, level, the majority of tateid taxa are difficult to identify using shells alone; anatomical features or molecular data are often needed for accurate identification. Most tateids have a dextral, conical to flattened or elongate shell, and in some species the horny operculum has pegs on the inner surface. One New Zealand species, *Potamopyrgus antipodarum,* is a parthenogenic brooder and has invaded Australia, Europe, and North America (Ponder, 1988; Gangloff, 1998; Levri et al., 2007). Other tateids have separate sexes and, although a few other New Zealand taxa brood, most lay egg capsules; all are microphagous feeders.

The genus *Tatea* occurs in temperate Australian estuaries (Ponder et al., 1991), but, with the exception of a few New Zealand species, most other tateids live in freshwater. Other Australian genera include *Fluvidona* Iredale, 1937 (Miller et al., 1999), *Posticobia* Iredale, 1943 (Clark, 2009), *Westrapyrgus* Ponder, Clark & Miller, 1999 (Ponder et al., 1999), and the artesian spring endemics *Caldicochlea* Ponder, 1997, *Fonscochlea* Ponder, Hershler & Jenkins, 1989, and *Trochidrobia*

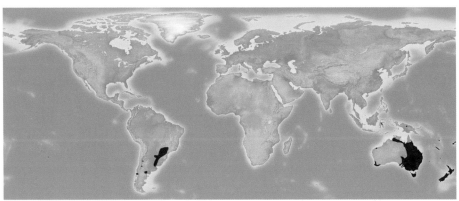

Distribution of Tateidae.

Austropyrgus petterdianus (Brazier, 1875). Australia: tributary of Brid River, off Upper Brid Rd., Tasmania. Photo by A. Hallan.

Ponder, Hershler & Jenkins, 1989 (Ponder et al., 1989, 1996). *Jardinella* Iredale & Whitley, 1938, occurs in coastal rivers and in artesian springs in Queensland (Ponder and Clark, 1990; Perez et al., 2005), while *Austropyrgus* Cotton, 1942, is the most speciose genus, with over 70 species (Clark et al., 2003). In Tasmania the seep and cave living *Nanocochlea* Ponder & Clark, 1993, and the cave living *Pseudotricula* Dang & Ho, 2006 (Ponder et al., 2005), are undoubted tateids, but the speciose genus *Beddomeia* Petterd, 1889, and its relatives *Phrantela* Iredale, 1943, and (in Victoria in mainland Australia) *Victodrobia* Iredale, 1943 (Ponder et al., 1993), are doubtfully included in the family (Zielske et al., 2011; Wilke et al., 2013), as is another taxon, the temperate Australian estuarine genus *Ascorhis* Ponder & Clark, 1988 (Ponder and Clark, 1988).

Population-level studies using allozymes (Ponder et al., 1994, 1995; Colgan and Ponder, 2000), DNA (Colgan et al., 2006; Murphy et al., 2012), and microsatellites (Worthington Wilmer et al., 2005, 2007, 2008, 2011) have been carried out on some Australian tateids, but published phylogenies of the Australian taxa are few and not comprehensive (Perez et al., 2005).

New Zealand has a diverse fauna, with 15 genera currently recognized, that include surface-living and subterranean taxa and a few estuarine species. The genera are *Catapyrgus* Climo, 1974, *Hadopyrgus* Climo, 1974, *Halopyrgus* Haase, 2008, *Kuschelita* Climo, 1974, *Leptopyrgus* Haase, 2008, *Meridiopyrgus* Haase, 2008, *Obtusopyrgus* Haase, 2008, *Opacuincola* Ponder, 1966,

Paxillostium Gardner, 1970, *Platypyrgus* Haase, 2008, *Potamopyrgus* Stimpson, 1865, *Rakipyrgus* Haase, 2008, *Rakiurapyrgus* Haase, 2008, *Sororipyrgus* Haase, 2008, and *Tongapyrgus* Haase, 2008 (Haase, 2008). Phylogenies of New Zealand tateids have been constructed using DNA data (Haase, 2005; Haase et al., 2007; Haase, 2008), and these taxa appear to have undertaken a significant monophyletic radiation.

Currently nine genera are recognized from New Caledonia: *Hemistomia* Crosse, 1872, *Heterocyclus* Crosse, 1872, *Kanakyella* Haase & Bouchet, 1998, *Pidaconomus* Haase & Bouchet, 1998, *Caledoconcha* Haase & Bouchet, 1998, *Leiorhagium* Haase & Bouchet, 1998, *Crosseana* Zielske & Haase, 2015, and *Novacaledonia* Zielske & Haase, 2015 (Starmühlner, 1970; Haase and Bouchet, 1998; Zielske and Haase, 2015). Species attributed to *Fluviopupa* also occur in New Caledonia, and other members of this genus are found on various Pacific Islands, including some in Fiji, Vanuatu, and French Polynesia and on Lord Howe Island (Ponder, 1982; Haase et al., 2005, 2006, 2010; Zielske and Haase, 2014; Zielske et al., 2017). One species has also been described from northeastern Australia (Ponder and Shea, 2013). The tateid fauna of New Guinea is largely unknown, but one cave-living taxon, the genus *Selmistomia* Bernasconi, 1995, has been named (Bernasconi, 1995).

Tateids occur on the island of Sulawesi (Indonesia), where two genera are recognized: *Sulawesidrobia* Ponder & Haase, 2005, and *Keindahan* Haase & Bouchet, 2006 (Ponder and Haase, 2005; Haase and Bouchet, 2006; Zielske et al., 2011). A radiation of the riverine genus *Potamolithus* occurs in South America (Pilsbry, 1911; Parodiz, 1965; Davis and Pons da Silva, 1984; López Armengol, 1996). To date this is the only tateid genus known from that continent, where it is one of the more diverse groups of freshwater gastropods; most live in streams and rivers, but at least two species live in caves (Bichuette and Trajano, 2003).

Some tateid snails occur in groundwater systems (caves, interstitial gravels, wells, etc.) in New Zealand (Climo, 1974, 1977; Haase, 2008). A radiation of the genus *Pseudotricula* has been described from a cave system in southwestern Tasmania (Ponder et al., 2005), and some undescribed taxa are known from wells and interstitial gravel habitats in Australia.

Many tateid species have very restricted ranges

(Colgan and Ponder, 1994; Ponder, 1994; Ponder and Colgan, 2002) and are thus of conservation concern because in unprotected areas their continued survival is often under threat from human activities. The arid zone artesian spring endemics in Australia have raised particular conservation concerns because of their threatened habitats (Ponder, 1986, 1995; Ponder and Walker, 2003).

LITERATURE CITED

Bernasconi, R. (1995). Two new cave prosobranch snails from Papua New Guinea: *Selmistomia beroni* n. gen. n. sp. (Caenogastropoda: Hydrobiidae) and *Georissa papuana* n. sp. (Archaeogastropoda: Hydrocenidae). Revue Suisse de Zoologie 102: 373-386.

Bichuette, M.E., and E. Trajano. 2003. A population study of epigean and subterranean *Potamolithus* snails from southeast Brazil (Mollusca: Gastropoda: Hydrobiidae). Hydrobiologia 505: 107-117.

Clark, S.A. 2009. The genus *Posticobia* (Mollusca: Caenogastropoda: Rissooidea: Hydrobiidae SL) from Australia and Norfolk Island. Malacologia 51: 319-341.

Clark, S.A., A.C. Miller, and W.F. Ponder. 2003. Revision of the snail genus *Austropyrgus* (Gastropoda: Hydrobiidae): A morphostatic radiation of freshwater gastropods in southeastern Australia. Records of the Australian Museum 28: 1-109.

Climo, F.M. 1974. Description and affinities of the subterranean molluscan fauna of New Zealand. New Zealand Journal of Zoology 1: 247-284.

Climo, F.M. 1977. Notes on the New Zealand Hydrobiid fauna (Mollusca: Gastropoda: Hydrobiidae). Journal of the Royal Society of New Zealand 7: 67-77.

Colgan, D.J., and W.F. Ponder. 1994. The evolutionary consequences of restrictions on gene flow: Examples from hydrobiid snails. Nautilus 108: 25-43.

Colgan, D.J., and W.F. Ponder. 2000. Incipient speciation in aquatic snails in an arid-zone spring complex. Biological Journal of the Linnean Society 71: 625-641.

Colgan, D.J., W.F. Ponder, and P. Da Costa. 2006. Mitochondrial DNA variation in an endemic aquatic snail genus, *Caldicochlea* (Hydrobiidae; Caenogastropoda) in Dalhousie Springs, an Australian arid-zone spring complex. Molluscan Research 26: 8-18.

Davis, G.M., and M.C. Pons da Silva. 1984. *Pomatolithus*: Morphology, convergence, and relationships among hydrobioid snails. Malacologia 25: 73-108.

Gangloff, M.M. 1998. The New Zealand mud snail in Western North America. Aquatic Nuisance Species Digest 2: 25-30.

Haase, M. 2005. Rapid and convergent evolution of parental care in hydrobiid gastropods from New Zealand. Journal of Evolutionary Biology 18: 1076-1086.

Haase, M. 2008. The radiation of hydrobiid gastropods in New Zealand: A revision including the description of new species based on morphology and mtDNA sequence information. Systematics and Biodiversity 6: 99-159.

Haase, M., and P. Bouchet. 1998. Radiation of crenobiontic gastropods on an ancient continental island: The *Hemistomia*-clade in New Caledonia (Gastropoda: Hydrobiidae). Hydrobiologia 367: 43-129.

Haase, M., and P. Bouchet. 2006. The radiation of hydrobioid gastropods (Caenogastropoda, Rissooidea) in ancient Lake Poso, Sulawesi. Hydrobiologia 556: 17-46.

Haase, M., O. Gargominy, and B. Fontaine. 2005. Rissooidean freshwater gastropods from the middle of the Pacific: The genus *Fluviopupa* on the Austral Islands (Caenogastropoda). Molluscan Research 25: 145-163.

Haase, M., W.F. Ponder, and P. Bouchet. 2006. The genus *Fluviopupa* Pilsbry, 1911 from Fiji (Caenogastropoda, Rissooidea). Journal of Molluscan Studies 72: 119-136.

Haase, M., B.A. Marshall, and I. Hogg. 2007. Disentangling causes of disjunction on the South Island of New Zealand: The Alpine fault hypothesis of vicariance revisited. Biological Journal of the Linnean Society 91: 361-374.

Haase, M., B. Fontaine, and O. Gargominy. 2010. Rissooidean freshwater gastropods from the Vanuatu archipelago. Hydrobiologia 637: 53-71.

Levri, E.P., A.A. Kelly, and E. Love. 2007. The invasive New Zealand mud snail (*Potamopyrgus antipodarum*) in Lake Erie. Journal of Great Lakes Research 33: 1-6.

López Armengol, M.F. 1996. Taxonomic revision of *Potamolithus agapetus* Pilsbry, 1911, and *Potamolithus buschii* (Frauenfeld, 1865) (Gastropoda: Hydrobiidae). Malacologia 38: 1-17.

Miller, A.C., W.F. Ponder, and S.A. Clark. 1999. Freshwater snails of the genera *Fluvidona* and *Austropyrgus* (Gastropoda, Hydrobiidae) from northern New South Wales and southern Queensland, Australia. Invertebrate Taxonomy 13: 461-493.

Murphy, N.P., M.F. Breed, M.T. Guzik, S.J.B. Cooper, and A.D. Austin. 2012. Trapped in desert springs: Phylogeography of Australian desert spring snails. Journal of Biogeography 39: 1573-1582.

Parodiz, J.J. 1965. The hydrobid snails of the genus *Potamolithus* (Mesogastropoda—Rissoacea). Sterkiana 20: 1-38.

Perez, K.E., W.F. Ponder, D.J. Colgan, S.A. Clark, and C. Lydeard. 2005. Molecular phylogeny and biogeography of spring-associated hydrobiid snails of the Great Artesian Basin, Australia. Molecular Phylogenetics and Evolution 34: 545-556.

Pilsbry, H.A. 1911. Non-marine mollusca of Patagonia. Princeton University Expedition to Patagonia 3: 513-633.

Ponder, W.F. 1982. Hydrobiidae of Lord Howe Island (Mollusca: Gastropoda: Prosobranchia). Australian Journal of Marine and Freshwater Research 33: 89-159.

Ponder, W.F. 1986. Mound springs of the Great Artesian Basin. Pp. 403-420 in Limnology in Australia (P. De Deckker and W. Williams, eds.). CSIRO, Melbourne.

Ponder, W.F. 1988. *Potamopyrgus antipodarum:* A molluscan colonizer of Europe and Australia. Journal of Molluscan Studies 54: 271-285.

Ponder, W.F. 1994. Australian freshwater mollusca: Conservation priorities and indicator species. Memoirs of the Queensland Museum 36: 191-196.

Ponder, W.F. 1995. Mound spring snails of the Australian Great Artesian Basin. Pp. 13-18 in The Conservation Biology of Mollucs (E.A. Kay, ed.). IUCN, Gland, Switzerland.

Ponder, W.F., and G.A. Clark. 1988. A morphological and electrophoretic examination of *Hydrobia buccinoides,* a variable brackish-water gastropod from temperate Australia (Mollusca: Hydrobiidae). Australian Journal of Zoology 36: 661-689.

Ponder, W.F., and G.A. Clark. 1990. A radiation of hydrobiid snails in threatened artesian springs in western Queensland. Records of the Australian Museum 42: 301-363.

Ponder, W.F., and D.J. Colgan. 2002. What makes a narrow range taxon? Insights from Australian freshwater snails. Invertebrate Systematics 16: 571-582.

Ponder, W.F., and M. Haase. 2005. A new genus of hydrobiid gastropods with Australian affinities from Lake Poso, Sulawesi (Gastropoda: Caenogastropoda: Rissooidea). Systematics 25: 27-36.

Ponder, W.F., and M.E. Shea. 2013. A new species of the *Fluviopupa* group (Caenogastropoda: Tateidae) from north-east Queensland, Australia. Molluscan Research 34: 71-78.

Ponder, W.F., and K.F. Walker. 2003. From mound springs to mighty rivers: The conservation status of freshwater molluscs in Australia. Aquatic Ecosystem Health and Management 6: 19-28.

Ponder, W.F., R. Hershler, and B. Jenkins. 1989. An endemic radiation of Hydrobiidae from artesian springs in northern South Australia: Their taxonomy, physiology, distribution and anatomy. Malacologia 31: 1-140.

Ponder, W.F., D.J. Colgan, and G.A. Clark. 1991. The morphology, taxonomy and genetic structure of *Tatea* (Mollusca: Gastropoda: Hydrobiidae), estuarine snails from temperate Australia. Australian Journal of Zoology 39: 447-497.

Ponder, W.F., G.A. Clark, A.C. Miller, and A. Toluzzi. 1993. On a major radiation of freshwater snails in Tasmania and eastern Victoria: A preliminary overview of the *Beddomeia* group (Mollusca: Gastropoda. Hydrobiidae). Invertebrate Taxonomy 5: 501-750.

Ponder, W.F., D.J. Colgan, G.A. Clark, A.C. Miller, and T. Terzis. 1994. Microgeographic, genetic and morphological differentiation of freshwater snails—the Hydrobiidae of Wilson's Promontory, Victoria, south-eastern Australia. Australian Journal of Zoology 42: 557-678.

Ponder, W.F., P.E. Eggler, and D.J. Colgan. 1995. Genetic differentiation of aquatic snails (Gastropoda: Hydrobiidae) from artesian springs in arid Australia. Biological Journal of the Linnean Society 56: 553-596.

Ponder, W.F., D.J. Colgan, T. Terzis, S.A. Clark, and A.C. Miller. 1996. Three new morphologically and genetically determined species of hydrobiid gastropods from Dalhousie Springs, northern South Australia, with the description of a new genus. Molluscan Research 17: 49-109.

Ponder, W.F., S.A. Clark, and A.C. Miller. 1999. A new genus and two new species of Hydrobiidae (Mollusca: Gastropoda: Caenogastropoda) from south Western Australia. Journal of the Royal Society of Western Australia 82: 109-120.

Ponder, W.F., S.A. Clark, S. Eberhard, and J.B. Studdert. 2005. A radiation of hydrobiid snails in the caves and streams at Precipitous Bluff, southwest Tasmania, Australia (Mollusca: Caenogastropoda: Rissooidea: Hydrobiidae s.l.). Zootaxa 1074: 1-66.

Ponder, W.F., T. Wilke, W.-C. Zhang, R.E. Golding, H. Fukuda, and R.A.B. Mason. 2008. *Edgbastonia alanwillsi* n. gen. and n. sp. (Tateinae: Hydrobiidae *s.l.:* Rissooidea: Caenogastropoda): A snail from an artesian spring group in western Queensland, Australia, convergent with some Asian Amnicolidae. Molluscan Research 28: 89-106.

Starmühlner, F. 1970. Etudes hydrobiologiques en Nouvelle Calédonie (Mission 1965 du Premier Institut de Zoologie de l'Université de Vienne): Die Mollusken der neukaledonischen Binnengewässer. Cahiers ORSTOM, Série Hydrobiologie 4: 3-127.

Wilke, T., M. Haase, R. Hershler, H.-P. Liu, B. Misof, and W.F. Ponder. 2013. Pushing short DNA fragments to the limit: Phylogenetic relationships of 'hydrobioid' gastropods (Caenogastropoda: Rissooidea). Molecular Phylogenetics and Evolution 66: 715-736.

Worthington Wilmer, J., and C. Wilcox. 2007. Fine scale patterns of migration and gene flow in the endangered mound spring snail, *Fonscochlea accepta* (Mollusca: Hydrobiidae) in arid Australia. Conservation Genetics 8: 617-628.

Worthington Wilmer, J., J.M. Hughes, J. Ma, and C. Wilcox. 2005. Characterization of microsatellite loci in the endemic mound spring snail *Fonscochlea accepta* and cross species amplification in four other hydrobiid snails. Molecular Ecology Notes 5: 205-207.

Worthington Wilmer, J., C. Elkin, C. Wilcox, L. Murray, D. Niejalke, and H. Possingham. 2008. The influence of multiple dispersal mechanisms and landscape structure on population clustering and connectivity in fragmented artesian spring snail populations. Molecular Ecology 17: 3733-3751.

Worthington Wilmer, J., L. Murray, C. Elkin, C. Wilcox, D. Niejalke, and H. Possingham. 2011. Catastrophic floods may pave the way for increased genetic diversity in endemic artesian spring snail populations. PLoS One 6: e28645.

Zielske, S., and M. Haase. 2014. When snails inform about geology: Pliocene emergence of islands of Vanuatu indi-cated by a radiation of truncatelloidean freshwater gastro-pods (Caenogastropoda: Tateidae). Journal of Zoological Systematics and Evolutionary Research 52: 217-236.

Zielske, S., and M. Haase. 2015. Molecular phylogeny and a modified approach of character-based barcoding refining the taxonomy of New Caledonian freshwater gastropods (Caenogastropoda, Truncatelloidea, Tateidae). Molecular Phylogenetics and Evolution 89: 171-181.

Zielske, S., M. Glaubrecht, and M. Haase. 2011. Origin and radiation of rissooidean gastropods (Caenogastropoda) in ancient lakes of Sulawesi. Zoologica Scripta 40: 221-237.

Zielske, S., W.F. Ponder, and M. Haase. 2017. The enigmatic pattern of long-distance dispersal of minute freshwater gastropods (Caenogastropoda, Truncatelloidea, Tateidae) across the South Pacific. Journal of Biogeography 44: 195-206.

24 Valvatidae Gray, 1840

CATHARINA CLEWING AND CHRISTIAN ALBRECHT

Valvatidae Gray, 1840 (= Borystheniinae Starobogatov, 1983), also referred to as valve snails, are nonpulmonate freshwater gastropods belonging to the informal group of "Lower Heterobranchia" (Bouchet and Rocroi, 2005). It represents the sole freshwater family of the Valvatoidea (other recent families are the marine Hyalogyrinidae Waren & Bouchet, 1993, and Cornirostridae Ponder, 1990). Systematic affinities of these prosobranchs have long been debated (Rath, 1988). The Valvatidae have then repeatedly been demonstrated to represent the earliest offshoot of Heterobranchia, based on morphological-anatomical (e.g., Rath, 1986; Haszprunar, 1988; Ponder, 1990) and molecular phylogenetic analyses (e.g., Dinapoli and Klussmann-Kolb, 2010; Jörger et al., 2010).

Valvatid snails have a specific feature, the pallial tentacle, which is a long, heavily ciliated extension of the mantle. This tentacle likely fulfils a cleaning function to avoid dirtying of the body (Dillon, 2000, and references therein). The shells of valvatids are small, dextral, discoid to ovate, with a large umbilicus. The operculum is circular, corneous, and multi-spiral (Brown, 1994), whereas the protoconchs are rather uniform (Bandel, 2010). Further characteristics include a featherlike gill. The central tooth does not have basal denticles (Brown, 1994). These gastropods are egg-laying and mostly hermaphrodites (Burch, 1989) and oviparous, with a notable exception, the ovoviviparous *Borysthenia naticina* (Glöer, 2002; but see Hawe et al., 2013 for a discussion on that).

Recently an extensive review of extant valvatid taxon names revealed 29 extant genus-level taxa; however, the taxonomic validity of these taxa is in need of revision (Haszprunar, 2014). There are different views of assigning genera vs. subgenera, and some taxa are even assigned to other families. The genus *Andrusovia* Brusina in Westerlund, 1902, might serve as an example. It was originally classified as a genus of Valvatidae, but currently it is also believed to belong to Hydrobiidae-Belgrandiinae (Kantor et al., 2009). In a global compilation of species diversity of freshwater gastropods, 71 extant species are listed (Strong et al., 2008), the majority of which are from the Palearctic (60 species), whereas 10 are from the

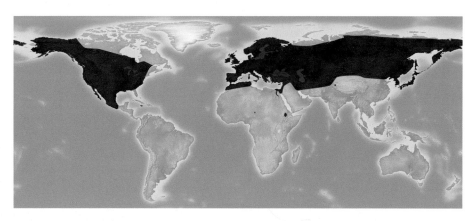

Distribution of Valvatidae.

Nearctic and 1 from the Afrotropical (Strong et al., 2008, table 2). Hawe et al. (2013) mentioned approximately 60 species spread over 16 supraspecific taxa. The extensive study of Haszprunar (2014) revealed more than 210 extant valvatid taxon names (including species, subspecies, and named varieties/synonyms).

Although quite some attention has been paid to studying the anatomy and morphology of *Valvata,* these attempts are remarkably restricted to very few selected species (Hawe et al., 2013). There is also no extensive family-level molecular phylogenetic study available to date (Hauswald et al., 2008; Clewing et al., 2014).

The oldest fossil records of valvatids apparently date back to Jurassic times (Tracey et al., 1993; Bandel, 1991; Taylor, 1988). However, there are several additional records (approximately 30) of valvatoid taxa described from the Mesozoic found in China, Germany, France, Scotland, England, and the United States (Haszprunar, 2014). The Neogene of Europe was also comparatively rich in valvatids (Harzhauser and Mandic, 2008; Neubauer et al., 2015). In Africa the Valvatidae were more widespread in the Holocene and Pliocene (Van Damme, 1984), with a special form occurring, e.g., in Lake Turkana (Scholz and Glaubrecht, 2010). Other species with peculiar shell shapes, such as keel-bearing species, lived in freshwater lakes of Rhodes and Kos, and in Lake Pannon in the Pontian of Hungary (Bandel, 2010). Extant keel-bearing species occur in North America or ancient lakes on the Balkans (Hauswald et al., 2008).

Members of the Valvatidae are distributed across the Northern Hemisphere. In the north, valvatids occur toward the margins of the continents except for the northernmost parts of Siberia (Bănărescu, 1990) and Canada (Clarke, 1973). There are apparently no records from either Iceland or Greenland. Valvatidae are present throughout the Palearctic, with the genus *Valvata* occurring in the entire range of the family. The situation is less clear on Mediterranean islands (Beckmann, 2007; Welter-Schultes, 2012; Seddon et al., 2014), with notable absence from some major Mediterranean islands, such as Cyprus or Crete (Bank, 2011a, b), and other islands being doubtful in terms of presence of valvatids (e.g., Corse). The southernmost limits of the continuous range of Valvatidae occurrences are at the southern end of the

Valvata tricarinata (Say, 1817). USA: Michigan. 4.3 mm. Poppe 812546.

Plateau of Mexico in America, northwestern Africa, and along the Palaearctic border in Asia (Bănărescu, 1990). In addition there are some isolated occurrences, such as in the highlands of Ethiopia or potentially Lake Chad, the current status of which require new fieldwork (Brown, 1994). The Pleistocene-Holocene distribution of *Valvata* in northern Africa was much more extended (Van Damme, 1984). In Asia, there is a probably isolated population thriving on the Tibetan Plateau at altitudes greater than 4000 m above sea level (see map; Clewing et al., 2014). The affinity of *Valvata pygmaea* C. B. Adams, 1849, from Jamaica has to be confirmed (Paul et al., 1993). This is also true for *V. mucronata* Menke, 1830, from the island of Madeira (Bank, 2011c).

Among the Valvatidae we find widespread species such as *Valvata piscinalis* (O. F. Müller, 1774) occurring throughout Europe and parts of Asia and other species that are restricted or even endemic to particular regions or settings such as ancient lakes. In Europe, Lake Ohrid is an example for an assemblage of endemic species (Hauswald et al., 2008), whereas Lake Baikal (Russia) might be considered to be a hotspot of valvatid diversity (4 nonendemic and 11 endemic species; see Sitnikova et al., 1993). Lake Biwa (Japan) hosts one endemic species (Kihira et al., 2003). Endemism on the one hand and invasive species on the other can be found among extant Valvatidae. For example, the European *Valvata piscinalis* has been established as an invasive species in the Laurentian lakes and potentially has an impact on native fauna there (Grigorovitch et al., 2005).

Valvatids prefer generally cold and clean lakes, rivers, and streams (Strong et al., 2008). Brown (1994) mentioned lakes, permanent pools, and slowly flowing streams as typical habitats. Valve snails feed on detritus and decaying organic material. *Valvata* is also known as the most important second intermediate host for the trematode *Echinoparyphium,* a parasite group that causes animal diseases (Huffman and Fried, 2012, in Dillon, 2000).

Relatively little is known about the conservation status of valvatids. Of the 26 species that have been assessed by the IUCN Red List to date (IUCN, 2017), 3 are listed as Critically Endangered or Endangered, 6 as Vulnerable or Near Threatened, 9 as Least Concern, and 8 as Data Deficient. *Valvata (Cincinna) kizakikoensis* (Fujita & Habe, 1991) is endemic to a few small lakes in Japan and is critically endangered (Van Damme and Vinarski, 2012). Vulnerable species include *Valvata hirsutecostata* Polinski, 1929 (Albrecht et al., 2010a), *V. utahensis* Call, 1884 (Cordeira and Perez, 2011), *V. relicta* (Polinski, 1929) (Albrecht et al., 2010b), and *V. virens* Tryon, 1863 (Bogan, 1996). Three other species being endangered and near threatened are known from the Balkans (Cuttelod et al., 2011). These assessments are certainly not representative of the global situation of the conservation status of Valvatidae.

LITERATURE CITED

Albrecht, C., T. Hauffe, and K. Schreiber. 2010a. *Valvata hirsutecostata.* The IUCN Red List of Threatened Species 2010: e.T155779A4841171. Accessed 19 April 2016.

Albrecht, C., T. Hauffe, and K. Schreiber. 2010b. *Valvata relicta.* The IUCN Red List of Threatened Species 2010: e.T155637A4812316. Accessed 19 April 2016.

Bănărescu P. 1990. Zoogeography of fresh waters. Vol. 1. General Distribution and Dispersal of Freshwater Animals. Aula-Verlag, Wiesbaden.

Bandel K. 1991. Gastropods from brackish and freshwater of the Jurassic-Cretaceous transition (a systematic reevaluation). Berliner Geowissenschaftliche Abhandlungen 135: 9-55.

Bandel K. 2010. Valvatiform Gastropoda (Heterostropha and Caenogastropoda) from the Paratethys Basin compared to living relatives, with description of several new genera and species. Paläontologie, Stratigraphie, Fazies (18), Freiberger Forschungshefte, C, 536: 91-155.

Bank, R.A. 2011a. Checklist of the land and freshwater Gastropoda of Cyprus. www.nmbe.ch/sites/default/files/uploads/pubinv/fauna_europaea_-_gastropoda_of_cyprus_0.pdf.

Bank, R.A. 2011b. Checklist of the land and freshwater Gastropoda of Greece. www.nmbe.ch/sites/default/files/uploads/pubinv/fauna_europaea_-_gastropoda_of_greece.pdf.

Bank, R.A. 2011c. Checklist of the land and freshwater Gastropoda of Macaronesia (Azores, Canary Islands, Madeira). www.nmbe.ch/sites/default/files/uploads/pubinv/fauna_europaea_-_gastropoda_of_macaronesia.pdf.

Beckmann, K.-H. 2007. Die Land- und Süßwassermollusken der balearischen Inseln. ConchBooks, Hackenheim, Germany.

Bogan, A.E. 1996. *Valvata virens.* The IUCN Red List of Threatened Species 1996: e.T22836A9394277. Accessed 19 April 2016.

Bouchet, P., and J.-P. Rocroi. 2005. Classification and nomenclator of gastropod families. Malacologia 47: 1-397.

Brown, D.S. 1994. Freshwater Snails of Africa and Their Medical Importance. 2nd ed. Taylor & Francis, London.

Burch, J.B. 1989. North American freshwater snails. Malacological Publications, Hamburg, Michigan.

Clarke, A.H. 1973. The freshwater molluscs of the Canadian Interior Basin. Malacologia 13: 1-509.

Clewing, C., P.V. von Oheimb, M. Vinarski, T. Wilke, and C. Albrecht. 2014. Freshwater mollusc diversity at the roof of the world: Phylogenetic and biogeographical affinities of Tibetan Plateau *Valvata.* Journal of Molluscan Studies 80: 452-455.

Cordeiro, J., and K. Perez. 2011. *Valvata utahensis.* The IUCN Red List of Threatened Species 2011: e.T22835A9394141. Accessed 19 April 2016.

Cuttelod, A., M. Seddon, and E. Neubert. 2011. European Red List of Non-Marine Molluscs. Publications Office of the European Union, Luxembourg.

Dillon, R.T., Jr. 2000. The Ecology of Freshwater Molluscs. Cambridge University Press, Cambridge, UK.

Dinapoli, A., and A. Klussmann-Kolb. 2010. The long way to diversity—phylogeny and evolution of the Heterobranchia (Mollusca; Gastropoda). Molecular Phylogenetics and Evolution 55: 60-76.

Glöer, P. 2002. Süsswassergastropoden Nord- und Mitteleuropas: Bestimmungsschlüssel, Lebensweise, Verbreitung. Die Tierwelt Deutschlands 73. ConchBooks, Hackenheim, Germany.

Grigorovich, I.A., E.L. Mills, C.B. Richards, D. Breneman,

and J.J.H. Ciborowski. 2005. European valve snail *Valvata piscinalis* (O. F. Müller, 1774) in the Laurentian Great Lakes basin. Journal of Great Lakes Research 31: 135-143.

Harzhauser, M., and O. Mandic. 2008. Neogene lake systems of central and south-eastern Europe: Faunal diversity, gradients and interrelations. Palaeogeography, Palaeoclimatology, Palaeoecology 260: 417-434.

Haszprunar, G. 1988. On the origin and evolution of major gastropod groups, with special reference to the Streptoneura. Journal of Molluscan Studies 54: 367-441.

Haszprunar, G. 2014. A nomenclator of extant and fossil taxa of the Valvatidae (Gastropoda, Ectobranchia). ZooKeys 377: 1-172.

Hauswald, A.-K., C. Albrecht, and T. Wilke. 2008. Testing two contrasting evolutionary patterns in ancient lakes: Species flock versus species scatter in valvatid gastropods of Lake Ohrid. Hydrobiologia 615: 169-179.

Hawe, A., M. Hess, and G. Haszprunar. 2013. 3D reconstruction of the anatomy of the ovoviviparous (?) freshwater gastropod *Borysthenia naticina* (Menke, 1845) (Ectobranchia: Valvatidae). Journal of Molluscan Studies 79: 1-14.

Huffman, J.E., and B. Fried. 2012. The biology of *Echinoparyphium* (Trematoda, Echinostomatidae). Acta Parasitologica 57: 199-210.

IUCN Red List of Threatened Species. 2017. Version 2017-1. www.iucnredlist.org. Accessed 26 July 2017.

Jörger, K.M., I. Stöger, Y. Kano, H. Fukuda, T. Knebelsberger, and M. Schrödl. 2010. On the origin of Acochlidia and other enigmatic euthyneuran gastropods, with implications for the systematics of Heterobranchia. BMC Evolutionary Biology 10: 323-342.

Kantor, Y.I., M.V. Vinarski, A.A. Schileyko, and A.V. Sysoev. 2009. Catalogue of the continental mollusks of Russia and adjacent territories. Available at www.ruthenica.com /documents/Continental_Russian_molluscs_ver2-2.pdf.

Kihira, H., M. Matsuda, and R. Uchiyama. 2003. Illustrated Guide to the Freshwater Molluscs of Japan 1. Freshwater Molluscs from Lake Biwa and the Yodo River. Pisces, Tokyo.

Neubauer, T.A., M. Harzhauser, A. Kroh, E. Georgopoulou, and O. Mandic. 2015. A gastropod-based biogeographic scheme for the European Neogene freshwater systems. Earth-Science Reviews 143: 98-116.

Paul, C.R.C., P. Hales, R.A. Perrott, and F.A. Street-Perrott. 1993. The freshwater Mollusca of Jamaica. Geological Society of America Memoirs 182: 363-370.

Ponder, W.F. 1990. The anatomy and relationships of the Orbitestellidae (Gastropoda: Heterobranchia. Journal of Molluscan Studies 56: 515-532.

Rath, E. 1986. Beiträge zur Anatomie und Ontogenie der Valvatidae (Mollusca: Gastropoda). PhD diss., University of Vienna, Austria.

Rath, E. 1988. Organization and systematic position of the Valvatidae. Malacological Review Supplement 4: 194-204.

Scholz, H., and M. Glaubrecht. 2010. A new and open coiled *Valvata* (Gastropoda) from the Pliocene Koobi Fora Formation of the Turkana Basin, northern Kenya. Journal of Paleontology 84: 996-1002.

Seddon, M.B., U. Kebapçi, and D. Van Damme. 2014. *Valvata piscinalis*. The IUCN Red List of Threatened Species: e.T156186A42435636. http://dx.doi.org/10.2305/IUCN.UK .2014-1.RLTS.T156186A42435636.en. Accessed 19 April 2016.

Sitnikova, T.Y., V.A. Fialkov, and Ya I. Starobogatov. 1993. Gastropoda from underwater hydrothermal vent of Baikal Lake. Ruthenica 3: 133-136.

Strong, E.E., O. Gargominy, W. F. Ponder, and P. Bouchet. 2008. Global diversity of gastropods (Gastropoda; Mollusca) in freshwater. Hydrobiologia 595: 149-166.

Taylor, D.W. 1988. Aspects of freshwater mollusc ecological biogeography. Palaeogeography, Palaeoclimatology, Palaeoecology 62: 511-576.

Tracey, S., J.A. Todd, and D.H. Erwin. 1993. Mollusca: Gastropoda. Pp. 131-167 in The Fossil Record (M.J. Benton, ed.). Chapman and Hall, London.

Van Damme, D. 1984. The freshwater Molluscs of northern Africa: Distribution, biogeography and palaeoecology. P. 176 in Developments in Hydrobiology, vol. 25 (H.J. Dumont, ed.). W. Junk, Dordrecht.

Van Damme, D., and M. Vinarski. 2012. *Cincinna kizakikoensis*. IUCN Red List of Threatened Species 2012: e.T188986A1916302. Accessed 19 April 2016.

Welter-Schultes, F. 2012. European Non-Marine Molluscs: A Guide for Species Identification. Planet Poster Editions, Göttingen.

25 Glacidorbidae Ponder, 1986

WINSTON PONDER

Glacidorbids are small to tiny freshwater snails that have a planispiral shell and an operculum. The family was named by Ponder (1986) on the basis of anatomical features of *Glacidorbis* Iredale, 1943, a genus originally named as a planorbid. Smith (1973) tentatively treated the group as valvatids, but it was then placed in the Hydrobiidae (Meier-Brook and Smith, 1975; Smith, 1979). Ponder (1986) moved the group from the caenogastropods to the pulmonates and created a new superfamily for them. Subsequently Haszprunar (1988) and Huber (1993) moved glacidorbids to the lower heterobranchs, mainly on the basis of neural characters, while Starobogatov (1988) treated them as *incertae sedis*. However, more recent molecular analyses (Holznagel et al., 2010; Schrödl et al., 2011; Golding, 2012) have shown that the group is correctly placed in the basal panpulmonates and is sister to the estuarine Amphiboloidea.

The first glacidorbid named was from southern New South Wales, Australia (Iredale, 1943), and Smith (1973) added an additional species from Tasmania. A species from southern Chile was then added (Meier-

Brook and Smith, 1975) and Bunn and Stoddart (1983) described an additional species from southwestern Australia. In a revision of the Australian members of the group, Ponder and Avern (2000) recognized 19 species, 15 of which were new, with most taxa from Tasmania. Rumi et al. (2015) described a new genus and three new species from southern Argentina.

The current understanding of the family includes four Australian genera (*Glacidorbis*, Iredale, 1943, by far the most diverse genus, *Striadorbis* Ponder & Avern, 2000, *Benthodorbis* Ponder & Avern, 2000, and *Tasmodorbis* Ponder & Avern, 2000), the latter two genera restricted to Tasmania, and two (*Gondwanorbis* Ponder, 1986, and *Patagonorbis* Rumi & Gutiérrez Gregoric, 2015) from South America. Most species live in Tasmania, with a few in temperate Australia and in the high country in New South Wales (Ponder and Avern, 2000). Four species have now been described from southern South America (Argentina and Chile) (Rumi et al., 2015). Thus, the distribution of the family is indicative of a relict Gondwanan distribution. The recent discovery of additional taxa in South America

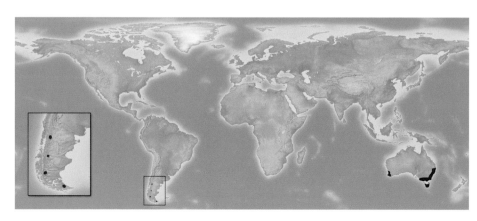

Distribution of Glacidorbidae.

suggests that additional new members of this family will be discovered there.

Some glacidorbids are vulnerable, having narrow to very narrow ranges, and the two species of *Benthodorbis* are confined to lakes on the central plateau of Tasmania (see Ponder and Avern, 2000, for details). One species (*Glacidorbis costata*), from a drained swamp in northern Tasmania, is apparently extinct (Ponder and Avern, 2000). Similarly, the South American taxa, so far as is known, have confined ranges.

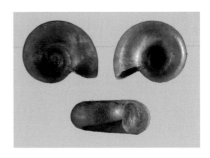

Glacidorbis hedleyi Iredale, 1943. Australia: Terrace Creek, Jenden State Forest, New South Wales. Photo by A. Hallan.

LITERATURE CITED

Bunn, S.E., and J.A. Stoddart. 1983. A new species of the prosobranch gastropod *Glacidorbis* and its implications for the biogeography of South Western Australia. Records of the Western Australian Museum 11: 49-57.

Golding, R.E. 2012. Molecular phylogenetic analysis of mudflat snails (Gastropoda: Euthyneura: Amphiboloidea) supports an Australasian centre of origin. Molecular Phylogenetics and Evolution 63: 72-81.

Haszprunar, G. 1988. On the origin and evolution of major gastropod groups, with special reference to the Streptoneura (Mollusca). Journal of Molluscan Studies 54: 367-441.

Holznagel, W.E., D.J. Colgan, and C. Lydeard. 2010. Pulmonate phylogeny based on 28S rRNA gene sequences: A framework for discussing habitat transitions and character transformation. Molecular Phylogenetics and Evolution 57: 1017-1025.

Huber, G. 1993. On the cerebral nervous system of marine Heterobranchia (Gastropoda). Journal of Molluscan Studies 59: 381-420.

Iredale, T. 1943. A basic list of the fresh water Mollusca of Australia. Australian Zoologist 10: 188-230.

Meier-Brook, C., and B.J. Smith. 1975. *Glacidorbis* Iredale 1943, a genus of freshwater prosobranchs with a Tasmanian-Southeast Australian-South Andean distribution. Archiv für Molluskenkunde 106: 191-198.

Ponder, W.F. 1986. Glacidorbidae (Glacidorbacea: Basommatophora), a new family and superfamily of operculate freshwater gastropods. Zoological Journal of the Linnean Society 87: 53-83.

Ponder, W.F., and G.J. Avern. 2000. The Glacidorbidae (Mollusca: Gastropoda: Heterobranchia) of Australia. Records of the Australian Museum 52: 307-353.

Rumi, A., D.E. Gutiérrez Gregoric, N. Landoni, J. Cárdenas Mancilla, S. Gordillo, J. Gonzalez, and D. Alvarez. 2015. Glacidorbidae (Gastropoda: Heterobranchia) in South America: Revision and description of a new genus and three new species from Patagonia. Molluscan Research 35: 143-152.

Schrödl, M., K.M. Jörger, and N.G. Wilson. 2011. Bye bye "Opisthobranchia"! A review on the contribution of mesopsammic sea slugs to euthyneuran systematics. Thalassas 27: 101-112.

Smith, B.J. 1973. A new species of snail from Lake Pedder, Tasmania, possibly belonging to the family Valvatidae. Journal of the Malacological Society of Australia 2: 429-434.

Smith, B.J. 1979. A new species of *Glacidorbis* (? Hydrobiidae, Gastropoda) from Great Lake, Tasmania. Journal of the Malacological Society of Australia 4: 121-127.

Starobogatov, I. 1988. On the systematic position of the genus *Glacidorbis* (Gastropoda *incertae sedis*) [in Russian, with English summary]. Proceedings of the Zoological Institute, Leningrad 187: 78-84.

26 Tantulidae Rankin, 1979, and Acochlidiidae Küthe, 1935

TIMEA P. NEUSSER AND MICHAEL SCHRÖDL

The Acochlidimorpha is a small group of mainly marine mesopsammic mollusks, which inhabit the interstices of coastal sands worldwide. These slugs have been regarded as enigmatic opisthobranchs, because they include the only gonochoristic opisthobranch species—and the only ones which succeeded to colonize freshwater systems (e.g., Schrödl and Neusser, 2010). Today, the Acochlidimorpha have been removed from the "Opisthobranchia" and are classified among panpulmonate gastropods (Bouchet et al., 2017)—this is a group of marine, intertidal, freshwater, and terrestrial slugs, snails, and limpets—based on molecular data (Jörger et al., 2010; Zapata et al., 2014; Teasdale, 2017). The remarkable flexibility of habitat choice united in the small clade of Acochlidimorpha was recently further underlined by the discovery of semiterrestrial (Swennen and Buatip, 2009; Neusser et al., 2011, 2015) and even fully terrestrial (Kano et al., 2015) Aitengidae, and the first marine-epibenthic acochlidian (Bathyhedylidae) from deep waters (Neusser et al., 2016).

Presenting the only known clade of freshwater slugs, the Acochlidimorpha form a fascinating group. Current phylogenetic hypotheses (Jörger et al., 2010; Schrödl and Neusser, 2010) suggest that the colonization of the limnic habitat from a marine ancestor occurred twice independently in Acochlidimorpha: once in the small, interstitial monotypic Tantulidae Rankin, 1979, with the single species *Tantulum elegans* Rankin, 1979, from St. Vincent Island in the Caribbean, and second, along the stemline of the family Acochlidiidae sensu Schrödl and Neusser (2010), large-sized (up to 3.5 cm) slugs living benthically in coastal rivers of different Indo-Pacific Islands (Schrödl and Neusser, 2010) and comprising currently seven species.

These unique, limnic slugs were first discovered in the end of the nineteenth century by two different expeditions to islands in the Indo-Pacific archipelago. Strubell (1892) presented a short notice on two distinct species which he collected on the Molucca Island Ambon, namely *Acochlidium amboinense* Strubell, 1892, and *Strubellia paradoxa* (Strubell, 1892; as *Acochlidium paradoxum*); later these species were investigated gross-anatomically and histologically by

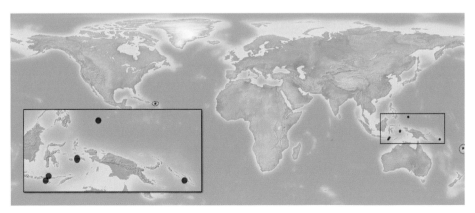

Distribution of Tantulidae from St. Vincent Island in the Caribbean and Acochlidiidae from coastal rivers of different Indo-Pacific Islands (see map inset for details).

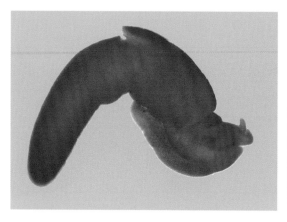

Tantulum elegans Rankin, 1979. Saint Vincent and the Grenadines: St. Vincent Island, ~1.5 mm. Preserved specimen stained with safranine.

Acochlidium bayerfehlmanni Wawra, 1980. Republic of Palau. 1.0–1.5 cm. Photo by Y. Kano.

Bücking (1933) and Küthe (1935), respectively. A few years after the first discoveries by Strubell, Bergh (1895) described *Palliohedyle weberi* (Bergh, 1895; as *Hedyle*), which was discovered in a freshwater stream on Flores Island by Max Weber in 1889.

The second phase of freshwater slug discoveries was in the middle of the twentieth century. During an expedition to the island Sumba in 1949, the Swiss ornithologist Ernst Sutter collected freshwater slugs (Bühler and Sutter, 1951), which were considered as *Palliohedyle weberi* by Benthem Jutting (1955) based on the external morphology. However, she pointed out that the specimens from Sumba differed in their body coloration from individuals described by Bergh (1895). A few years later, some specimens were collected during the course of an ecological survey on Palau Island and, based on gross dissections of the radula and penial armature, assigned to *Acochlidium amboinense* (see Bayer and Fehlmann, 1960).

In 1979 the discovery of the first and solely limnic interstitial acochlidian slug was published. Based on paraffin sections, Rankin (1979) described *Tantulum elegans* collected from the muddy interstices of a mountain marsh far away from the sea on St. Vincent Island in the Caribbean.

The Austrian naturalist Erhard Wawra was the first to study the comparative anatomy of limnic Acochlidiidae based on histological paraffin sections. Wawra (1979) analyzed the originally collected material from Sumba and redescribed these slugs as *Acochlidium sutteri* Wawra, 1979. Specimens from Palau collected

by Greg Bright were described as the new species *Acochlidium bayerfehlmanni* Wawra, 1980, rejecting a previous assignment to *A. amboinense* suggested by Bayer and Fehlmann (1960). *Acochlidium bayerfehlmanni* differs from *A. sutteri* in slight details of the penial armature (Wawra, 1980). He also reexamined specimens collected by Ferdinand Starmühlner on the Solomon Islands and assigned them to *Strubellia paradoxa* (see Wawra, 1974, 1988).

The last species of the genus *Acochlidium* Strubell, 1892, described was *Acochlidium fijiense* (Haynes & Kenchington, 1991; as *A. fijiensis*) from Fiji based on radula features and the penial armature (Haynes and Kenchington, 1991). For this species, Haase and Wawra (1996) provided for the first time a detailed microanatomical description of the reproductive system based on histological sections and scanning electron microscopy and explained the functionality of the complex male copulatory organs.

Traditionally, the large-sized limnic acochlidian species were examined by their external morphology and traditional dissection, describing mainly features of the radula and the gross anatomy of the penial armature. Detailed, reliable, and especially comparable data about the micro-anatomy of all organ systems were not available until recently. Nowadays, modern imaging techniques, including computer-based three-dimensional reconstructions (for example, with the software Amira) based on histological semithin section series, allow for in-depth examination of the micro-anatomy of small mesopsammic slugs, but also of complex organ systems of large-sized specimens. They also facilitate establishing large data sets of comparative anatomical features for future phylogenetic analyses. The 3D reconstruction of paratypes of the limnic interstitial *Tantulum elegans* corrected and complemented the original description mainly of the central nervous and the circulatory systems and added new micro-anatomical data about tiny nervous features and the special androdiaulic reproductive system (Neusser and Schrödl, 2007). Brenzinger et al. (2011a) were the first to examine the micro-anatomy of a representative of the limnic family Acochlidiidae. The 3D reconstruction of a paratype and of newly collected specimens of the large-sized limnic *Strubellia paradoxa* from near the type locality on Ambon (collected by Matthias Glaubrecht, CeNak, Hamburg) supplemented and corrected the original description by Küthe (1935), providing new anatomical data on all organ systems (Brenzinger et al., 2011a). Within an integrative approach combining 3D micro-anatomical and molecular data, Brenzinger et al. (2011b) examined specimens of the genus *Strubellia* Odhner, 1937, from Vanuatu and the Solomon Islands and compared the results with *Strubellia paradoxa* from the Solomon Islands (Wawra 1974, 1988). Molecular analyses argued for establishing the new species *Strubellia wawrai* Brenzinger, Neusser, Jörger & Schrödl, 2011, supported by slight morphological differences concerning the male copulatory organs. When the feeding habits were recorded for the first time in limnic Acochlidimorpha, *S. wawrai* was shown to feed on eggs (with calcareous capsules) of freshwater neritids inhabiting the same streams. An amphidromous lifestyle was suggested for limnic Acochlidiidae (Brenzinger et al., 2011b), similar to neritid snails (Kano et al., 2011; see this volume, p. 31), explaining the dispersal of limnic slugs between different Indo-Pacific Islands.

Rankin (1979) provided the first classification of marine and limnic Acochlidimorpha, creating 19 genera with numerous families for just 25 nominal species. This unnecessary inflation of taxon names on the species, genus, and even family level was heavily criticized by several authors (Arnaud et al., 1986; Wawra, 1987; Schrödl and Neusser, 2010). Rankin's system was replaced by a different classification proposed by Wawra (1987), subdividing the Acochlidiida (as Aco-

chlidia) into the two major clades, Microhedylacea and Hedylopsacea. Recently, Wawra's classification was largely confirmed both by cladistic analyses based on morphological characters (Schrödl and Neusser, 2010) and by molecular analyses (Jörger et al., 2010, 2014). According to Schrödl and Neusser (2010), limnic acochlidian species are classified within the clade Hedylopsacea in two families. The monotypic Tantulidae includes the small, interstitial *Tantulum elegans* from St. Vincent Island. The family Acochlidiidae sensu Schrödl & Neusser (2010) comprises three genera of large-sized and benthic slugs. The genus *Acochlidium* Strubell, 1892, exhibits the broadest distribution, with the four recognized species *Acochlidium amboinense* from Ambon, *A. sutteri* from Sumba, *A. bayerfehlmanni* from Palau, and *A. fijiense* from Fiji. The genus *Palliohedyle* Rankin, 1979, includes *Palliohedyle weberi* from Flores, and the genus *Strubellia* Odhner, 1937, includes *Strubellia paradoxa* from Ambon and *S. wawrai* from the Solomon Islands and Vanuatu.

In morphological cladistic analyses, the Tantulidae resulted as the first basal offshoot within the clade Hedylopsacea, being the sister group to a combined clade of marine Hedylopsidae, Pseudunelidae, and the limnic Acochlidiidae (Schrödl and Neusser, 2010). The Acochlidiidae were recovered sister to marine and (temporary) brackish Pseudunelidae in morphological (Schrödl and Neusser, 2010) and molecular analyses (Jörger et al., 2010, 2014). Molecular clock analyses estimated the transition to the freshwater habitat in Acochlidiidae to approximately 30 mya in the Paleogene (Jörger et al., 2014).

Despite recent expeditions to the Indo-Pacific Archipelago dedicated to surveying the limnic slug diversity, our knowledge on distribution is still limited and the slug's occurrence is patchy. Without reliable data on distribution ranges of the individual species, the risk of extinction or endangerment of species remains speculative. However, some type localities have changed considerably due to human impact (e.g., straightening of watercourses and pollution); we repeatedly were unable to find acochlidiids in highly polluted rivers in city centers, despite their occurrence on nearby less polluted streams—a potential indication of their vulnerability to human impact. Moreover, the up- and downstream migration of amphidromous Acochlidiidae may be interfered with by human impact, such as dams. The type locality of the enigmatic *Tantulum elegans* on St. Vincent Island (Harrison and Rankin, 1976; Rankin, 1979) was also considerably altered by anthropogenic influence, i.e., agriculture, and recollection attempts at the type locality were unsuccessful (K. Jörger, LMU, Munich, pers. comm.).

To explore distribution patterns and the biogeographic history of freshwater slugs, future research will concentrate on the radiation of Acochlidiidae, including newly collected specimens from Ambon, Sulawesi, Solomon Islands, and Papua New Guinea. Delineation of the recognized and newly collected *Acochlidium* populations based on external and morphological characters only is problematic and ambiguous. Preliminary molecular analyses support clusters of *Acochlidium* not entirely congruent with those recognized by traditional morphology. An integrative molecular and micro-anatomical approach is thus indispensable to reevaluate the morphological characters traditionally used for species description. Moreover, little-studied organ systems, such as the circulatory and excretory systems, should be analyzed in micro-anatomical and ultrastructural detail to provide information on the unique capability to inhabit freshwater. Last but not least, an updated classification including the new molecular and morphological evidences needs to be provided.

LITERATURE CITED

Arnaud, P.M., C. Poizat, and L. Salvini-Plawen. 1986. Marine-interstitial Gastropoda (including one freshwater interstitial species): A faunistic, distributional, and ecological synthesis of the world fauna inhabiting subterranean waters (including the marine interstitial). Pp. 153-161 in Stygofauna Mundi (L. Botosaneanu, ed.). Brill/Backhuys, Leiden.

Bayer, F.M., and H.A. Fehlmann. 1960. The discovery of a freshwater opisthobranchiate mollusk, *Acochlidium amboinense* Strubell, in the Palau Islands. Proceedings of the Biological Society of Washington 73: 183-194.

Benthem Jutting, W.S.S. van. 1955. Süsswassermollusken von Sumba. Verhandlungen der Naturforschenden Gesellschaft in Basel 66: 49-60.

Bergh, R. 1895. Die Hedyliden, eine Familie kladohepatischer Nudibranchien. Verhandlungen der kaiserlich königlichen zoologisch-botanischen Gesellschaft Wien 45: 4-12.

Bouchet, P., J.-P. Rocroi, B. Hausdorf, A. Kaim, Y. Kano, A. Nützel, P. Parkhaev. M. Schrödl, and E. Strong. 2017. Revised classification, nomenclator and typification of gastropod and monoplacophoran families. Malacologia 61: 1-526.

Brenzinger, B., T.P. Neusser, M. Glaubrecht, G. Haszprunar, and M. Schrödl. 2011a. Redescription and three-dimensional reconstruction of the limnic acochlidian gastropod *Strubellia paradoxa* (Strubell, 1892) (Gastropoda: Euthyneura) from Ambon, Indonesia. Journal of Natural History 45: 183-209.

Brenzinger, B., T.P. Neusser, K.M. Jörger, and M. Schrödl. 2011b. Integrating 3D microanatomy and molecules: Natural history of the Pacific freshwater slug *Strubellia* Odhner, 1937 (Heterobranchia: Acochlidia), with description of a new species. Journal of Molluscan Studies 77: 351-374.

Bücking, G. 1933. *Hedyle amboinensis* (Strubell). Zoologische Jahrbücher (Systematik) 64: 549-582.

Bühler, A., and E. Sutter. 1951. Wissenschaftliche Ergebnisse der Sumba-Expedition des Museums für Völkerkunde und des Naturhistorischen Museums in Basel, 1949: Reisebericht und allgemeine Einführung. Verhandlungen der Naturforschenden Gesellschaft in Basel 62: 181-217.

Haase, M., and E. Wawra. 1996. The genital system of *Acochlidium fijiensis* (Opisthobranchia: Acochlidioidea) and its inferred function. Malacologia 38: 143-151.

Harrison, A.D., and J.J. Rankin. 1976. Hydrobiological studies of Easter Lesser Antillean Islands I. St. Vincent: Freshwater fauna—its distribution, tropical river zonation and biogeography. Archiv für Hydrobiologie (Monographische Beiträge) 50: 275-311.

Haynes, A., and W. Kenchington. 1991. *Acochlidium fijiensis* sp. nov. (Gastropoda: Opisthobranchia: Acochlidiacea) from Fiji. Veliger 34: 166-171.

Jörger, K.M., I. Stöger, Y. Kano, H. Fukuda, T. Knebelsberger, and M. Schrödl. 2010. On the origin of Acochlidia and other enigmatic euthyneuran gastropods, with implications for the systematics of Heterobranchia. BMC Evolutionary Biology 10: 323-342.

Jörger, K.M., B. Brenzinger, T.P. Neusser, A.V. Martynov, N.G. Wilson, and M. Schrödl. 2014. Panpulmonate habitat transitions: Tracing the evolution of Acochlidia (Heterobranchia, Gastropoda). bioRxiv. http://dx.doi.org/10.1101/010322.

Kano, Y., E.E. Strong, B. Fontaine, O. Gargominy, M. Glaubrecht, and P. Bouchet. 2011. Focus on freshwater snails. Pp. 257-264 in The Natural History of Santo (P. Bouchet, H. Le Guyader, and O. Pascal, eds.). Patrimoines Naturels, vol. 69. Muséum National d'Histoire Naturelle, Paris.

Kano, Y., T.P. Neusser, H. Fukumori, K.M. Jörger, and

M. Schrödl. 2015. Sea-slug invasion of the land. Biological Journal of the Linnean Society 116: 253-259.

Küthe, P. 1935. Organisation und systematische Stellung des Acochlidium paradoxum Strubell. Zoologische Jahrbücher der Abteilung für Systematik 66: 513-540.

Neusser, T.P., and M. Schrödl. 2007. Tantulum elegans reloaded: A computer-based 3D-visualization of the anatomy of a Caribbean freshwater acochlidian gastropod. Invertebrate Biology 126: 18-39.

Neusser, T.P., H. Fukuda, K.M. Jörger, Y. Kano, and M. Schrödl. 2011. Sacoglossa or Acochlidia? 3D reconstruction, molecular phylogeny and evolution of Aitengidae (Gastropoda: Heterobranchia). Journal of Molluscan Studies 77: 332-350.

Neusser, T.P., A.J. Bourke, K. Metcalfe, and R.C. Willan. 2015. First record of Aitengidae (Mollusca: Panpulmonata: Acochlidia) for Australia. Northern Territory Naturalist 26: 27-31.

Neusser, T.P., K.M. Jörger, E. Lodde-Bensch, E.E. Strong, and M. Schrödl. 2016. The unique deep sea-land connection: Interactive 3D visualization and molecular phylogeny of Bathyhedyle boucheti n. sp. (Bathyhedylidae n. fam.)—the first panpulmonate slug from bathyal zones. PeerJ 4:e2738.

Rankin, J.J. 1979. A freshwater shell-less Mollusc from the Caribbean: Structure, biotics and contribution to a new understanding of the Acochlidioidea. Royal Ontario Museum Life Sciences Contributions 116: 1-123.

Schrödl, M., and T.P. Neusser. 2010. Towards a phylogeny and evolution of Acochlidia (Mollusca: Gastropoda: Opisthobranchia). Zoological Journal of the Linnean Society 158: 124-154.

Strubell, A.D. 1892. [no title]. In Verhandlungen des Naturhistorischen Vereins der Preussischen Rheinlande, Westphalens, 49. Jg, Sitzung der niederrheinischen Gesellschaft, 62.

Swennen, C.K., and S. Buatip. 2009. Aiteng ater, new genus, new species, an amphibious and insectivorous sea slug that is difficult to classify (Mollusca: Gastropoda: Opisthobranchia: Sacoglossa(?): Aitengidae, new family). Raffles Bulletin of Zoology 57: 495-500.

Teasdale, L.C. 2017. Phylogenomics of the pulmonate land snails. PhD diss., University of Melbourne. 199 pp. http://hdl.handle.net/11343/128240.

Wawra, E. 1974. The rediscovery of Strubellia paradoxa (Strubell) (Gastropoda: Euthyneura: Acochlidiacea) on the Solomon Islands. Veliger 17: 8-10.

Wawra, E. 1979. Acochlidium sutteri nov. spec. (Gastropoda, Opisthobranchia, Acochlidiacea) von Sumba, Indonesien. Annalen des Naturhistorischen Museums in Wien, Serie B, Botanik und Zoologie 82: 595-604.

Wawra, E. 1980. Acochlidium bayerfehlmanni spec. nov. (Gastropoda: Opisthobranchia: Acochlidiacea) from Palau Islands. Veliger 22: 215-218.

Wawra, E. 1987. Zur Anatomie einiger Acochlidia (Gastropoda, Opisthobranchia) mit einer vorläufigen Revision des Systems und einem Anhang über Platyhedylidae (Opisthobranchia, Ascoglossa). PhD diss., Universität Wien.

Wawra, E. 1988. Strubellia paradoxa (Strubell 1892) (Gastropoda: Opisthobranchia) von den Salomon Inseln. Zoologischer Anzeiger 220: 163-172.

Zapata, F., N.G. Wilson, M. Howison, S.C.S. Andrade, K.M. Jörger, M. Schrödl, F.E. Goetz, G. Giribet, and C.W. Dunn. 2014. Phylogenomic analyses of deep gastropod relationships reject Orthogastropoda. Proceedings of the Royal Society of London B 281: 20141739.

27 Chilinidae Dall, 1870

DIEGO E. GUTIÉRREZ GREGORIC

The Chilinidae is comprised of only one genus, *Chilina* Gray, 1828, which has almost 50 species. The family is endemic in southern South America, including Argentina, Chile, Uruguay, Brazil, Paraguay, and the Malvinas Islands (Pilsbry, 1911; Castellanos and Gaillard, 1981; Brown and Pullan, 1987). Some records also mention the family in Peru, but that has not been confirmed. Originally the representatives of this genus were within the Auriculidae and Lymnaeidae as subgenus and genus, respectively (Gray, 1828, 1847). Dall (1870) created the family Chilinidae, including only the genus *Chilina*.

The Chilinidae is one of the oldest freshwater families currently known. Such primitive features as the presence of one chiastoneuric nervous system, horizontal lamellar tentacles, a noncontractile pneumostome, and an incomplete division of male and female ducts (Haeckel, 1911; Harry, 1964) have indicated a relationship to different groups of gastropods (Hubendick, 1945, 1978). These authors assign Chilinidae to the "Lower Basommatophora" classification, joining to Ellobiidae and Otinidae,

and consider Chilinidae as a possible sister group of Hygrophila and Stylommatophora. Dayrat et al. (2001) published a molecular phylogeny of Euthyneura that argued for the monophyly of Hygrophila and proposed the Chilinidae as a basal group. Klussmann-Kolb et al. (2008) distinguished two clades within the Hygrophila, the first including *Chilina* and *Latia* Gray, 1850, and the second comprising higher limnic Basommatophora. More recently, Dayrat et al. (2011), on the basis of molecular analyses, concluded that the Hygrophila was not a monophyletic group and proposed a relationship between the Chilinoidea and the Amphiboiloidea, although this contention was not well supported.

The genus and family have an oval (oblong to ventricose) shell with expanded last whorl; a nervous system with vestigial chiastoneury; pulmonary roof pigmented with kidney occupying almost entire length; kidney inner wall with numerous transverse trabeculae of irregular contour; rectum on right side of mantle cavity, anus near pneumostome; incomplete division of male and female ducts; common

Distribution of Chilinidae.

Chilina fulgurata Pilsbry, 1911. Argentina: Near Comodoro Rivadavia, Chubut Province. 12.1 mm. Poppe 207295.

duct opening to hermaphrodite duct, with irregular contours on both sides; proximal portion of uterus with glandular walls; calcareous granules in vaginal lumen and secondary bursa copulatrix or accessory seminal receptacle present; penial terminal portion with cuticularized toothlike structures; radula with rows of teeth arranged in V (Castellanos and Gaillard, 1981; Ovando and Gutiérrez Gregoric, 2012; Gutiérrez Gregoric et al., 2014).

Castellanos and Miquel (1980) and Castellanos and Gaillard (1981) carried out the first review of Chilinidae species present in Argentina, Brazil, and Uruguay, and with reference to the ones in Chile. These authors concluded that there are 16 species in Argentina and 23 in Chile, presenting 3 species in common. Valdovinos Zarges (1999, 2006) provides a list of present species in Chile, about 30 species, of which 5 would be found in Argentina. However, Valdovinos Zarges (1999, 2006) did not use the synonymic list by Castellanos and Gaillard (1981). In Brazil there is a subspecies called *Chilina fluminea parva,* which some authors consider as a species, *C. parva* (Olazarri, 1968; Pereira, 1997; Simões, 2002; Simone, 2006). However, Lanzer (1997) treats it as a subspecies, because the differences founded in the shell and radula are not enough for the specific differentiation. Simone (2006) and Agudo (2008), mentioned *C. globosa* for the southern of Brazil; however, for Castellanos and Gaillard (1981) it is a synonym of *C. fluminea*. In Uruguay two species are present, which are also found in Argentina (Scarabino, 2004). In Paraguay only one species is cited in the Parana River, bordering with Argentina, which is also found in Argentina (Quintana, 1982).

Historically, the descriptions of the species of the family were based almost exclusively on shell characters. Due to the huge polymorphism that some species can show, anatomical and molecular studies are needed in most species. Recently in Argentina, a review of species has begun, with descriptions of the shell, radula, reproductive and nervous systems, and genetics analyses (Ituarte, 1997; Gutiérrez Gregoric and Rumi 2008; Gutiérrez Gregoric, 2008, 2010; Ovando and Gutiérrez Gregoric, 2012; Gutiérrez Gregoric et al., 2014; Gutiérrez Gregoric and de Lucía, 2016). These works have described seven new species and redescribed seven other species, including the reclassification of a subspecies into a species. That means that now there are 24 species of Chilinidae in Argentina. However, it is necessary to study the species that are present in southern Argentina and Chile.

About their distribution, Castellanos and Gaillard (1981) organized the morphological diversity of Chilinidae into three "groups of species": (1) *Fluminea* group: species with thick shell, short or very short spire, and two columellar teeth, distributed in Del Plata Basin and northwestern Argentina; (2) *Parchappii* group: species with large-size shell, slender, elongated, conical spire, moderately high, generally with one tooth on the edge columellar, distributed in central and southern Argentina; (3) *Gibbosa* group: species with the last whorl large and angular, spire usually well developed, whorls straight, narrow opening and one or two teeth columellar. It is distributed in southern Argentina and Chile. However, there is an overlap of shell and distributional characters that do not allow us to follow with the classification. In relation to global distribution, the relationship between Chilinidae (South America) and Latidae (Australasia) is not unique among freshwater gastropods. Glacidorbidae is represented by 4 species in South America (Argentina and Chile) and 19 in Australasia (Ponder and Avern, 2000; Rumi et al., 2015), and Tateidae presents a native genus in South America (*Potamolithus*) and several in Australasia (Wilke et al., 2013). These records confirm the Gondwana distribution in freshwater organisms.

In contrast to other families, representatives of this family reach larger sizes in colder temperatures than in hot ones. So, sizes up to 5 cm are recorded in the Argentine and Chilean south, while species near the tropics do not exceed 2 cm (Castellanos and Miquel, 1991). *Chilina* species can be found in quite different types of habitats, such as lakes, lagoons, dams, waterfalls, streams, rivers, canals, and estuaries; all generally with clean oxygenated water and variable temperature ranges (Gutiérrez Gregoric, 2008; Cuezzo, 2009).

From the point of view of health, Chilinidae species are among the group of freshwater gastropods which act as intermediate hosts of Schistosomatidae (Trematoda: Digenea), which produce outbreaks of "swimmer's itch" or cercarial dermatitis (Acha and Szyfres, 2003; Flores et al., 2015). In Argentina, the first report of schistosome cercariae were in *C. fluviatilis* (= *C. fluminea*) from the Paraná River (Szidat, 1951). In southern Argentina, cercariae of the family Schistosomatidae in *C. dombeiana, C. gibbosa, C. neuquenensis,* and *C. perrieri* (Martorelli, 1984; Flores, 2005; Flores and Semenas, 2008; Flores et al., 2015) have been recorded and in areas of the Río de la Plata River in *C. fluminea* (Miquel, 1984). In Chile, cercariae that produce dermatitis in people emerging out of the *C. dombeiana* have been recorded (Valdovinos Zarges and Balboa, 2008).

Representatives of Chilinidae are one of the main foods of some native and exotic fishes that have commercial or sports interests, and they are distributed in Argentinean and Chilean Patagonia (*Percichthys trucha, Patagonina hatcheri, Salmo trutta,* and *Oncorhynchus mykiss*) (Ferriz, 1993–1994; Figueroa et al., 2010). When fishes eat the snails, they receive some parasites that cause, among other things, ocular Diplostomiasis (parasite cataract in fish or eye fluke disease), reducing their productive value. In this case, cercariae and their subsequent maturation cause damage in the vitreous and retina, cataracts, and eventually blindness (Semenas, 1998).

Biology and ecology of Chilinidae have received little study. A few years ago, some studies were conducted in Argentina: *C. fluminea* (Miquel, 1986; Gutiérrez Gregoric et al., 2012); *C. gibbosa* (Bosnia et al., 1990); *C. parchappii* (Estebenet et al., 2002; Martín, 2003); and *C. megastoma* (Gutiérrez Gregoric et al., 2010). In Chile, work has been done on *C. ovalis* (Quijón and Jaramillo, 1999; Quijón et al., 2001)

and *C. patagonica* (Landler and von Oheimb, 2013). This knowledge is basic and essential to define and evaluate possible measures to control parasites or dermatitis.

Of the 24 species of *Chilina* found in Argentina, 18 are endemic and 11 vulnerable. The vulnerability can be (1) only known from the type locality (3 species); (2) species occurring in protected areas (4 species); (3) without recent record (4 species); (4) continuous restricted distribution (9 species); (5) discontinuous restricted distribution (3 species) (Rumi et al., 2006; Gutiérrez Gregoric et al., 2014; Gutiérrez Gregoric and de Lucía, 2016). Of the 12 species assessed by the IUCN Red List of Threatened Species, only 1 species has been assessed as Vulnerable (*C. angusta* for Chile), 7 as Data Deficient, and 4 as Least Concern (IUCN, 2017). The presence of Chilinidae endemic species in highly oxygenated freshwater environments (waterfalls, jumps, and rapids) (almost five species) should be taken into account by government agencies before construction of dams that modify these types of environments. These environments are the most vulnerable continental environments.

LITERATURE CITED

Acha, P.N., and B. Szyfres. 2003. Zoonoses and Communicable Diseases Common to Man and Animals, vol. 3, Parasitoses. PAHO Scientific and Technical Publication, no. 580. Pan American Health Organization, Washington, D.C.

Agudo, A.I. 2008. Listagem sistemática dos moluscos continentais ocorrentes no estado de Santa Catarina, Brasil. Comunicaciones de la Sociedad Malacológica del Uruguay 9: 147–179.

Bosnia, A., F. Kaisin, and A. Tablado. 1990. Population dynamics and production of the freshwater snail *Chilina gibbosa* Sowerby 1841 (Chilinidae, Pulmonata) in a North-Patagonian Reservoir. Hydrobiología 190: 97–110.

Brown, D.S., and N.B. Pullan. 1987. Notes on shell, radula and habitat of *Chilina* (Basommatophora) from the Falkland Island. Journal of Molluscan Studies 53: 105–108.

Castellanos, Z.A. de, and M.C. Gaillard. 1981. Chilinidae. Vol. 15 (4) in Fauna de agua dulce de la República Argentina (R.A. Ringuelet, ed.). FECIC-CONICET, Buenos Aires, Argentina.

Castellanos, Z.A. de, and S. Miquel. 1980. Notas complementarias al género *Chilina* Gray (Mollusca Pulmonata). Neotropica 26: 171–178.

Castellanos, Z.A. de, and S. Miquel. 1991. Distribución de los Pulmonata Basommatophora. Vol. 15 (9) in Fauna de agua dulce de la República Argentina (Z.A. Castellanos, ed.). PROFADU-CONICET, Buenos Aires, Argentina.

Cuezzo, M.G. 2009. Mollusca Gastropoda. Pp. 595-654 in Macroinvertebrados bentónicos sudamericanos, sistemática y biología (E. Domínguez and H. Fernández, eds.). Fundación Miguel Lillo, Tucumán, Argentina.

Dall, W.H. 1870. On the genus *Pompholyx* and its allies, with a revision of the Limnaeidae of authors. Annals of the Lyceum of Natural History of New York 9: 333-361.

Dayrat, B., A. Tillier, G. Lecointre, and S. Tillier. 2001. New clades of Euthyneuran gastropods (Mollusca) from 28S rRNA sequences. Molecular Phylogenetics and Evolution 19: 225-235.

Dayrat, B., M. Conrad, S. Balayan, T.R. White, C. Albrecht, R. Golding, S.R. Gomes, M.G. Harasewych, and A.M. de Frias Martins. 2011. Phylogenetic relationships and evolution of Pulmonate gastropods (Mollusca): New insights from increased taxon sampling. Molecular Phylogenetics and Evolution 59: 425-437.

Estebenet, A.L., N.J. Cazzaniga, and N.V. Pizani. 2002. The natural diet of the Argentinean endemic snail *Chilina parchappii* (Basommatophora: Chilinidae) and two other coexisting pulmonate gastropods. Veliger 45: 71-78.

Ferriz, R.A. 1993-1994. Algunos aspectos de la dieta de cuatro especies ícticas del río Limay (Argentina). Revista de Ictiologia 2-3: 1-7.

Figueroa, R., V.H. Ruiz, P. Berrios, A. Palma, P. Villegas, and A. Andreu-Soler. 2010. Trophic ecology of native and introduced fish species from the Chillán River, South-Central Chile. Journal of Applied Ichthyology 26: 78-83.

Flores, V. 2005. Estructura comunitaria de digeneos larvales en *Chilina dombeiana* y *Heleobia hatcheri* (Mollusca, Gastropoda) de la región andino patagónica. PhD diss., Universidad Nacional del Comahue, Argentina.

Flores, V., and L. Semenas. 2008. Larval digenean community parasitizing the freshwater snail, *Chilina dombeyana* (Pulmonata: Chilinidae) in Patagonia, Argentina, with special reference to the Notocotylid *Catatropis chilinae*. Journal of Parasitology 94: 305-313.

Flores, V., S. Brant, and E.S. Loker. 2015. Avian schistosomes from the South American endemic gastropod genus *Chilina* (Pulmonata: Chilinidae), with a brief review of South American schistosome species. Journal of Parasitology 101: 565-576.

Gray, J.E. 1828. Spicilegia Zoologica, or Original Figures and Short Systematic Descriptions of New and Unfigured Animals. Part 1. Treuttel, Würtz, London.

Gray, J.E. 1847. A list of the genera of recent Mollusca, their synonyms and types. Proceedings of the Zoological Society of London 15: 129-219.

Gutiérrez Gregoric, D.E. 2008. Estudios morfoanatómicos y tendencias poblacionales en especies de la familia Chilinidae Dall, 1870 (Mollusca: Gastropoda) en la Cuenca del Plata. PhD diss., Universidad Nacional de La Plata, Argentina.

Gutiérrez Gregoric, D.E. 2010. Redescription of two endemic species of Chilinidae (Gastropoda: Basommatophora) from Del Plata Basin (South America). Journal of Conchology 40: 321-332.

Gutiérrez Gregoric, D.E., and A. Rumi. 2008. *Chilina iguazuensis* (Gastropoda: Chilinidae), new species from Iguazú National Park, Argentina. Malacologia 50: 321-330.

Gutiérrez Gregoric, D.E., and M. de Lucía. 2016. Diversity hotspots of freshwater gastropods: Three new species from Uruguay River (South America). PeerJ 4e2138.

Gutiérrez Gregoric, D.E., V. Núñez, and A. Rumi. 2010. Populational studies in an endemic gastropod species of waterfalls environment. American Malacological Bulletin 28: 159-165.

Gutiérrez Gregoric, D.E., V. Núñez, and A. Rumi. 2012. Population dynamics of the freshwater gastropod *Chilina fluminea* (Chilinidae), of temperate area from Argentina. Veliger 51: 109-116.

Gutiérrez Gregoric, D.E., N.F. Ciocco, and A. Rumi. 2014. Two new species of *Chilina* Gray from Cuyo Malacological Province, Argentina. (Gastropoda: Hygrophila: Chilinidae). Molluscan Research 34: 84-97.

Haeckel, W. 1911. Beiträge zur Anatome der Gattung *Chilina*. Zoologische Jahrbücher 13: 89-136.

Harry, W.H. 1964. The anatomy of *Chilina fluctuosa* Gray reexamined, with prolegomena on the phylogeny of the higher limnic Basommatophora (Gastropoda: Pulmonata). Malacologia 1: 355-385.

Hubendick, B. 1945. Phylogenie und Tiergeographie der Siphonariidae zur Kenntnis der Phylogenie in der Ordnung Basommatophora und des Ursprungs der pulmonaten Gruppe. Zoologiska Bidrag fran Uppsala 24: 1-216.

Hubendick, B. 1978. Systematics and comparative morphology of the Basommatophora. Pp. 1-47 in Pulmonates, vol. 2A, Systematics, Evolution and Ecology (V. Fretter and J. Peake, eds.). Academic Press, London.

Ituarte, C.F. 1997. *Chilina megastoma* Hylton Scott, 1958 (Pulmonata: Basommatophora): A study of topotypic specimens. American Malacological Bulletin 14: 9-15.

IUCN Red List of Threatened Species. 2017. Version 2017-1. www.iucnredlist.org. Accessed 26 July 2017.

Klussmann-Kolb, A., A. Dinapoli, K. Kuhn, B. Streit, and C. Albrecht. 2008. From sea to land and beyond—new insights into the evolution of euthyneuran Gastropoda (Mollusca). BMC Evolutionary Biology 8:57.

Landler, L., and P.V. von Oheimb. 2013. Y Axis orientation in the South American freshwater snail species *Chilina patagonica* (Gastropoda : Chilinidae). Molluscan Research 33: 98-103.

Lanzer, R. 1997. *Chilina* (Basommatophora; Chilinidae) nas lagoas costeiras do Río Grande do Sul, Brasil: Concha, rádula, habitat e distribuição. Iheringia 82: 93-106.

Martín, P.R. 2003. Allometric growth and inter-population morphological variation of the freshwater snail *Chilina parchappii* (Gastropoda: Chilinidae) in the Napostá Grande stream, Southern Pampas, Argentina. Studies on Neotropical Fauna and Environment 38: 71-78.

Martorelli, S.R. 1984. Sobre una cercaria de la familia Schistosomatidae (Digenea) parásita de *Chilina gibbosa* Sowerby, 1841 en el lago Pellegrini, provincia de Río Negro, República Argentina. Neotropica 30: 97-106.

Miquel, S.E. 1984. Contribución al conocimiento biológico de gasterópodos pulmonados del área rioplatense, con especial referencia a *Chilina fluminea* (Maton). PhD diss., Universidad Nacional de La Plata, Argentina.

Miquel, S.E. 1986. El ciclo de vida y la evolución gonadal de *Chilina fluminea fluminea* (Maton, 1809) (Gastropoda; Basommatophora; Chilinidae). Neotropica 32: 23-34.

Olazarri, J. 1968. Hallazgo del holotipo y "status" de *Chilina parva* Martens, 1868 (Moll. Gastr.). Comunicaciones Zoológicas del Museo de Historia Natural de Montevideo 9: 1-5.

Ovando, X.M.C., and D.E. Gutiérrez Gregoric. 2012. Systematic revision of *Chilina* Gray (Gastropoda: Pulmonata) from northwestern Argentina and description of a new species. Malacologia 55: 117-134.

Pereira, P.C. 1997. Estudo da forma da concha nas espécies *Chilina fluminea* (Maton, 1811) e *Chilina parva* Martens, 1868 (Mollusca; Pulmonata; Chilinidae): Uma análise multivariada. Master's thesis, Pontifica Universidade Católica do Rio Grande do Sul, Porto Alegre, Brazil.

Pilsbry, H.A. 1911. Non-marine Mollusca of Patagonia. Princeton University Expedition to Patagonia 3: 513-633.

Ponder, W.F., and G.J. Avern. 2000. The Glacidorbidae (Mollusca: Gastropoda: Heterobranchia) of Australia. Records of the Australian Museum 52: 307-353.

Quijón, P., and E. Jaramillo. 1999. Gastropods and intertidal soft-sediments: The case of *Chilina ovalis* Sowerby (Pulmonata. Basommathopora) in South-Central Chile. Veliger 42: 72-84.

Quijón, P., H. Contreras, and E. Jaramillo. 2001. Population biology of the intertidal snail *Chilina ovalis* Sowerby (Pulmonata) in the Queule River Estuary, South-Central Chile. Estuaries 24: 69-77.

Quintana, M.G. 1982. Catálogo preliminar de la malacofauna del Paraguay. Revista del Museo Argentino de Ciencias Naturales Bernardino Rivadavia 21 (3): 61-158.

Rumi, A., D.E. Gutiérrez Gregoric, M.V. Núñez, I.I. César, M.A. Roche, M.P. Tassara, S.M. Martín, and M.F. López Armengol. 2006. Freshwater Gastropoda from Argentina: Species richness, distribution patterns, and an evaluation of endangered species. Malacologia 49: 189-208.

Rumi, A., D.E. Gutiérrez Gregoric, N. Landoni, J. Cárdenas Mancilla, S. Gordillo, J. Gonzalez, and D. Alvarez. 2015. Glacidorbidae in South America: Revision and description of a new genus and new species from Argentinean Patagonia and Chile. Molluscan Research 35: 143-153.

Scarabino, F. 2004. Lista sistemática de los gastrópoda dulceacuícolas vivientes de Uruguay. Comunicaciones de la Sociedad Malacológica del Uruguay 8: 347-356.

Semenas, L. 1998. Primer registro de diplostomiasis ocular en trucha arco iris cultivada en Patagonia (Argentina). Archivos de Medicina Veterinaria 30: 165-170.

Simões, R.I. 2002. Comunidade de moluscos bentônicos na área de abrangência da usina hidrelétrica de Dona Francisca, rio Jacuí, Rio Grande do Sul, Brasil: Fase de pré e pós-enchimento do reservatório. Master's thesis, Universidad Federal do Rio Grande do Sul Brazil. www.lume .ufrgs.br. Accessed 28 May 2018.

Simone, L.R.L. 2006. Land and Freshwater Molluscs of Brazil: An Illustrated Inventory on the Brazilian Malacofauna, Including Neighbor Regions of South America, Respect to the Terrestrial and Freshwater Ecosystems. EGB, Sao Paulo, Brazil.

Szidat, L. 1951. Cercarias schistosómicas y dermatitis schistosómica humana en la República Argentina. Comunicaciones del Instituto Nacional de Investigación de las Ciencias Naturales, Ciencias Zoológicas 2: 129-150.

Valdovinos Zarges, C. 1999. Biodiversidad de moluscos chilenos: Base de datos taxonómica y distribucional. Gayana 63: 111-64.

Valdovinos Zarges, C. 2006. Estado de conocimiento de los gastrópodos dulceacuícolas de Chile. Gayana 70: 88-95.

Valdovinos Zarges, C., and C. Balboa. 2008. Cercarial dermatitis and lake eutrophication in South-Central Chile. Epidemiology and Infection 136: 391-394.

Wilke, T., M. Haase, R. Hershler, H.-P. Liu, B. Misof, and W.F. Ponder. 2013. Pushing short DNA fragments to the limit: Phylogenetic relationships of "hydrobioid" gastropods (Caenogastropoda: Rissooidea). Molecular Phylogenetics and Evolution 66: 715-736.

28 Latiidae Hutton, 1882

CHRISTIAN ALBRECHT

The monogeneric Latiidae are a limpetlike group of freshwater pulmonate snails restricted to New Zealand. There has been debate on the phylogenetic status of these presumably basal Hygrophila (e.g., Hubendick, 1978; Meier-Brook, 1984; Nordsieck, 1992). Even phylogenetic affinities to the Holarctic Acroloxidae were suggested (Hubendick, 1978). Recently, molecular studies included *Latia* and crucial other taxa such as Chilinidae from South America (Dayrat et al., 2011; Holznagel et al., 2010, Jörger et al., 2010, Klussmann-Kolb et al., 2008). These studies consistently revealed that Hygrophila, the former so-called higher limnic Basommatophora, comprise two clades; on the one hand the Chilinoidea (sensu Boss [1982], comprising Chilinidae and Latiidae), and on the other the higher limnic pulmonate groups. *Chilina* and *Latia* are sister taxa, a relationship that has been mentioned earlier (e.g., Hubendick 1978; Nordsieck 1990). This relation is supported by morphological characteristics such as the position of the eyes at the exterior bases of the tentacles. They also share the incomplete division of the male and female ducts (Hubendick, 1978).

Currently, Latiidae are found exclusively on the North Island of New Zealand (Starobogatov, 1986; Taylor, 1988). However, a recent study reported a fossil species, *Latia manuheriki* Marshall, 2011, from the southern part of the South Island (Marshall, 2011). The current checklist of recent Mollusca of New Zealand lists three species: *L. climoi* Starobogatov, 1986, *L. lateralis* (Gould, 1852), and *L. neritoides* Gray, 1850 (Spencer et al., 2014). Two additional taxa have been synonymized with *L. neritoides* (Marshall, 2011). Doubts have been expressed as to the actual validity of three separate species, given the absence of shell morphological differences across the North island, which amounts to the whole range of *Latia* (Marshall, 2011). This taxonomic situation has been addressed by Wakerly (2014) in a M.Sc. thesis recently finished at the University of Otago. Wakerley (2014) concluded that instead of representing a species complex, all *Latia* taxa are conspecific. However, pronounced phy-

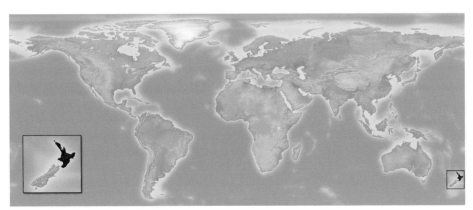

Distribution of Latiidae.

logeographic structuring was found throughout its range, according to Wakerly's interpretation of only a single valid species, *Latia neritoides*.

Latiidae are primarily occurring in streams with gravel or stony grounds flowing through both open and covered lands. The animals prefer hard substrates but can occasionally be found on macrophytes, too. Members of Latiidae represent the only bioluminescent freshwater gastropods—a phenomenon that has attracted some scientific attention (Meyer-Rochow and Moore, 1988, 2009). Studies continue to focus on the chemical nature of the substances secreted by these enigmatic animals (e.g., Oba et al., 2017; Ohmiya et al., 2005; Waldenmaier et al., 2012).

Latia neritoides is a widespread and abundant species for which a stable population trend has been predicted (Moore, 2013).

Latia neritoides Gray, 1850. New Zealand: Waikato River, North Island. 7.3 mm. Poppe 496366.

LITERATURE CITED

Boss, K.J. 1982. Mollusca. Pp. 945-1166 in Synopsis and Classification of Living Organisms, vol. 1 (Parker, S.P., ed.). McGraw Hill, New York.

Dayrat, B., M. Conrad, S. Balayan, T.R. White, C. Albrecht, R. Golding, S.R. Gomes, M.G. Harasewych, and A.M. de Frias Martins. 2011. Phylogenetic relationships and evolution of pulmonate gastropods (Mollusca): New insights from increased taxon sampling. Molecular Phylogenetics and Evolution 59: 425-437.

Holznagel, W.E., D.J. Colgan, and C. Lydeard. 2010. Pulmonate phylogeny based on 28S rRNA gene sequences: A framework for discussing habitat transitions and character transformation. Molecular Phylogenetics and Evolution 57: 1017-1025.

Hubendick, B. 1978. Systematics and comparative morphology of the Basommatophora. Pp. 1-47 in Pulmonates, vol. 2A, Systematics, Evolution and Ecology (V. Fretter and J. Peake, eds.). Academic Press, London.

Jörger, K.M., I. Stöger, Y. Kano, H. Fukuda, T. Knebelsberger, and M. Schrödl. 2010. On the origin of Acochlidia and other enigmatic euthyneuran gastropods, with implications for the systematics of Heterobranchia. BMC Evolutionary Biology 10: 323.

Klussmann-Kolb, A., A. Dinapoli, K. Kuhn, B. Streit, and C. Albrecht. 2008. From sea to land and beyond—new insights into the evolution of euthyneuran Gastropoda (Mollusca). BMC Evolutionary Biology 8: 57.

Marshall, B.A. 2011. A new species of *Latia* Gray, 1850 (Gastropoda: Pulmonata: Hygrophila: Chilinoidea: Latiidae) from Miocene Palaeo-lake Manuherikia, southern New Zealand, and biogeographic implications. Molluscan Research 31: 47-52.

Meier-Brook, C. 1984. A preliminary biogeography of freshwater pulmonate gastropods. Pp. 23-37 in World-Wide Snails: Biogeographical Studies on Non-Marine Mollusca (A. Solem and A.C. van Bruggen, eds.). Brill, Leiden.

Meyer-Rochow, V.B., and S. Moore. 1988. Biology of *Latia neritoides* Gray 1850 (Gastropoda, Pulmonata, Basommatophora): The only light-producing freshwater snail in the world. Internationale Revue der gesamten Hydrobiologie und Hydrographie 73: 21-42.

Meyer-Rochow, V.B., and S. Moore. 2009. Hitherto unreported aspects of the ecology and anatomy of a unique gastropod: The bioluminescent freshwater pulmonate Latia neritoides. Pp. 85-104 in Bioluminescence in Focus—A Collection of Illuminating Essays (V.B. Meyer-Rochow, ed.). Research Signpost, Kerala, India.

Moore, S. 2013. *Latia neritoides*. The IUCN Red List of Threatened Species 2013: e.T198906A2545381. T198906A2545381.en. Accessed 7 January 2016.

Nordsieck, H. 1992. Phylogeny and system of the Pulmonata (Gastropoda). Archiv für Molluskenkunde 121: 31-52.

Oba, Y., C.V. Stevani, A.G. Oliveira, A.S. Tsarkova, T.V. Chepurnykh, and I.V. Yampolsky. 2017. Selected least studied but not forgotten bioluminescent systems. Photochemistry and Photobiology 93: 405-415.

Ohmiya, Y., S. Kojima, M. Nakamura, and H. Niwa. 2005.

Bioluminescence in the limpet-like snail, Latia neritoides. Bulletin of the Chemical Society of Japan 78: 1197–1205.

Spencer, H.G., R.C. Willan, B. Marshall, and T.J. Murray. 2014. Checklist of the Recent Mollusca recorded from the New Zealand Exclusive Economic Zone. www.molluscs .otago.ac.nz/index.html.

Starobogatov, Ya.I. 1986. On the taxonomy of the gastropod molluscan genus *Latia* (Gastropoda Pulmonata Latiidae). Trudy Zoologicheskogo Instituta Akademii Nauk SSSR 148: 93–96.

Taylor, D.W. 1988 Aspects of freshwater mollusc ecological biogeography. Palaeogeography, Palaeoclimatology, Palaeoecology 62: 511–576.

Wakerley, G.L.J. 2014. Taxonomy of an unusual limpet: *Latia neritoides* (Gastropoda: Pulmonata: Hygrophila: Chilinoidea: Latiidae). Master's thesis, University of Otago. Retrieved from http://hdl.handle.net/10523/4913.

Waldenmaier, H.E., A.G. Oliveira, and C.V. Stevani. 2012. Thoughts on the diversity of convergent evolution of bioluminescence on earth. International Journal of Astrobiology 11: 335–343.

29 Lymnaeidae Rafinesque, 1815

MAXIM V. VINARSKI, CATHARINA CLEWING, AND
CHRISTIAN ALBRECHT

Lymnaeidae is a large and diverse family of freshwater pulmonates widely distributed on all continents except Antarctica (Hubendick, 1951; Starobogatov, 1970; Bănărescu, 1990; Ashworth and Preece, 2003). This family is well recognized, as some of its members have a tremendous biomedical and veterinary importance because they serve as intermediate hosts for trematode species that affect both livestock and humans (Tolan, 2011). *Fasciola hepatica* L., 1758 (Digenea: Fasciolidae), causing fascioliasis in vertebrate definite hosts is probably the best known example (Fürst et al., 2012; Cabada and White, 2012). Fascioliasis is a major human parasitic disease affecting around 17 million people and putting more than 90 million at risk worldwide (Tolan, 2011).

As shell shapes tend to be very plastic and anatomical characters are largely uniform within the family, it turns out to be challenging to estimate how many extant species and genera of Lymnaeidae there are. Although second in species richness, Lymnaeidae is probably the most controversially discussed family of the limnic pulmonates. Recent classification systems recognized a single genus, *Lymnaea* Lamarck, 1799 (e.g., Walter, 1968), two genera (Hubendick, 1951; Kruglov, 2005), or far more than two genera (e.g., Burch, 1989; Ponder & Waterhouse, 1997; Correa et al., 2010; Vinarski, 2013). Estimates of global species richness of Lymnaeidae vary from only 40 (Hubendick, 1951), to about 100 (Strong et al., 2008), to more than 200 species (Kruglov, 2005). Species richness numbers clearly depend on methodological grounds followed by a particular author. Most nominal species in the family were delineated on a strictly conchological basis, and their validity has not been checked by anatomical or molecular taxonomic methods. Although it seems that cryptic speciation is uncommon for lymnaeids, some examples have been found within the genus *Galba* Schrank, 1803 (Bargues et al., 2007, 2011).

Some members of Lymnaeidae are among the most ecologically flexible freshwater gastropods occurring in extreme habitats such as remote desert pools (Brown, 1994), geothermal springs (Vinarski et al., 2016), or high-elevational areas (e.g., Tibetan Plateau;

Distribution of
Lymnaeidae.

Lymnaea stagnalis (Linnaeus, 1758). France: Tang de Lindre, Foret Domaniale de Romersberg, Moselle Department. 48.9 mm. Poppe 759803.

Oheimb et al., 2011; Clewing et al., 2016). Lymnaeids have even occupied areas in the past that are uninhabitable today, e.g., Antarctica (Ashworth and Preece, 2003). The vast lymnaeid fossil record goes back to early Cretaceous times for taxa such as *Lanx* Clessin, 1880, and *Stagnicola* Jeffreys, 1830, or even early Jurassic for *Galba* Schrank, 1803 (Zilch, 1959–1960; Taylor, 1988; Tracey et al., 1993).

The phylogenetic relationships within the family are still not resolved completely, though some more or less comprehensive hypotheses have been proposed (e.g., Inaba, 1969; Remigio and Blair, 1997; Kruglov, 2005; Bargues et al., 2003; Correa et al., 2010; Vinarski, 2013). Vinarski (2013) proposed to delineate two subfamilies of living Lymnaeidae, Lymnaeinae *s. str.,* and Radicinae Vinarski, 2013 (= Amphipepleinae Pini, 1877), whereas the allocation of extinct species and genera among subfamilies remains highly problematic. The only extant representatives of limpet lymnaeids, the genera *Lanx* Clessin, 1880, and *Fisherola* Hannibal, 1912, occur in the Nearctic, both being endemic to North America (Burch, 1989). The phylogenetic relationships and taxonomic position of these limpets are not well understood. Sometimes the two genera have been separated into a family of their own (Lancidae Hannibal, 1914; see Staroboga-

tov, 1967), though Bouchet et al. (2017) rank it as a subfamily. The North American pateliform *Lanx* were not separated distinctly from the remaining lymnaeids in an unpublished phylogeny (Albrecht, pers. comm.), which would imply that a family status is unsupported. In the absence of a comprehensive phylogeny, Lancinae Hannibal, 1914, should be treated as a subfamily of Lymnaeidae (but see Vinarski, 2013), comprising the genera *Lanx, Fisherola,* and *Idaholanx* Clarke, Campbell & Lydeyard, 2017 (Campbell et al., 2017).

Though the distribution range of the family is virtually cosmopolitan, the species richness of lymnaeid snails does vary substantially in different zoogeographic regions. The highest number of species is observed in the Palearctic (Kruglov, 2005; Vinarski and Kantor, 2016) and Nearctic (Burch, 1989; Johnson et al., 2013), whereas lymnaeid diversity in other zoogeographic regions seems to be impoverished (Hubendick, 1951; Brown, 1994). The widespread global latitudinal gradient of increasing species richness toward lower latitudes (Hillebrand, 2004), is not observed for Lymnaeidae, whose species diversity is highest in the temperate belt of the Northern Hemisphere (Starobogatov, 1970). This also holds for all freshwater pulmonates (Hubendick, 1962).

The Palearctic maintains dozens of lymnaeid species as well as a series of endemic genera and subgenera (for example, *Aenigmomphiscola* Kruglov & Starobogatov, 1981, *Myxas* Sowerby, 1822, *Omphiscola* Rafinesque, 1819, *Peregriana* Servain, 1881). Most species of Lymnaeidae of the Palearctic have rather wide ranges, and some have been introduced by humans to other continents [for example, *Galba truncatula* (O.F. Müller, 1774) introduced to New Zealand and South America; see Hubendick, 1951; Bargues et al., 2011]. On the other hand, there are a handful of endemic lymnaeids in Europe and Asia, especially in the ancient lakes [*Radix pinteri* Schütt, 1974, and *R. relicta* Poliński, 1929, from lakes Ohrid and Prespa, Balkans, and *R. onychia* (Westerlund, 1887) from lake Biwa, Japan] or from mountain lakes such as Lake Teletskoye, Altay Mountains, Siberia [*Radix teletzkiana* (Kruglov & Starobogatov, 1983)]. Lake Baikal, with a plethora of unique species, genera, and even families of freshwater snails, has no endemic lymnaeid taxon (Kozhov, 1962), except for *Radix intercisa* (Lindholm,

1909) (Vinarski and Kantor, 2016). Some Palearctic lymnaeid species are currently invading the littoral of the lake (*Radix auricularia;* see Stift et al., 2004).

Northern Africa maintains few lymnaeid species of Palearctic origin (Van Damme, 1984; Brown, 1994), whereas the Middle East fauna includes a series of nominal endemic species belonging mostly to the subgenus *Radix* s. str. (Annandale and Prashad, 1919; Glöer & Pešić, 2012). However, the validity of these numerous taxa is still not clear. The lymnaeid species of Palearctic origin are also known in the region (Neubert, 1998; Glöer & Pešić, 2012). Two species in the genus *Stagnicola* Jeffreys, 1830, endemic to Turkey, were recently described (Glöer and Yildirim, 2006).

The lymnaeid fauna of central, Southeast, and South Asia is characterized by the dominance of the genus *Radix* Montfort, 1810 (subgenus *Radix s. str.*). Some species of this group represent true endemics to these regions, e.g., *Radix quadrasi* (Mőllendorff, 1898), endemic to Southeast Asia. The genera *Bullastra* Bergh, 1901, and *Orientogalba* Kruglov & Starobogatov, 1985, are South Asian in their origin and distribution, though some species of *Orientogalba* Kruglov & Starobogatov, 1985, reach northern Asia as well as Hawaii and Guam (see Kruglov and Starobogatov, 1993).

Sub-Saharan Africa has only one native species of Lymnaeidae, *Radix natalensis* (Krauss, 1848), widely distributed in tropical Africa as well as in Madagascar and Arabia (Brown, 1994; Neubert, 1998). Also, the Palearctic *Galba truncatula* is known from eastern Africa (Brown, 1994; note that Kruglov and Staroboga-tov, 1985, treat African records of this snail as belong-ing to a separate species classified by these authors within the genus *Orientogalba*). Recently, at least three more species of lymnaeid snails have been registered in Africa as invaders from Europe, Asia, and North America, respectively: *Lymnaea stagnalis* (L., 1758) in Cameroon (Tchakonté et al., 2014), *Radix rubiginosa* (Michelin, 1831) (Appleton and Miranda, 2015) in South Africa, and *Pseudosuccinea columella* (Say, 1817) in countries such as Zimbabwe, Mozambique, Zam-bia, Kenya, and Egypt (Appleton and Miranda, 2015). A conchologically peculiar species, *Lantzia carinata* Jousseaume, 1872, constituting a monotypic genus, is endemic to Réunion (Brown, 1994).

The Nearctic holds a very diverse fauna of Lym-naeidae, including several endemic genera: *Acella*

Haldeman, 1841; *Bulimnea* Haldeman, 1841, *Hinkleyia* F. C. Baker, 1928, and some others (see Kruglov and Starobogatov, 1993; Vinarski, 2013). The overall spe-cies richness of the Nearctic Lymnaeidae is about 60 (Johnson et al., 2013), though future research based on the integrative taxonomy approach may reduce this number. The core of the Nearctic lymnaeid fauna constitutes the species of the genera *Galba* and *Stagnicola* s. lato. The genus *Ladislavella* Dybowski, 1913 (= *Stagnicola,* partim; = *Catascopia* Meier-Brook & Bargues, 2002) is also distributed in northern Asia and even in central and eastern Europe (Vinarski, 2012; Vinarski et al., 2016). In Europe, it is represented by a single species, *L. terebra* (Westerlund, 1885), formerly known as *Stagnicola occultus* (Jackiewicz, 1959).

The South American fauna of Lymnaeidae consists of few species belonging to four genera only (*Galba,* *Limnobulla* Kruglov & Starobogatov, 1985; *Pectinidens* Pilsbry, 1911; *Pseudosuccinea* Say, 1817). The genus *Galba* dominates this continent. For example, in Venezuela there are five species of this genus with full absence of other lymnaeid genera (Pointier, 2015). The genus *Pectinidens* is restricted to the southern half of South America, whereas *Limnobulla* is endemic to the Falkland Islands.

The Australian and Pacific islands maintain a rela-tively small number of lymnaeid species, but the fauna of these regions is marked by the presence of several endemic genera, each containing few species (Ponder and Waterhouse, 1997; Puslednik et al., 2009). These are *Austropeplea* Cotton, 1942 (Australia, New Zealand, New Guinea), *Erinna* H. Adams & A. Adams, 1855 (Hawaii), and *Pseudoisidora* Thiele, 1931 (Hawaii). The phylogenetic position of the Australasian pond snails was reviewed by Puslednik et al. (2009), although a great deal of additional work should be done to resolve all questions concerning this group in Australia and adjacent regions. More dedicated sur-veys are also needed to assess the conservation status of the restricted or even endemic species, especially the species on oceanic islands.

The current picture of the lymnaeid biogeography is complicated by active human-mediated dispersal of highly invasive species into other continents, which can potentially lead to further homogenization of native malacofaunas. The world's most effective lymnaeid invaders are *Lymnaea stagnalis, Galba trun-*

catula, and especially *Pseudosuccinea columella,* which has become established in Europe, Australia, Pacific Islands, South Africa, and elsewhere.

LITERATURE CITED

Annandale, N., and B. Prashad. 1919. Report on the freshwater gastropod molluscs of Lower Mesopotamia. Part 1. The genus *Limnaea.* Records of the Indian Museum 18: 103-115.

Appleton, C.C., and N.A.F. Miranda. 2015. Two Asian freshwater snails newly introduced into South Africa and an analysis of alien species reported to date. African Invertebrates 56: 1-17.

Ashworth, A.C., and R.C. Preece. 2003. The first freshwater molluscs from Antarctica. Journal of Molluscan Studies 69: 89-92.

Bănărescu, P. 1990. Zoogeography of fresh waters. Vol. 1. General Distribution and Dispersal of Freshwater Animals. Aula-Verlag, Wiesbaden.

Bargues, M.D., P. Horák, R.A. Patzner, J.-P. Pointier, M. Jackiewicz, C. Meier-Brook, and S. Mas-Coma. 2003. Insights into relationships of Palearctic and Nearctic lymnaeids (Mollusca: Gastropoda) by rDNA ITS-2 sequencing and phylogeny of stagnicoline intermediate host species of *Fasciola hepatica.* Parasite 10: 243-255.

Bargues, M.D., P. Artigas, R.L. Mera y Sierra, J.-P. Pointier, and S. Mas-Coma. 2007. Characterisation of *Lymnaea cubensis, L. viatrix* and *L. neotropica* n. sp., the main vectors of *Fasciola hepatica* in Latin America, by analysis of their ribosomal and mitochondrial DNA. Annals of Tropical Medicine and Parasitology 101: 621-641.

Bargues, M.D., P. Artigas, M. Khoubbane, R. Flores, P. Glöer, R. Rojas-Garcia, K. Ashrafi, G. Falkner, and S. Mas-Coma. 2011. *Lymnaea schirazensis,* an overlooked snail distorting fascioliasis data: Genotype, phenotype, ecology, worldwide spread, susceptibility, applicability. PLoS ONE 6: e24567.

Bouchet, P., J.-P. Rocroi, B. Hausdorf, A. Kaim, Y. Kano, A. Nützel, P. Parkhaev, M. Schrödl, and E.E. Strong. 2017. Revised classification, nomenclator and typification of gastropod and monoplacophoran families. Malacologia 61: 1-526.

Brown, D.S. 1994. Freshwater Snails of Africa and Their Medical Importance. 2nd ed. Taylor & Francis, London.

Burch, J.B. 1989. North American Freshwater Snails. Malacological Publications, Hamburg, Michigan.

Cabada, M.M., and A.C. White Jr. 2012. New developments in epidemiology, diagnosis, and treatment of fascioliasis. Current Opinion Infectious Diseases 25: 518-22.

Campbell, S.C., S.A. Clark, and C. Lydeard. 2017. Phylogenetic analysis of the Lancinae (Gastropoda, Lymnaeidae) with a description of the U.S. federally endangered Banbury Springs lanx. *ZooKeys* 663:107-132.

Clewing, C., C. Albrecht, and T. Wilke. 2016. A complex system of glacial sub-refugia drives endemic freshwater biodiversity on the Tibetan Plateau. PLoS ONE 11: e0160286.

Correa, A.C., J.C. Escobar, P. Durand, F. Renaud, P. David, P. Jarne, J.-P. Pointier, and S. Hurtrez-Boussèz. 2010. Bridging gaps in the molecular phylogeny of the Lymnaeidae (Gastropoda: Pulmonata), vectors of Fascioliasis. BMC Evolutionary Biology 10: 381.

Fürst, T., U. Duthaler, B. Sripa, J. Utzinger, and J. Keiser. 2012. Trematode infections: Liver and lung flukes. Infectious Disease Clinics North America 26: 399-419.

Glöer, P., and V. Pešić. 2012. The freshwater snails (Gastropoda) of Iran, with descriptions of two new genera and eight new species. ZooKeys 219: 11-61.

Glöer, P., and M.Z. Yildirim. 2006. *Stagnicola* records from Turkey with the description of two new species, *Stagnicola tekecus* n. sp. and *S. kayseris* n. sp. (Gastropoda: Lymnaeidae). Journal of Conchology 39: 85-89.

Hillebrand, H. 2004. On the generality of the latitudinal diversity gradient. American Naturalist 163: 192-211.

Hubendick, B. 1951. Recent Lymnaeidae: Their variation, morphology, taxonomy, nomenclature, and distribution. Küngliga Svenska Vetenskapsakademiens Handlingar 3: 1-223.

Hubendick, B. 1962. *Aspects* of the *diversity* of the freshwater fauna. Oikos 13:249-261.

Inaba, A. 1969. Cytotaxonomic studies of lymnaeid snails. Malacologia 7: 143-168.

Johnson, P.D., A.E. Bogan, K.M. Brown, N.M. Burkhead, J.R. Cordeiro, J.T. Garner, P.D. Hartfield, et al. 2013. Conservation status of freshwater gastropods of Canada and the United States. Fisheries 38: 247-282.

Kozhov, M.M. 1962. Biology of Lake Baikal [in Russian]. Soviet Academy of Sciences Press, Moscow.

Kruglov, N.D. 2005. Molluscs of the family Lymnaeidae (Gastropoda, Pulmonata) in Europe and northern Asia [in Russian]. Smolensk State Pedagogical University Publishing, Smolensk, Russia.

Kruglov, N.D., and Ya.I. Starobogatov. 1985. The volume of the subgenus *Galba* and of other similar subgenera of the genus *Lymnaea* (Gastropoda, Pulmonata) [in Russian]. Zoologicheskij Zhurnal, 64: 24-35.

Kruglov, N.D., and Ya.I. Starobogatov. 1993. Annotated and illustrated catalogue of species of the family Lymnaeidae (Gastropoda Pulmonata Lymnaeiformes) of Palaearctic and adjacent river drainage areas. Part I. Ruthenica 3: 65-92.

Neubert, E. 1998. Annotated checklist of the terrestrial

and freshwater molluscs of the Arabian Peninsula with description of new species. Fauna of Arabia 17: 333-461.

Oheimb, P.V.V., C. Albrecht, F. Riedel, L. Du, J. Yang, D.C. Aldridge, U. Bößneck, H. Zhang, and T. Wilke. 2011. Freshwater biogeography and limnological evolution of the Tibetan Plateau—insights from a plateau-wide distributed gastropod taxon (*Radix* spp.). PLoS ONE 6: e26307.

Pointier, J.-P. 2015. Freshwater Molluscs of Venezuela and Their Medical and Veterinary Importance. ConchBooks, Harxheim, Germany.

Ponder, W.F., and J. Waterhouse. 1997 A new genus and species of Lymnaeidae from the lower Franklin River, south western Tasmania. Journal of Molluscan Studies 63: 441-468.

Puslednik, L., W.F. Ponder, M. Dowton, and A.R. Davis. 2009. Examining the phylogeny of the Australasian Lymnaeidae (Heterobranchia: Pulmonata: Gastropoda) using mitochondrial, nuclear and morphological markers. Molecular Phylogenetics and Evolution 52: 643-659.

Remigio, E.A., and D. Blair. 1997. Molecular systematics of the freshwater snail family Lymnaeidae (Pulmonata: Basommatophora) utilising mitochondrial ribosomal DNA sequences. Journal of Molluscan Studies 63: 173-185.

Starobogatov, Ya.I. 1967. On the systematization of freshwater pulmonate molluscs [in Russian]. Trudy Zoologicheskogo Instituta AN SSSR 42: 280-304.

Starobogatov, Ya.I. 1970. Molluscan Fauna and Zoogeographic Zonation of Continental Freshwater Bodies of the World [in Russian]. Nauka, Leningrad.

Stift, M., E. Michel, T.Ya. Sitnikova, E.Yu. Mamonova, and D. Yu Sherbakov. 2004. Palaearctic gastropod gains a foothold in the dominion of endemics: Range expansion and morphological change of *Lymnaea* (*Radix*) *auricularia* in Lake Baikal. Hydrobiologia 513: 101-108.

Strong, E.E., O. Gargominy, W.F. Ponder, and P. Bouchet. 2008. Global diversity of gastropods (Gastropoda: Mollusca) in freshwater. Hydrobiologia 595: 149-166.

Taylor, D.W. 1988. Aspects of freshwater mollusc ecological biogeography. Palaeogeography, Palaeoclimatology, Palaeoecology 62: 511-576.

Tchakonté, S., G.A. Ajeagah, D. Diomandé, A.I. Camara, and P. Ngassam. 2014. Diversity, dynamic and ecology of freshwater snails related to environmental factors in urban and suburban streams in Douala-Cameroon (Central Africa). Aquatic Ecology 48: 379-395.

Tolan, R.W., Jr. 2011. Fascioliasis due to *Fasciola hepatica* and *Fasciola gigantica* infection: An update on this 'neglected' neglected tropical disease. *Laboratory Medicine* 42: 107-117.

Tracey, S., J.A. Todd, and D.H. Erwin. 1993. Mollusca: Gastropoda. Pp. 131-167 in The Fossil Record (M.J. Benton, ed.). Chapman and Hall, London.

Van Damme, D. 1984. The freshwater molluscs of Northern Africa: Distribution, biogeography and paleoecology. Pp. xii, 1-163 in Developments in Hydrobiology, vol. 25 (H.J. Dumont, ed.). W. Junk, Dordrecht, Netherlands.

Vinarski, M.V. 2012. The lymnaeid genus *Catascopia* Meier-Brook et Bargues, 2002 (Mollusca: Gastropoda: Lymnaeidae), its synonymy and species composition. Invertebrate Zoology (Moscow) 9: 91-104.

Vinarski, M.V. 2013. One, two, or several? How many lymnaeid genera are there? Ruthenica 23: 41-58.

Vinarski, M.V., O.V. Aksenova, Yu.V. Bespalaya, I.N. Bolotov, M.Yu. Gofarov, and A.V. Kondakov. 2016. *Ladislavella tumrokensis:* The first molecular evidence of a Nearctic clade of lymnaeid snails inhabiting Eurasia. Systematics and Biodiversity 14: 276-287.

Vinarski, M.V., and Yu. I. Kantor. 2016. Analytical Catalogue of Fresh and Brackish Water Molluscs of Russia and Adjacent Countries. KMK Scientific Press, Moscow.

Walter, H.J. 1968. Evolution, taxonomic revolution, and zoogeography of the Lymnaeidae. Bulletin of the American Malacological Union 34: 18-20.

Zilch, A. 1959-1960. Gastropoda (Euthyneura). Pp. 91-102 In: Handbuch der Paläozoologie, vol. 6 (O.H. Schindewolf, ed.). Gebrüder Bornträger, Berlin.

30 Acroloxidae Thiele, 1931

BJÖRN STELBRINK, ALENA A. SHIROKAYA, AND
CHRISTIAN ALBRECHT

Acroloxidae are a Holarctic group of freshwater pulmonate snails with a strongly disjunct distribution across the Northern Hemisphere. The family consists of five genera: *Acroloxus* Beck, 1838 (with *Dinarancylus* Starobogatov, 1991, and *Costovelletia* Starobogatov, 1991, sensu Kruglov & Starobogatov, 1991 synonymized here), and the Lake Baikal endemic genera *Baicalancylus* Starobogatov, 1967, *Frolikhiancylus* Sitnikova & Starobogatov, 1993 (originally considered as a subgenus of *Pseudancylastrum*), *Gerstfeldtiancylus* Starobogatov, 1989, and *Pseudancylastrum* Lindholm, 1909. According to Strong et al. (2008) and Kantor et al. (2010), ca. 50 species are known to science, including two new species recently described from the Anatolian Lake Eğirdir (*A. egirdirensis;* Shirokaya et al., 2012) and northern Iran (*A. pseudolacustris;* Glöer & Pešić, 2012). Acroloxidae mostly, but not exclusively, inhabit stagnant or slowly flowing water bodies of various nature, ranging from lowland swamps to glacial lakes. However, several species described from the Russian Far East also occur in river systems (Prozorova, 1996; Kantor et al., 2010).

Acroloxid species live mostly on hard substrates or macrophytes and are found down to incredible depths of 912 m in Lake Baikal. In fact, that is the deepest known record of any freshwater pulmonate gastropods (Sitnikova, 2006; Sitnikova and Shirokaya, 2013).

Acroloxidae are considered an ancient gastropod family that may have originated in the Cretaceous (Starobogatov, 1970; Taylor, 1988). However, the few distinguishable and unique shell characters in both extant and fossil species often make the species assignment difficult. Several fossil species have been assigned to the potentially older and morphologically different genus *Pseudancylastrum* (Clarke, 1970), while fossils attributed to the genus *Acroloxus* found in central Europe and Anatolia are mainly from the Pliocene (Starobogatov, 1970; Schütt and Kavuşan, 1984).

Acroloxidae build a well-supported monophyletic freshwater limpet family, which is characterized by quite a few morphological features, such as dextral anatomy (Hubendick, 1972), reduced lung, a semidiaulic genital system, and the lack of an anterior

Distribution of Acroloxidae.

gizzard (Nordsieck, 1992). Members of that family also have an apomorphic general appearance of their radula, which is typically heavy and narrow. Absence of a pallial cavity, presence of a pseudobranch, a protoconch with pores, and absence of columns of palisade type in the teleoconch inner structure are additional autapomorphies of the Acroloxidae. Their phylogenetic distinctness within the limnic pulmonates has recently been clarified by molecular studies (Albrecht et al., 2004; Walther et al., 2006). This is not the sister family to Latiidae from New Zealand as previously proposed (e.g., Hubendick, 1978). Instead, its members form a highly supported group within the Hygrophila, with closer affinities to Planorbidae, Physidae, and Lymnaeidae rather than to the Chilinoidea (Latiidae and Chilinidae) (Klussmann-Kolb et al., 2008; Dayrat et al., 2011).

The most speciose genus (*Acroloxus:* 26 species) also includes the most widespread species and is found in both parts of the Holarctic. In the Nearctic, the only known species is *Acroloxus coloradensis,* which is recorded from several glacial lakes in eastern Ontario and southern Quebec and the Rocky Mountains (Canada: Alberta, British Columbia; United States: Colorado and Montana) (Clarke, 1973; Anderson, 2006; Hossack and Newell, 2013). However, hardly anything is known about population genetics and phylogeographic patterns (but see Ellis et al., 2005), which may explain the potential relict distribution of this widespread but rare species.

In the Euro-Mediterranean subregion (sensu Bănărescu, 1992), the most widespread species is *Acroloxus lacustris,* which occurs from the Iberian peninsula to the Caspian Sea, where its distribution may overlap with other described species (Hubendick, 1978; Bănărescu, 1990; Kruglov and Staroboga-tov, 1991; Kerney, 1999; Kantor et al., 2010; Welter-Schultes, 2012). Further species have been described from Europe (e.g., *A. shadini, A. oblongus*); however, their validity is questionable (see, e.g., Vinarski and Kramarenko, 2015), and only 3 further species are listed for this area in Fauna Europaea (www.fauna-eu.org) and other literature: *A. improvisus, A. macedonicus,* and *A. tetensi.* These 3 species are true endemics, with the first 2 species inhabiting the Balkan ancient Lake Ohrid (see Stelbrink et al., 2016 for phylogeographic patterns within the lake), while the third

Acroloxus lacustris (Linnaeus, 1758). Belgium: Canal Bruxelles (Charleroi Basin). Seneffe, Hainaut Province. 4 mm. Poppe 805728.

species has been described from caves in Slovenia (Kuščer, 1932; Hubendick, 1960). The remaining 18 *Acroloxus* species are scattered across continental Russia, with a comparatively high diversity found in the Amur region of the Russian Far East (Starobogatov et al., 2004; Kantor et al., 2010; Shirokaya et al., 2011). Given the patchy distribution, plus the number of recognized species, a revision of that genus across the Holarctic using molecular data seems to be necessary. However, a first attempt to disentangle interspecific relationships within the Euro-Mediterranean subregion has been recently made using a molecular phylogenetic framework (Stelbrink et al., 2016).

The oldest and largest ancient lake in the world, Lake Baikal, represents a hotspot of molluscan biodiversity. This lake hosts a total of 148 gastropod species, 78% of which are endemic to the lake (Sitnikova, 2006), and it also houses four genera of acroloxids comprising a total of 25 endemic species (Shirokaya et al., 2011). Thus, the endemic Baikalian fauna accounts for about 50% of the world's acroloxid diversity. A recent time-calibrated molecular phylogenetic study using two mitochondrial (16S and COI) and two nuclear (28S and H3) markers revealed that the four genera (*Baicalancylus, Frolikhiancylus, Gerstfeldtiancylus,* and *Pseudancylastrum*) are reciprocally monophyletic and that the species flock most likely diverged in the Pliocene and thus much later than the lake's assumed origin of c. 30 mya (Stelbrink et al., 2015). Another major finding emerging from this phylogenetic study is the placement of the abyssal species *Frolikhiancylus frolikhae* within the Baikal flock. This genus was originally considered as a subgenus of *Pseudancylastrum* (Sitnikova et al., 1993) but seems to be reciprocally monophyletic and is rather closely related to *Gerstfeldtiancylus* based on molecular markers (Stelbrink et al., 2015). More important, however, is its astonishing ecology, as this species was found in the abyssal of Lake Baikal at depths of 340–430 and 912 m in oil-seeps and hydrothermal vents (Sitnikova et al., 1993; Sitnikova and Shirokaya, 2013).

There is disparity in the degree of conservation assessments regarding Acroloxidae. Whereas the Lake Baikal endemics are less completely assessed and considered to be either Data Deficient or Least Concern, with many species not listed at all, most other endemics are found in categories of higher conservation relevance. *A. coloradensis* from North America, the lake endemics *A. improvisus* and *A. egirdirensis,* and the cave species *A. tetensi,* are assessed as Vulnerable, *A. macedonicus* from ancient Lake Ohrid even as Critically Endangered.

LITERATURE CITED

Albrecht, C., Z. Fehér, T. Hauffe, and K. Schreiber. 2010. *Acroloxus improvisus.* The IUCN Red List of Threatened Species 2010: e.T155993A4880775. Accessed 22 October 2018.

Albrecht, C., T. Hauffe, and K. Schreiber. 2010. *Acroloxus macedonicus.* The IUCN Red List of Threatened Species 2010: e.T156051A4901391. Aceessed 22 October 2018.

Albrecht, C., T. Wilke, K. Kuhn, and B. Streit. 2004. Convergent evolution of shell shape in freshwater limpets: The African genus *Burnupia.* Zoological Journal of the Linnean Society 140: 577–586.

Anderson, T. 2006. Rocky Mountain Capshell snail (*Acroloxus coloradensis*): A technical conservation assessment. USDA Forest Service, Rocky Mountain Region.

Bănărescu, P. 1990. Zoogeography of fresh waters. Vol. 1. General Distribution and Dispersal of Freshwater Animals. Aula-Verlag, Wiesbaden.

Bănărescu, P. 1992. Zoogeography of fresh waters. Vol. 2. Distribution and Dispersal of Freshwater Animals in North America and Eurasia. Aula-Verlag, Wiesbaden.

Clarke, A.H. 1970. On *Acroloxus coloradensis* (Henderson) (Gastropoda: Basommatophora) in eastern Canada. Publications in Zoology (National Museums of Canada, National Museum of Natural Sciences) 2: 1-13.

Clarke, A.H. 1973. The freshwater molluscs of the Canadian Interior Basin. Malacologia 13: 1-509.

Dayrat, B., M. Conrad, S. Balayan, T.R. White, C. Albrecht, R. Golding, S.R. Gomes, M.G. Harasewych, and A.M. de Frias Martins. 2011. Phylogenetic relationships and evolution of pulmonate gastropods (Mollusca): New insights from increased taxon sampling. Molecular Phylogenetics and Evolution 59: 425-437.

Ellis, B., L. Marnell, M. Anderson, J. Stanford, A. Albrecht, T. Wilke, and C. Relyea. 2005. Natural treasure at Glacier National Park: A rare and tiny freshwater limpet. Flathead Lake Journal 2005: 1-3.

Glöer, P., and V. Pešić. 2012. The freshwater snails (Gastropoda) of Iran, with descriptions of two new genera and eight new species. ZooKeys 219: 11-61.

Hossack, B.R., and R.L. Newell. 2013. New distribution record for the rare limpet *Acroloxus coloradensis* (Henderson, 1930) (Gastropoda: Acroloxidae) from Montana. Nautilus 127: 40-41.

Hubendick, B. 1960. The Ancylidae of Lake Ochrid and their bearing on intralacustrine speciation. Proceedings of the Zoological Society of London 133: 497-529.

Hubendick, B. 1972. The European fresh-water limpets (Ancylidae and Acroloxidae). Informations de la Sociètè belge de Malacologie 1: 109-126.

Hubendick, B. 1978. Systematics and comparative morphology of the Basommatophora. Pp. 1-47 in Pulmonates, vol. 2A, Systematics, Evolution and Ecology (V. Fretter and J. Peake, eds.). Academic Press, London.

Kantor, Y., M. Vinarski, A. Schileyko, and A. Sysoev. 2010. Catalogue of the continental mollusks of Russia and adjacent territories, v.2.3.1. Available at www.ruthenica.com/documents/Continental_Russian_molluscs_ver2-3-1.pdf.

Kebapçı, U., & Seddon, M.B. 2014. *Acroloxus egirdirensis.* The IUCN Red List of Threatened Species 2014: e.T22483644A42418070. Accessed 22 October 2018.

Kerney, M. 1999. Atlas of the Land and Freshwater Molluscs of Britain and Ireland. Harley Books, Colchester.

Klussmann-Kolb, A., A. Dinapoli, K. Kuhn, B. Streit, and C. Albrecht. 2008. From sea to land and beyond—new insights into the evolution of euthyneuran Gastropoda (Mollusca). BMC Evolutionary Biology 8: 57.

Kruglov, N.D., and Ya.I. Starobogatov. 1991. Generic composition of the family Acroloxidae (Gastropoda, Pulmonata) and the species of the genus *Acroloxus* found in the USSR [in Russian]. Zoologicheskij Zhurnal 70: 66-80.

Kuščer, L. 1932. Höhlen- und Quellenschnecken aus dem

Flußgebiet der Ljubljanica. Archiv für Molluskenkunde 64: 48-61.

Nordsieck, H. 1992. Phylogeny and system of the Pulmonata (Gastropoda). Archiv für Molluskenkunde 121: 31-52.

Ormes, M., and J. Cordeiro. 2017. *Acroloxus coloradensis*. The IUCN Red List of Threatened Species 2017: e.T315A62821344. Accessed 22 October 2018.

Prozorova, L.A. 1996. On the species composition of the Acroloxidae (Gastropoda, Pulmonata) family in the Russian Far East [in Russian]. Zoologicheskij Zhurnal 75: 494-498.

Schütt, H., and G. Kavuşan. 1984. Mollusken der miozänen Süßwasserablagerungen in der Umgebung vom Harmancık bei Kütahya-Bursa in Nordwestanatolien. Archiv für Molluskenkunde 114: 217-229.

Shirokaya, A.A., L.A. Prozorova, T.Y. Sitnikova, D.V. Matafonov, and C. Albrecht. 2011. Fauna and shell morphology of limpets of the genus *Acroloxus* Beck (Gastropoda: Pulmonata: Acroloxidae), living in Lake Baikal (with notes on Transbaikalia limpets). Ruthenica 21: 73-80.

Shirokaya, A., Ü. Kebapçi, T. Hauffe, and C. Albrecht. 2012. Unrecognized biodiversity in an old lake: A new species of *Acroloxus* Beck, 1837 (Pulmonata, Hygrophila, Acroloxidae) from Lake Eğirdir, Turkey. Zoosystematics and Evolution 88: 159-170.

Sitnikova, T.Y. 2006. Endemic gastropod distribution in Baikal. Hydrobiologia 568: 207-211.

Sitnikova, T.Y., and A.A. Shirokaya. 2013. New data in deepwater Baikal limpets found in hydrothermal vents and oil-seeps. Archiv für Molluskenkunde 142: 257-278.

Sitnikova, T.Y., V.A. Fialkov, and Ya.I. Starobogatov. 1993. Gastropoda from underwater hydrothermal vent of Baikal Lake. Ruthenica 3: 133-136.

Slapnik, R. 2011. *Acroloxus tetensi*. The IUCN Red List of Threatened Species 2011: e.T155811A4846657. Accessed 22 October 2018.

Starobogatov, Ya.I. 1970. Molluscan Fauna and Zoogeographic Zonation of Continental Freshwater Bodies of the World [in Russian]. Nauka, Leningrad.

Starobogatov, Ya.I., L.A. Prozorova, V.V. Bogatov, and E.M. Sayenko. 2004. Molluscs. Pp. 11-251 in Key to the Freshwater Invertebrates of Russia and Adjacent Countries [in Russian] (S.J. Tsalolikhin, ed.). Nauka, St. Petersburg.

Stelbrink, B., A.A. Shirokaya, C. Clewing, T.Y. Sitnikova, L.A. Prozorova, and C. Albrecht. 2015. Conquest of the deep, old and cold: An exceptional limpet radiation in Lake Baikal. Biology Letters 11: 20150321.

Stelbrink, B., A.A. Shirokaya, K. Föller, T. Wilke, and C. Albrecht. 2016. Origin and diversification of Lake Ohrid's endemic acroloxid limpets: The role of geography and ecology. BMC Evolutionary Biology 16: 273.

Strong, E.E., O. Gargominy, W.F. Ponder, and P. Bouchet. 2008. Global diversity of gastropods (Gastropoda: Mollusca) in freshwater. Hydrobiologia 595: 149-166.

Taylor, D.W. 1988. Aspects of freshwater mollusc ecological biogeography. Palaeogeography, Palaeoclimatology, Palaeoecology 62: 511-576.

Vinarski, M.V., and S.S. Kramarenko. 2015. How does [*sic*] the discrepancies among taxonomists affect macroecological patterns? A case study of freshwater snails of Western Siberia. Biodiversity and Conservation 24: 2097-2091.

Walther, A.C., T. Lee, J.B. Burch, and D. Ó Foighil. 2006. *Acroloxus lacustris* is not an ancylid: A case of misidentification involving the cryptic invader *Ferrissia fragilis* (Mollusca: Pulmonata: Hygrophila). Molecular Phylogenetics and Evolution 39: 271-275.

Welter-Schultes, F.W. 2012. European Non-Marine Molluscs: A Guide for Species Identification. Planet Poster Editions, Göttingen.

31 Bulinidae P. Fischer & Crosse, 1880

CHRISTIAN ALBRECHT, BJÖRN STELBRINK, AND
CATHARINA CLEWING

Bulinidae comprise small to medium-sized planorboid gastropods, reaching up to 25 mm in height or diameter. They are sinistral and either high-spired (e.g., *Bulinus* Müller, 1781) or discoid (e.g., *Indoplanorbis* Annandale & Prashad, 1921) and possess a large pseudobranch that is deeply folded and vascularized. Buliniforme pulmonate gastropods have traditionally been considered to be the second major branch of planorboid gastropods (e.g., Hubendick, 1955; Burch, 1965; Starobogatov, 1967). Hubendick (1978), in his account, even listed a total of 16 genera in his "Bulininae," including limpet-shaped (ancylid), planispiral, and buliniform taxa. Recent molecular phylogenetic analyses, however, have suggested a very different scenario for planorboid gastropods, in which the bulinine forms would be reduced to be represented by *Bulinus* and *Indoplanorbis* only (Albrecht et al., 2007; Jørgensen et al., 2011). These phylogenetic suggestions were followed in the most recent classification of the worldwide gastropods (Bouchet et al., 2017), in which the family Bulinidae, Fischer & Crosse, 1880, is proposed, comprising the subfamilies Bulininae

Fischer & Crosse, 1880, and Plesiophysinae Bequaert & Clench, 1939. The latter group is only tentatively included because the affinities of *Plesiophysa* Fisher, 1883, the sole genus in that subfamily, have never been tested in a phylogenetic framework. Meier-Brook (1984) suggested the center of origin for his Bulininae (i.e., *Bulinus* and *Indoplanorbis*) to be in the Madagascar–eastern Africa–India part of Gondwanaland during Triassic-Jurassic periods. Then, after the collision of India with Asia, adjacent areas were colonized. Meier-Brook argued further that the Neotropics or even Australia would have been entered via the eastern Antarctica connection. A second scenario would have been the migration via Europe and southwestern Asia or, alternatively, a later migration from Africa to India after its collision (Meier-Brook, 1984). The fossil record is not too helpful in this context: the oldest *Bulinus* fossils belong to the *B. africanus* species group from the early Miocene (ca. 20 mya) deposits in Sperrgebiet, Namibia (Pickford, 2008). However, Liu et al. (2010) pointed out that further sampling is needed to better understand the origin of the group.

Distribution of Bulinidae.

Gastropods of the genus *Bulinus* O.F. Müller, 1781, are well known as intermediate hosts for parasites causing one of the most debilitating—but nowadays neglected—tropical diseases called schistosomiasis (bilharzia). This group of pulmonate snails is ecologically flexible and occurs throughout Africa and some adjacent regions, including the circum-Mediterranean countries, Arabia, and southwestern Asia (Madsen, 2017). *Bulinus* occurs throughout the African continent, except for the arid desert region, and on most tropical continental islands (Brown, 1994) and Madagascar (Stothard et al., 2001), La Reunion, Mauritius (Griffith and Florens, 2006), and Sao Tome (Brown, 1981). Whereas most of the species inhabit sub-Saharan Africa, a single species, *Bulinus truncatus* (Audouin, 1827), is known from parts of Spain, France, the Maghreb states, and Mediterranean islands such as Crete, Corse, Sardinia, and Sicily. It is *B. truncatus* that is known to have patchy occurrences in the Near East at the Iranian Caspian Sea, in Iraq, Iran, and Syria and Arabia, where it has been recorded from western regions in scattered localities (Brown, 1994, and references therein).

The taxonomic history of *Bulinus* is long; in fact, it goes back to Adanson (1757), who described "Le Bulin"; later, O. F. Müller (1781) formally named the genus after *Bulinus senegalensis.* More than 100 species had been described by 1915, almost entirely based on shells, when the life cycle of the parasite *Schistosoma haematobium* became known (Brown, 1994). Today, the classification still largely relies on the early accounts of Mandahl-Barth (1957, 1965), and the system is based on both shell and anatomical characters; however, the definition of the majority of the more than 30 species currently recognized is still unsatisfactory (Brown, 1994). A variety of taxonomic characters have been employed in *Bulinus,* ranging from (shell) morphology to genital anatomy and radulae, chromosome numbers, and data from electrophoresis and immunodiffusion (summarized in Brown, 1994). Conchological characters are of restricted value in a planorboid snail genus such as *Bulinus* that is characterized by a rather uniform shell shape largely lacking specific characters such as keels. For matters of practicality, four species groups of *Bulinus* have been defined (e.g., Brown, 1981): the *B. africanus* group, *B. forskalii* group, *B. reticulatus* group, and the *B. trunca-*

Bulinus tropicus (Krauss, 1848). Democratic Republic of Congo: Sange, Katanga Province. 17 mm. Poppe 826007.

tus/tropicus complex. These species groups have been basically confirmed by phylogenetic studies based on mitochondrial and nuclear markers (e.g., Kane et al., 2008; Jørgensen et al., 2011) that all, unfortunately, suffer from unresolved or poorly supported relationships within and between the proposed species groups. However, these studies have also revealed a great extent of unrecognized and cryptic diversity in *Bulinus* spp. A severe underestimation of genetic diversity in *Bulinus* has been proposed based on RAPD (Stothard and Rollinson, 1996), microsatellites (e.g., Emery et al., 2003), mitochondrial (Kane et al., 2008), and nuclear DNA data sets (Stothard et al., 1996, 2001; Jørgensen et al., 2011, 2013; Zein-Eddine et al., 2014). In unison, these studies concluded that much more genetic study is needed to identify cryptic species (complexes) and to study the role of hybridization in *Bulinus.*

Indoplanorbis exustus is an intermediate host for the *Schistosoma indicum* species group, and the role of this snail in the transmission of several other medically and veterinary important parasites has been emphasized repeatedly (Liu et al., 2010; Devkota et al., 2015). The species is rather ecologically flexible and thrives in unspecific freshwater habitats that are not flowing, but it requires warm climates (Raut et al., 1992). In Africa, it was found in a rice estate in Nigeria and at margins of artificial lakes in Ivory Coast (Brown, 1994). Originally described from India, *Indoplanorbis exustus* (Deshayes, 1834) is widely distributed in southern Asia, ranging from Iran to China eastward, including India, and from the southeastern Himalayas to Southeast Asia southward. Its range extends from Muscat in southeastern Arabia (Wright and Brown, 1980), Socotra Island (Wright, 1971), Oman to Iran (Liu et al., 2010) in the west, and throughout tropical Southeast Asia, including islands such as Sri Lanka, Borneo, and the Philippines (e.g., Devkota, et al. 2015). It is also known from Java (e.g., Brown, 1994; Liu et al., 2010; Devkota et al., 2015), Sulawesi, Sumatra, and even Japan (Brandt, 1974; Kihira et al., 2003). A recent range expansion throughout the Indo-Australian Archipelago is suggested by a new phylogeographic study (Gauffre-Autelin et al., 2017). Records from western Africa—Nigeria near Lagos (Kristensen and Ogunnowo, 1987), Cameroon (Campell et al., 2017), and Ivory Coast (Mouchet et al., 1987)—likely represent recent introductions. However, *Indoplanorbis exustus* is easily confused, based on shell characteristics, with planorbids such as *Biomphalaria* and *Helisoma/Planorbella,* and thus the real distribution of this species in some regions might be masked. *Indoplanorbis exustus* seems to be prone to human-mediated dispersal, as it is has been found in remote places such as the Hawaiian archipelago, French Polynesia, and New Guinea (Cowie, 2001). It has the potential to become a globally invasive species, as records from the Lesser Antilles document (Pointier et al., 2005).

This species has also been genetically investigated in a phylogeographic framework. Based on specimens from 10 countries and two mitochondrial markers, Liu et al. (2010) proposed that the ancestor of this group might have originated in the northern Indian Assam region and subsequently colonized the subcontinent. Colonization and diversification was assumed to be triggered by the Himalayan uplift and the onset of monsoon intensification. The study of Liu et al. (2010) further revealed a high genetic diversity within this species, suggesting that the four (Devkota et al., 2015) to five (Gauffre-Autelin et al., 2017) distinct mitochondrial lineages found might represent a cryptic species complex.

A total of six nominal species of *Plesiophysa* Fischer,

1883, have been described for this Neotropical genus that are conchologically very similar. *Plesiophysa dolichomastix* Paraense, 2002, and *Plesiophysa guadeloupensis* (Fischer, 1883) are known from different regions in southeastern Brazil (Haas, 1838; Paraense 2002a, b; Simone, 2006). Other species and records of *P. guadeloupensis* and the remaining species are known only from a few Caribbean islands, such as Guadeloupe, St. Martin, Puerto Rico, and Barbados (Paraense, 2002a, b, 2008). A phylogenetic study of *Plesiophysa* would be very helpful for enlightening the affinities to the Bulininae and origin of this group.

The conservation status of the *Plesiophysa* species is unknown, whereas *Indoplanorbis exustus* was assessed least concern (LC, Budha et al., 2012). Of the 38 species of *Bulinus* in the Red List database of the IUCN, 23 were Least Concern and 8 were Data Deficient. Two endemic *Bulinus* species are assessed as Endangered; 3 further species had been listed as Near Threatened and 2 others as Vulnerable.

LITERATURE CITED

Adanson, M. 1757. Histoire naturelle du Sénégal: Coquillages; avec la relation abrégée d'un voyage fait en ce pays, pendant les années 1749-53. Chez Claude-Jean-Baptiste Bauche, Paris.

Albrecht, C., K. Kuhn, and B. Streit. 2007. A molecular phylogeny of Planorboidea (Gastropoda, Pulmonata): Insights from enhanced taxon sampling. Zoologica Scripta 36: 27-39.

Bouchet, P., J.P. Rocroi, B. Hausdorf, A. Kaim, Y. Kano, A. Nützel, P. Parkhaev, M. Schrödl, and E.E. Strong. 2017. Revised classification, nomenclator and typification of gastropod and monoplacophoran families. Malacologia 61: 1-526.

Brandt, R.A.M. 1974. The non-marine aquatic Mollusca of Thailand. Archiv für Molluskenkunde 105: i-iv, 1-423.

Brown, D.S. 1981. Generic nomenclature of freshwater snails commonly classified in the genus *Bulinus* (Mollusca: Basommatophora). Journal of Natural History 15: 909-915.

Brown, D. 1994. Freshwater Snails of Africa and Their Medical Importance. 2nd ed. Taylor & Francis, London.

Budha, P.B., J. Dutta, and B.A. Daniel. 2012. *Indoplanorbis exustus*. The IUCN Red List of Threatened Species 2012: e.T165594A17211568. http://dx.doi.org/10.2305/IUCN.UK.2012.RLTS.T165594A17211568.en. Accessed 30 September 2016.

Burch, J.B. 1965. Chromosome number and systematics in

euthyneuran snails. Proceedings of the First European Malacological Congress 1962: 215-241.

Campbell, S.J., J.R. Stothard, F. O'Halloran, D. Sankey, T. Durant, D.E. Ombede, G.D. Chuinteu, B.L. Webster, L. Cunningham, E.J. LaCourse, and L.S. Tchuem-Tchuenté. 2017. Urogenital schistosomiasis and soil-transmitted helminthiasis (STH) in Cameroon: an epidemiological update at Barombi Mbo and Barombi Kotto crater lakes assessing prospects for intensified control interventions. Infectious Diseases of Poverty 6: 49.

Cowie, R.H. 2001. Invertebrate invasions on Pacific Islands and the replacement of unique native faunas: A synthesis of the land and freshwater snails. Biological Invasions 3: 119-136.

Devkota, R., S.V. Brant, and E.S. Loker. 2015. The *Schistosoma indicum* species group in Nepal: Presence of a new lineage of schistosome and use of the *Indoplanorbis exustus* species complex of snail hosts. International Journal for Parasitology 45: 857-870.

Emery, A.M., N.J. Loxton, J.R. Stothard, C.S. Jones, J. Spinks, J. Llewellyn-Hughes, L.R. Noble, and D. Rollinson. 2003. Microsatellites in the freshwater snail *Bulinus globosus* (Gastropoda: Planorbidae) from Zanzibar. Molecular Ecology Notes 3: 108-110.

Gauffre-Autelin, P., T. von Rintelen, B. Stelbrink, and C. Albrecht. 2017. Recent range expansion of an intermediate host for animal schistosome parasites in the Indo-Australian Archipelago: Phylogeography of the freshwater gastropod *Indoplanorbis exustus* in South and Southeast Asia. Parasites & Vectors 10: 126.

Haas, F. 1938. Neue Binnen-Mollusken aus Nordost-Brasilien. Archiv für Molluskenkunde 70: 46-51.

Hubendick, B. 1955. Phylogeny in the Planorbidae. Transactions of the Zoological Society of London 28: 453-542.

Hubendick, B. 1978. Systematics and comparative morphology of the Basommatophora. Pp. 1-47 in Pulmonates, vol. 2A, Systematics, Evolution and Ecology (V. Fretter and J. Peake, eds.). Academic Press, London.

Jørgensen, A., H. Madsen, A. Nalugwa, S. Nyakaana, D. Rollinson, J.R. Stothard, and T.K. Cristensen. 2011. A molecular phylogenetic analysis of *Bulinus* (Gastropoda: Planorbidae) with conserved nuclear genes. Zoologica Scripta 40: 126-136.

Jørgensen, A., J.R. Stothard, H. Madsen, A. Nalugwa, S. Nyakaana, and D. Rollinson. 2013. The ITS2 of the genus Bulinus: Novel secondary structure among freshwater snails and potential new taxonomic markers. Acta Tropica 128: 218-225.

Kane, R.A., J.R. Stothard, A.M. Emery, and D. Rollinson. 2008. Molecular characterization of freshwater snails in

the genus *Bulinus:* A role for barcodes? Parasites & Vectors 1: 15.

Kihira, H., M. Matsuda, and R. Uchiyama. 2003. Illustrated Guide to the Freshwater Molluscs of Japan. 1. Freshwater Molluscs from Lake Biwa and the Yodo River. Pisces, Tokyo. 1-159.

Kristensen, T.K., and O. Ogunnowo. 1987. Indoplanorbis exustus (Deshayes, 1834), a freshwater snail new for Africa, found in Nigeria (Pulmonata: Planorbidae). Journal of Molluscan Studies 53: 245-246.

Liu, L., M.M. Mondal, M.A. Idris, H.S. Lokman, P.J. Rajapakse, F. Satrija, J.L. Diaz, E.S. Upatham, and S.W. Attwood. 2010. The phylogeography of *Indoplanorbis exustus* (Gastropoda: planorbidae) in Asia. Parasites & Vectors 3: 57.

Madsen, H. 2017. *Schistosoma* intermediate host snails. Pp. 38-55 in *Schistosoma:* Biology, Pathology and Control (B.G.M. Jamieson, ed.). CRC Press, Boca Raton, Florida.

Mandahl-Barth, G. 1957. Intermediate hosts of *Schistosoma:* African *Biomphalaria* and *Bulinus:* 2. Bulletin of the World Health Organization 17: 1-65.

Mandahl-Barth, G. 1965. The species of the genus *Bulinus,* intermediate hosts of *Schistosoma.* Bulletin of the World Health Organization 33: 33-44.

Meier-Brook, C. 1984. A preliminary biogeography of freshwater pulmonate gastropods. Pp. 23-37 in World-Wide Snails: Biogeographical Studies on Non-Marine Mollusca (A. Solem and A.C. van Bruggen, eds.). Brill, Leiden.

Mouchet, F., J.L. Rey, and P Cunin. 1987. Découverte d'Indoplanorbis exustus (Planorbidae; Bulininae) à Yamoussokro, Côte d'Ivoire. Bulletin de la Société de Pathologie Exotique 80: 811-812.

Paraense, W.L. 2002a. *Plesiophysa dolichomastix* sp. n. (Gastropoda: Planorbidae). Memórias do Instituto Oswaldo Cruz 97: 505-508.

Paraense, W.L. 2002b. The genus *Plesiophysa,* with a redescription of *P. ornata* (Haas, 1938) (Gastropoda: Planorbidae). Brazilian Journal of Biology 62: 333-338.

Pickford, M. 2008. Freshwater and terrestrial Mollusca from Early Miocene deposits of the northern Sperrgebiet, Namibia. Memoir of the Geological Survey of Namibia 20: 65-74.

Pointier, J.-P. 2008. Guide to the Freshwater Molluscs of the Lesser Antilles. Conchbooks, Hackenheim, Germany.

Pointier, J.P., P. David, and P. Jarne. 2005. Biological invasions: The case of planorbid snails. Journal of Helminthology 79: 249-256.

Raut, S.K., M.S. Rahman, and S.K. Samanta. 1992. Influence of temperature on survival, growth and fecundity of the freshwater snail *Indoplanorbis exustus* (Deshayes). Memorias do Instituto Oswaldo Cruz 87: 15-19.

Simone, L.R.L. 2006. Land and freshwater molluscs of Brazil. EGB, Fapesp, São Paulo.

Starobogatov, Ya.I. 1967. On the Systematization of Freshwater Pulmonate Molluscs [in Russian]. Trudy Zoologicheskogo Instituta AN SSSR 42: 280-304.

Stothard, J.R., and D. Rollinson. 1996. An evaluation of random amplified polymorphic DNA (RAPD) for the identification and phylogeny of freshwater snails of the genus Bulinus (Gastropoda: Planorbidae). Journal of Molluscan Studies 62: 165-176.

Stothard, J.R., S. Hughes, and D. Rollinson. 1996. Variation within the Internal Transcribed Spacer (ITS) of ribosomal DNA genes of intermediate snail hosts within the genus Bulinus (Gastropoda: Planorbidae). Acta Tropica 61: 19-29.

Stothard, J.R., P. Brémond, L. Andriamaro, B. Sellin, E. Sellin, and D. Rollinson. 2001. Bulinus species on Madagascar: Molecular evolution, genetic markers and compatibility with Schistosoma haematobium. Parasitology 123: 261-275.

Wright, C.A. 1971. Bulinus on Aldabra and the subfamily Bulininae in the Indian Occan area. Philosophical Transactions of the Royal Society B 260: 299-313.

Wright, C.A., and D.S. Brown. 1980. The freshwater Mollusca of Dhofar. Journal of Oman Studies, Special Report No. 2:97-102.

Zein-Eddine, R., F.F. Djuikwo-Teukeng, M. Al-Jawhari, B. Senghor, T. Huyse, and G. Dreyfuss. 2014. Phylogeny of seven Bulinus species originating from endemic areas in three African countries, in relation to the human blood fluke Schistosoma haematobium. BMC Evolutionary Biology 14: 271.

32 Burnupiidae Albrecht, 2017

CHRISTIAN ALBRECHT AND CATHARINA CLEWING

The monogeneric Burnupiidae are a limpetlike group of freshwater pulmonate snails predominantly occurring in Africa. The genus *Burnupia* Walker, 1912, has traditionally been seen as member of either a freshwater limpet family Ancylidae (Boss, 1982) or as a member of the Planorbidae (Bouchet and Rocroi, 2005). A phylogenetic analysis of Planorboidea that included two *Burnupia* species from South Africa resulted in a striking position of these taxa being very distinct from all other limpetlike pulmonate gastropods (Albrecht et al., 2004). This phylogenetic distinctness was found again in a larger phylogenetic analysis (Albrecht et al., 2007) and an examination of the available information on shell morphology (Walker, 1912), anatomy (Hubendick, 1964; Hodgson and Healy, 1998), cytogenetics (Burch, 1965), and a molecular apomorphy (Albrecht et al., 2004) eventually led to the formal description of the family Burnupiidae Albrecht, 2017, in Bouchet et al. (2017).

The majority of species of *Burnupia* occur in sub-Saharan Africa, particularly in eastern and southern parts, from the isolated Ethiopian highlands down to the Cape region. The arid southwest and large coastal stretches are free of *Burnupia* records; however, isolated populations in Namibia and subfossil records in Botswana exist (Walker, 1923; Pilsbry and Bequaert, 1927; Mandahl-Barth, 1968). The current African distribution has been summarized in Brown (1994).

A single small-sized species was more recently described from South America, specifically Brazil (*Burnupia ingae* Lanzer, 1991). Its presence is now confirmed from all Brazilian states (Lacerda and Santos, 2011), but records are still scarce (Ovando et al., 2016). The phylogenetic affinity of the Brazilian species remains untested. An African–South American distribution pattern is biogeographically interesting as outline for the case of the planorbid genus *Biomphalaria,* for which various hypotheses had been put forward, ranging from a Gondwanan origin hypothesis to recent human-mediated transatlantic dispersal out of Africa (e.g., Morgan et al., 2001). A phylogenetic study of *Burnupia* would also provide interesting insights into the origin, age, and relationship of the New World species and the African species.

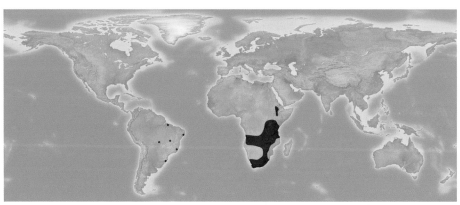

Distribution of Burnupiidae.

Burnupia capensis (Walker, 1912). South Africa: Umnlatuzani River, Malvern, Natal. ~ 5mm. UMMZ 102479.

The number of species existing is unknown, and probably too many names/synonyms exist. Phylogenetic studies of the freshwater limpet *Ancylus* in Europe has shown that both taxonomic splitting and lumping have masked the actual diversity in that group (e.g., Albrecht et al., 2006). The situation in *Burnupia* might be quite similar. Brown (1994) followed earlier taxonomic accounts and grouped species according to their geographic distribution (in Africa). He listed 14 species for southern Africa, including the type species *B. caffra* (*Ancylus caffer* Krauss, 1848). Three species are named for the southeastern Congo region and finally eastern African species, numbering a total of four. The validity of all these nominal taxa has to be tested in a continentwide phylogeographic framework.

Burnupiidae are primarily occurring in perennial streams and rivers with gravel or stony grounds in lowland and high-mountain ranges. A few species are also known from lake edges, such as in the lakes Edward, Victoria, Kivu, and Tanganyika. All species seem to have a high oxygen demand, live on hard substrata, and not be drought resistant. Therefore, *Burnupia* species were thought to be of potential bioindication value. Consequently, a few *Burnupia* species have been used in ecotoxicological monitoring studies (e.g., Gerhardt and Palmer, 1998; Davies-Coleman and Palmer, 2004).

Of the 21 species assessed for the IUCN red list by 2017, 18 were Data Deficient and *Burnupia caffra* as Least Concern. The range-restricted eastern African *B. stuhlmanni* Martens, 1897, and *B. crassistriata* (Preston, 1911) have been assessed as Near Threatened (Lange, 2010a, b; Van Damme and Lange, 2016). A general lack of distributional data in concert with the unresolved taxonomic situation is responsible for a pronounced lack of knowledge of the conservation status, even of widespread taxa (e.g., de Kock and Wolmarans, 2009, 2017).

LITERATURE CITED

Albrecht, C., T. Wilke, K. Kuhn, and B. Streit. 2004. Convergent evolution of shell shape in freshwater limpets: The African genus *Burnupia*. Zoological Journal of the Linnean Society 140: 577–586.

Albrecht, C., S. Trajanovski, K. Kuhn, B. Streit, and T. Wilke. 2006. Rapid evolution of an ancient lake species flock: Freshwater limpets (Gastropoda: Ancylidae) in the Balkan Lake Ohrid. Organisms, Diversity & Evolution 6: 294–307.

Albrecht, C., K. Kuhn, and B. Streit. 2007. A molecular phylogeny of Planorboidea (Gastropoda: Pulmonata): Insights from enhanced taxon sampling. Zoologia Scripta 36: 27–39.

Boss, K.J. 1982. Mollusca. Pp. 945–1166 in Synopsis and Classification of Living Organisms, vol. 1 (Parker, S.P., ed.). McGraw Hill, New York.

Bouchet, P., and J.-P. Rocroi. 2005. Classification and nomenclator of gastropod families. Malacologia 47: 1–397.

Bouchet, P., J.P. Rocroi, B. Hausdorf, A. Kaim, Y. Kano, A. Nützel, P. Parkhaev, M. Schrödl, and E.E. Strong. 2017. Revised classification, nomenclator and typification of gastropod and monoplacophoran families. Malacologia 61: 1–526.

Brown, D.S. 1994. Freshwater Snails of Africa and Their Medical Importance. 2nd ed. Taylor & Francis, London.

Burch, J.B. 1965. Chromosome numbers and systematics in euthyneuran snails. Proceedings of the First European Malacological Congress 1962: 215–241.

Davies-Coleman, H.D., and C.G. Palmer. 2004. The use of a freshwater mollusc, *Burnupia stenochorias* (Ancylidae) as an ecotoxicological indicator in whole effluent toxicity testing. Pp. 309–315 in Proceedings of the 2004 Water Institute of Southern Africa Biennial Conference. Cape Town, South Africa.

De Kock, K.N., and C.T. Wolmarans. 2009. Distribution of *Burnupia capensis* (Walker, 1912) and *Burnupia stenochorias* (Melvill & Ponsonby, 1903) (Gastropoda: Ancylidae) in South Africa. Suid-Afrikaanse Tydskrif vir Natuurwetenskap en Tegnologie, 28: 220–236.

De Kock, K.N., and C.T. Wolmarans. 2017. Distribution and habitats of *Burnupia trapezoidea* (Boettger, 1910) (Gastropoda: Ancylidae) in South Africa. Water SA, 43: 258–263.

Gerhardt, A., and C. Palmer. 1998. Copper tolerances of *Adenophlebia auriculata* (Eaton) 1884 (Insecta: Ephemeroptera) and *Burnupia stenochorias* Cawston 1932 (Gastropoda: Ancylidae) in indoor artificial streams. Science of the Total Environment 215: 217-229.

Hodgson, A.N., and J.M. Healy. 1998. Comparative sperm morphology of the pulmonate limpets *Trimusculus costatus, T. reticulatus* (Trimusculidae) and *Burnupia stenochorias* and *Ancylus fluviatilis* (Ancylidae). Journal of Molluscan Studies 64: 447-460.

Hubendick, B. 1964. Studies on Ancylidae: The Subgroups. Meddelanden från Göteborgs Musei Zoologiska Avdelning 137: 1-71.

Lacerda, L.E.M., and S.B. Santos. 2011. *Burnupia ingae* Lanzer, 1991 (Gastropoda: Ancylidae): Current distribution in Brazil. Check List 7: 862-864.

Lange, C. 2010a. *Burnupia stuhlmanni.* The IUCN Red List of Threatened Species 2010: e.T44269A10885369. http://dx.doi.org/10.2305/IUCN.UK.2010-3.RLTS.T44269A10885369.en. Accessed 27 September 2016.

Lange, C. 2010b. *Burnupia crassistriata.* The IUCN Red List of Threatened Species 2010: e.T44268A10885139. http://dx.doi.org/10.2305/IUCN.UK.2010-3.RLTS.T44268A10885139.en. Accessed 27 September 2016.

Mandahl-Barth, G. 1968. Freshwater molluscs. Exploration Hydrobiologique Bangweulu-Lualupa 12: 1-68.

Morgan, J.A.T., R.J. Dejong, S.D. Snyder, G.M. Mkoji, and E.S. Loker. 2001. *Schistosoma mansoni* and *Biomphalaria:* Past history and future trends. Parasitology 123: 211-228.

Ovando, X.M.C., L.E.M. Lacerda, R. Roza Dutra, and S.B. Santos. 2016. *Burnupia ingae* Lanzer, 1991 (Gastropoda: Hygrophila): Contributions to the knowledge on its morphology and first records in Rio de Janeiro state, Brazil. Malacologia 59: 333-339.

Pilsbry, H.A., and J. Bequaert. 1927. The aquatic mollusks of the Belgian Congo. With a geographical and ecological account of Congo malacology. *Bulletin of the American Museum of Natural History* 53: 69-602.

Van Damme, D., and C. Lange. 2016. *Burnupia caffra.* IUCN Red List of Threatened Species 2016: e.T165802A84309445. http://dx.doi.org/10.2305/IUCN.UK.2016-3.RLTS.T165802A84309445.en.

Walker, B. 1912. A revision of Ancylini of South Africa. Nautilus 25: 139-144.

Walker, B. 1923. The Ancylidae of South Africa. Privately published.

33 Physidae Fitzinger, 1833

AMY R. WETHINGTON AND CHARLES LYDEARD

The freshwater gastropod family Physidae has a Holarctic distribution with extensions into Central and South America, achieving its greatest diversity in North America (Burch, 1982; Taylor, 2003) and Central America (Taylor, 2003). Over the past two centuries the family's distribution has increased worldwide through the introduction of a small number of species, particularly *Physa acuta* Draparnaud 1805 (e.g., see Dillon et al., 2002; Dillon, 2000), which, based on phylogenetic, fossil, distribution, and breeding studies, is thought to be a native of North America, not Europe, where it was described (Lydeard et al., 2016), and subsequently spread across the western Palearctic (Vinarski, 2017). Indeed, it is possible that nominal species of described species in Chile are actually introduced populations of *P. acuta* (Collado, 2017). *Physa acuta* is often referred to as *Physella acuta* (Burch, 1988; Burch and Tottenham, 1980; Turgeon et al., 1998, among many others) and less often as *Haitia acuta* (Taylor, 2003), although the generic-level nomenclature needs to be evaluated further.

Recent work concerning physid systematics has synonymized many previously known physid species (e.g., *Physa integra, P. heterostropha, P. virgata,* and *P. cubensis*) under the species name *acuta* based on morphological, behavioral, and molecular studies (Dillon et al., 2002; Paraense and Pointier, 2003; Taylor, 2003; Dillon and Wethington, 2006; Wethington and Lydeard, 2007; and Dillon et al., 2011). While there seems to be a growing congruence with regard to species names, generic names remain in flux (Te, 1978; Taylor, 2003; Wethington and Lydeard, 2007). For instance, IUCN and Thompson (2011) both recognize the Central and South American genera *Mayabina* Taylor, 2003, and *Mexinauta* Taylor, 2003, listed under Taylor's new tribe, Amecanautini. The tribe Amecanautini is largely based on slight variations in penial and mantle morphology yet to be included in any phylogenetic or genetic studies.

Burch's guides (Burch and Tottenham, 1980; Burch, 1988) for Physidae were based on an unpublished dissertation by George Te (1978). According to Te (1978), the genus *Physella* differed from the genus *Physa* by a variation in a mantle character. *Physa*'s

Distribution of Physidae.

mantle covered the shell on both sides, while *Physella*'s mantle was more fingerlike and mostly one-sided as it projected around the shell. This mantle character is much easier to see when a snail is alive, and upon closer study, this character can vary within a population. The original genus name for both *heterostropha* and *pomilia* is *Physa*. *Physa heterostropha* was named by Thomas Say (1817) from Philadelphia and *Physa pomilia* was named by Conrad (1834) from Randons Creek near Claiborne, Alabama. A later study would show that *P. heterostropha* and *P. pomilia* are each separate species (Wethington and Lydeard, 2007) and that *P. heterostropha* is best known as *Physa acuta*, a snail with an almost worldwide distribution (Dillon et al., 2002; Paraense and Pointier, 2003; Dillon and Wethington, 2006; Wethington and Lydeard, 2007; Dillon et al., 2011).

Taylor (2003) agreed that *P. heterostropha* should be synonymized under the name *acuta,* but he refers to this species as *Haitia acuta. Haitia* Clench & Aguayo, 1932, was described as a subgenus for their new species: *Physa* (*Haitia*) *elegans,* which Taylor (2003) elevated to the rank of genus, despite the lack of available material for the type species. To quote Taylor (2003), "Lack of characters, as well as of material, make this summary of *Haitia* among the least satisfactory of all genera in the family" (p. 129). *Physa pomilia,* a name synonymized under *acuta* by Taylor (2003), seems to be more restricted geographically and is most common along the southeastern seaboard (Wethington and Lydeard, 2007; Wethington et al., 2009).

Despite recent systematic efforts to organize the family (see above), Turgeon et al. (1998) still dominates. For simplicity, Turgeon et al.'s (1998) taxonomy will be used in this chapter. Turgeon et al. (1998) is largely based on the Burch guides to North American freshwater snails (Burch and Tottenham, 1980; Burch, 1988), which, as mentioned earlier, rely heavily on George Te's PhD dissertation (Te, 1978). As such, the family is divided into two main groups based on penial morphology, mantle morphology, and shell characteristics: Aplexinae (which includes two main groups: *Aplexa* Fleming, 1820 [e.g., *Aplexa hypnorum* (Linnaeus 1758) and *Stenophysa* Martens, 1898 (e.g., *Stenophysa marmorata* Guilding 1828)]), and Physinae (which includes two main groups: *Physa* Draparnaud, 1801, having penial type "a" and mantle extending

Physella gyrina (Say, 1821). Canada: Boundary Bay, Centennial Beach, British Columbia. 10.4 mm. Poppe 927952.

over the shell on both sides nearly equally [e.g., *Physa jennessi* Dall 1919] and *Physella* Haldeman, 1842 having fingerlike mantle projections, mostly on one side of the shell [e.g., *Physella gyrina* (Say 1921)]). *Physella* is further divided into three main groups: *Physella* having penial type "b" (e.g., *Physella gyrina*), and *Costatella* Dall, 1870, having penial type "c" (*Alampetista* Zilch, 1956, e.g., *Physella acuta*) or having penial type "bc," which includes an odd paraphyletic assemblage (*Costatella* Dall, 1870, e.g., *Physella zionis* Pilsbry 1926, and *Physella pomilia* (Conrad, 1834)); see Wethington and Lydeard (2007) and Wethington et al. (2009).

How the evolution of reproductive isolating mechanisms corresponds with phylogeny is considered to be an important unresolved question within families formerly placed in the now polyphyletic Basommatophora by Jarne et al. (2010). The only exploration involving the evolution of reproductive isolating mechanisms within freshwater snails is by Dillon and colleagues (2011) for the family Physidae. As suggested by Coyne and Orr (2004), reproductive incompatibility in the form of postmating reproductive isolating mechanisms seems to have evolved first in physids, most likely in allopatry, which was reinforced in sympatry by the development of premating reproductive isolating mechanisms (Dillon et al., 2011). This was inferred by examining the results of no-choice tests as well as reproductive behavior noted in choice tests. There was no premating isolation between *P. acuta* with *P. carolinae, P. pomilia,* or *P. gyrina.* F1 progeny resulted in crosses of *P. acuta* with *P. carolinae* and *P. pomilia,* but not with the more distantly related *P. gyrina.* F1 progeny were of normal viability in the more closely related *P. acuta–P. carolinae* crosses as compared with the more distant related *P. acuta–P. pomilia* crosses, where the F1 progeny had reduced viability. While viable F1 resulted in crosses between *P. acuta* and *P. carolinae,* they were sterile. Reproductive anatomy of penial structures corresponded with phylogeny, with *P. acuta* having a "c" penial morphology, *P. carolinae* and *P. pomilia* having a "bc" penial morphology, and *P. gyrina* having a "b" penial morphology (Wethington and Lydeard, 2007; Wethington et al., 2009).

Profound ecophenotypic plasticity has been found in physids (e.g., DeWitt, 1998; Britton and McMahon, 2004; Wethington and Guralnick, 2004; Auld and Relyea, 2011), as well as in other freshwater pulmonates (for lymnaids: Lam and Calow, 1988; Correa et al., 2010; Brönmark et al., 2011; for ancylids: Basch, 1963; and for planorbids: Hoverman et al., 2005; Hoverman and Relyea, 2007). Wave action can influence shell morphology. The shells of *P. gyrina* found in slow-moving waters such as ditches and small ponds are often thin, while the shells of *P. gyrina* found in large lakes such as the Great Lakes are often globose, some with shoulders (Dillon and Wethington, 2005; Wethington and Lydeard, 2007).

Elevated temperatures can affect shell shape in physids (Britton and McMahon, 2004; Wethington and Guralnick, 2004). Hot-water physids phenotypically resemble each other regardless of species. *Physella acuta, P. spelunca,* and *P. gyrina* (which includes *P. wrighti, P. johnsoni,* and *P. aurea*) living in hot springs are generally small as adults and have globose shells (Wethington and Guralnick, 2004). Britton and McMahon's (2004) study showed that physids grown in warmer water for five generations developed more globose shells. Warmer waters, where surface temperatures may be as high as 37 to 40° C, allowed Texan physids to reproduce more often, as many as three times a year, compared to the once- or twice-a-year reproductive cycles found in higher latitudes (McMahon, 1975).

Physids exhibit a remarkable plasticity with regard to life history shifts (Crowl and Covich, 1990), antipredator behavior (Snyder, 1967; Alexander and Covich, 1991a, b; Covich et al., 1994; Turner, 1996; McCarthy and Fisher, 2000; Bernot and Turner, 2001; Turner et al., 2000), and shell morphology (DeWitt, 1998; DeWitt et al., 2000), but this comes at an assortment of possible costs (DeWitt et al., 1999).

Physids can delay reproduction in order to reach larger sizes more quickly in the face of crayfish predators (Crowl and Covich, 1990). In DeWitt et al.'s (1999) study, smaller physids were much more active than larger physids when exposed to crayfish plus snail extract cue. This is interesting since *Planorbella trivolvis* are immune to crayfish predation by the time they reach 9 mm. Larger planorbids have stronger shells (Alexander and Covich, 1991b). Increased inbreeding depression correlates to increased waiting-to-self time in basommatophorans (Escobar et al., 2011). Crayfish (*Procambarus acutus*) predators caused an increase in inbreeding depression in *Physella acuta* via an increased wait-to-self-fertilize time and lowered survival among selfed eggs (Auld, 2010).

Freshwater gastropods, especially basommatophorans, have thin shells due to low calcium levels in their habitats (Vermeij and Covich, 1978; Covich, 2010). Induced shell response in freshwater pulmonates by predators is known to occur in physids (DeWitt, 1998; DeWitt et al., 2000). Physids reared in crayfish plus crushed snail cue water have shells that tend to be more elongate, hence harder for crayfish to enter, while physids reared in fish plus crushed snail cue water have rounder shells, hard for fish to grab (DeWitt, 1998; DeWitt et al., 2000; Langerhans and DeWitt, 2002). Auld and Relyea (2011) discovered that predator-induced physids grown with caged crayfish fed conspecific snails were less likely to be crushed and/or extracted from their shell by crayfish predators than controls. Predator-induced physids weighed more, had greater shell thickness, longer shells, and larger apertures than uninduced physids. These induced shell responses to predators may come at a cost to the affected snails in terms of growth rate (DeWitt 1998).

Physids are found in still, well aerated, shallow waters in fairly sunny areas where there are few predators or debris (Dawson, 1911). Some members require permanent sources of water, such as *P. acuta,* which cannot handle desiccation well, while others can live in temporary ponds, such as *P. gyrina,* and use their slight burrowing tendency to survive periodic droughts (Dawson, 1911; Clampitt, 1970) or to escape predators (Snyder, 1967).

The ecology of *P. zionis* is quite striking and different from all other physids, as it crawls vertically on the actual rock face where seeping occurs along the narrows in Zion National Park, Utah. These physids obtain maturity at a very small size, less than 5 mm in length, and lay correspondingly small egg masses, between 1 and possibly 4 eggs per capsule (Pilsbry, 1925; Christopher Rogers, pers. obs.) as compared to as many as 200+ eggs in some individual physids (pers. obs.). There are physids (*P. gyrina*) that are nearly sympatric to *P. zionis* but are found in the ponds, swampy areas, and Virgin River which runs through the narrows.

Physella spelunca has a unique ecology, as it lives in a heated spring within a cave filled with toxic sulfuric gas and feeds primarily on bacteria (Turner and Clench, 1974).

The *gyrina* group seems to be divided into two distinct ecological groups: the heat tolerant group (S after Dillon, 2000) and the heat intolerant group. Many members of the heat intolerant group are large (U after Dillon, 2000) as adults compared to the heat tolerant group and have a more globose shell, some with pronounced shoulders. The more globose shelled *gyrinas* can live as long as three years, as compared to the more stress driven heat tolerant *gyrinas,* which live a maximum of only one year (Wethington and Guralnick, 2004; Wethington and Lydeard, 2007). Depending on climate, *P. gyrina* can have one (in high latitudes) to three (in tropical areas) generations per year (Dillon, 2000).

There are 24 physid species that are listed as critically imperiled (G1), imperiled (G2), or vulnerable (G3), and one is presumed extinct in North America (Johnson et al., 2013). However, most physid species are secure or apparently secure, with a few regarded to be invasive in Eurasia and Africa (e.g., *P. acuta*). It should be noted that some taxa included in Johnson et al. (2013), such as *Physella johnsoni* (listed as G1) have been synonymized under more common taxonomic names such as *P. gyrina* (listed as G5).

Physids are not typically known to be vectors for human diseases (Brown, 1994), but there are a few reports where *P. acuta* has been introduced that suggest otherwise. For instance, there is an unconfirmed experimental infection of *S. haematobium* reported in the Middle East (Magzoub and Kasim, 1980), but there is no other evidence that *P. acuta* could be a host. Another medically important parasite carried by *P. acuta* as a second intermediate host is *Echinostoma liei* in Egypt (Christensen et al., 1980).

LITERATURE CITED

Alexander, J.E., Jr., and A.P. Covich. 1991a. Predator avoidance in the freshwater snail *Physella* virgata to the crayfish *Procambarus simulans.* Oecologia 87: 435-442.

Alexander, J.E., Jr., and A.P. Covich. 1991b. Predation risk and avoidance behavior in two freshwater snails. Biological Bulletin 180: 387-393.

Auld, J.R. 2010. The effects of predation risk on mating system expression in a freshwater snail. Evolution 64: 3476-3494.

Auld, J., and R. Relyea. 2011. Adaptive plasticity in predator-induced defenses in a common freshwater snail: Altered selection and mode of predation due to prey phenotype. Evolutionary Ecology 25: 189-202.

Basch, P.F. 1963. Environmentally influenced shell distortion in a fresh-water limpet. Ecology 44: 193-194.

Bernot, R.J., and A.M. Turner. 2001. Predator identity and trait mediated indirect effects in a littoral food web. Oecologia 129: 139-146.

Britton, D.K., and R.F. McMahon. 2004. Environmental and genetically induced shell shape variation in the freshwater pond snail *Physa* (*Physella*) *virgata* (Gould, 1855). American Malacological Bulletin 19: 93-100.

Brönmark, C., T. Lakowitz, and J. Hollander. 2011. Predator-induced morphological plasticity across local populations of a freshwater snail. PLoS ONE 6: e21773.

Brown, D.S. 1994. Freshwater Snails of Africa and Their Medical Importance. 2nd ed. Taylor & Francis, London.

Burch, J.B. 1982. North American freshwater snails: Identification keys, generic synonymy, supplemental notes, glossary, references, index. Walkerana 1: 217-365.

Burch, J.B. 1988. North American freshwater snails: Introduction, systematics, nomenclature, identification, morphology, habitats, distribution. Walkerana 2: 1-80.

Burch, J.B., and J. Tottenham. 1980. North American freshwater snails: Species list, ranges, and illustrations. Walkerana 1: 81-215.

Christensen, N.O., F. Frandsen, and M.Z. Roushdy. 1980.

The influence of environmental conditions and parasite-intermediate host-related factors on the transmission of *Echinostoma liei*. Zeitschrift für Parasitenkunde 63: 47-63.

Clampitt, P.T. 1970. Comparative ecology of the snails *Physa gyrina* and *Physa integra*. Malacologia 10: 113-151.

Collado, G.A. 2017. Unraveling cryptic invasion of a freshwater snail in Chile based on molecular and morphological data. Biodiversity and Conservation 26: 567-578.

Correa, A.C., J.S. Escobar, P. Durand, F. Renaud, P. David, P. Jarne, J-P Pointier, and S. Hurtrez-Boussès. 2010. Bridging gaps in the molecular phylogeny of the Lymnaeidae (Gastropoda: Pulmonata), vectors of Fascioliasis. BMC Evolutionary Biology 10: 381-393.

Covich, A.P. 2010. Winning the biodiversity arms race among freshwater gastropods: Competition and coexistence through shell variability and predator avoidance. Hydrobiologia 653: 191-215.

Covich, A.P., T.A. Crowl, J.E. Alexander Jr., and C.C. Vaughn. 1994. Predator-avoidance responses in freshwater decapods-gastropod interactions mediated by chemical stimuli. Journal of the North American Benthological Society 13: 291-298.

Coyne, J., and H. Orr. 2004. Speciation. Sinauer Associates, Sunderland, Massachusetts.

Crowl, T.A., and A.P. Covich. 1990. Predator-induced life history shifts in a freshwater snail. Science 247: 949-951.

Dawson, J. 1911. The biology of *Physa*. Behavior Monographs 1: 1-120.

DeWitt, T.J. 1998. Costs and limits of phenotypic plasticity: Tests with predator-induced morphology and life history in a freshwater snail. Journal of Evolutionary Biology 11: 465-480.

DeWitt, T.J., A. Sih, and J.A. Hucko. 1999. Trait compensation and cospecialization: Size, shape, and antipredator behavior. Animal Behaviour 58: 397-407.

DeWitt, T.J., B.W. Robinson, and D.S. Wilson. 2000. Functional diversity among predators of a freshwater snail imposes an adaptive trade-off for shell morphology. Evolutionary Ecology Research 2: 129-148.

Dillon, R.T., Jr. 2000. The Ecology of Freshwater Molluscs. Cambridge University Press, Cambridge, UK.

Dillon, R.T., Jr., and A.R. Wethington. 2005. The Michigan Physidae revisited: A population genetic survey. Malacologia 48: 133-142.

Dillon, R.T., Jr., and A.R. Wethington. 2006. No-choice mating experiments among six nominal taxa of the subgenus *Physella* (Basommatophora: Physidae). Heldia 6: 41-50.

Dillon, R.T., Jr., A.R. Wethington, J.M. Rhett, and T.P. Smith. 2002. Populations of the European freshwater pulmonate *Physa acuta* are not reproductively isolated from American *Physa heterostropha* or *Physa integra*. Invertebrate Biology 121: 226-234.

Dillon, R.T., Jr., A.R. Wethington, and C. Lydeard. 2011. The evolution of reproductive isolation in a simultaneous hermaphrodite, the freshwater snail *Physa*. BMC Evolutionary Biology 11: 144.

Escobar, J.S., J.R. Auld, A.C. Correa, J.M. Alonso, Y.K. Bony, M-A Coutellec, J.M. Koene, J.-P. Pointier, P. Jarne, and P. David. 2011. Patterns of mating-system evolution in hermaphroditic animals: Correlations among selfing rate, inbreeding depression, and the timing of reproduction. Evolution 65: 1233-1253.

Hovermand, J.T., and R.A. Relyea.2007. How flexible is phenotypic plasticity? Developmental windows for trait induction and reversal. Ecology 88: 693-705.

Hovermand, J.T., J.R. Auld, and R.A. Relyea. 2005. Putting prey back together again: Integrating predator-induced behavior, morphology, and life history. Oecologia 144: 481-491.

Jarne, P., P. David, J.-P. Pointier, and J.M. Koene. 2010. Basommatophoran gastropods. Pp. 173-196 in The Evolution of Sexual Characters in Animals (A. Cordoba-Aguilar and J.L. Leonard, eds.). Oxford University Press, New York.

Johnson, P.D., A.E. Bogan, K.M. Brown, N.M. Burkhead, J.R. Cordeiro, J.T. Garner, P.D. Hartfield, et al. 2013. Conservation status of freshwater gastropods of Canada and the United States. Fisheries 38: 247-282.

Lam, P., and P. Calow. 1988. Differences in the shell shape of *Lymnaea peregra* (Müller) (Gastropoda: Pulmonata) from lotic and lentic habitats; environmental or genetic variance? Journal of Molluscan Studies 54: 197-207.

Langerhans, R.B., and T.J. DeWitt. 2002. Plasticity constrained: Over-generalized induction cues cause maladaptive phenotypes. Evolutionary Ecology Research 4: 857-870.

Lydeard, C., D. Campbell, and M. Golz. 2016. *Physa acuta* Draparnaud, 1805 should be treated as a native of North America, not Europe. Malacologia 59: 347-350.

Magzoub, M., and A.A. Kasim, 1980. Schistosomiasis in Saudi Arabia. Annals of Tropical Medicine and Hygiene 74: 511-513.

McCarthy, T.M., and W.A. Fisher. 2000. Multiple predator avoidance behaviours of the freshwater snail *Physella heterostropha pomilia*: Responses vary with risk. Freshwater Biology 44: 387-397.

McMahon, R.F. 1975. Effects of artificially elevated water temperatures on the growth, reproduction and life cycle of a natural population of *Physa virgata* Gould. Ecology 56: 1167-1175.

Paraense, W.L., and J.-P. Pointier. 2003. *Physa acuta* Draparnaud, 1805 (Gastropoda: Physidae): A study of topotypic specimens. Memórias do Instituto Oswaldo Cruz, Rio de Janeiro, 98: 513-517.

Pilsbry, H.A. 1925. A fresh-water snail, *Physa zionis,* living under unusual conditions. Proceedings of the Academy of Natural Sciences of Philadelphia 77: 325-328.

Snyder, N.F.R. 1967. An alarm reaction of aquatic gastropods to intraspecific extract. Cornell University Agricultural Experiment Station, Memoir 403, Ithaca, New York.

Taylor, D.W. 2003. Introduction to Physidae (Gastropoda: Hygrophilia) biogeography, classification, morphology. Revista de Biologia Tropical 51, Supplement 1: 1-287.

Te, G.A. 1978. The systematics of the family Physidae (Basommatophora: Pulmonata). PhD diss., University of Michigan. 324 pages.

Thompson, F.G. 2011. An annotated checklist and bibliography of the land and freshwater snails of Mexico and Central America. Bulletin of the Florida Museum of Natural History 50: 1-299.

Turgeon, D.D., J.F. Quinn Jr., A.E. Bogan, E.V. Coan. F.G. Hochberg, W.G. Lyons, P.M. Mikkelsen, et al. 1998. Common and Scientific Names of Aquatic Invertebrates from the United States and Canada: Mollusks. 2nd ed. American Fisheries Society, Special Publication 26, Bethesda, Maryland.

Turner, A.M. 1996. Freshwater snails alter habitat use in response to predators. Animal Behaviour 51: 747-756.

Turner, R.D., and W.J. Clench. 1974. A new blind *Physa* from Wyoming with notes on its adaptations to the cave environment. Nautilus 88: 80-85.

Turner, A.M., R.J. Bernot, and C.M. Boes. 2000. Chemical cues modify species interactions: The ecological consequences of predator avoidance by freshwater snail. Oikos 88: 148-158.

Vermeij, G.J, and A.P. Covich. 1978. Coevolution of freshwater gastropods and their predators. American Naturalist 112: 833-843.

Vinarski, M.V. 2017. The history of an invasion: Phases of the explosive spread of the physid snail *Physella acuta* through Europe, Transcaucasia and Central Asia. Biological Invasions 19: 1299-1314.

Wethington, A.R., and R. Guralnick. 2004. Are hot spring physids distinctive lineages? A molecular systematic perspective. American Malacological Bulletin 19: 135-144.

Wethington, A.R., and C. Lydeard. 2007. A molecular phylogeny of Physidae (Gastropoda: Basommatophora) based on mitochondrial DNA sequences. Journal of Molluscan Studies 73: 241-257.

Wethington, A.R., J. Wise, and R.T. Dillon Jr. 2009. Genetic and morphological characterization of the Physidae of South Carolina (Pulmonata: Basommatophroa), with description of a new species. Nautilus 123: 282-292.

34 Planorbidae Rafinesque, 1815

CHRISTIAN ALBRECHT, BJÖRN STELBRINK, AND
CATHARINA CLEWING

Planorbidae represent the most diverse taxon of freshwater pulmonate gastropods on earth that has an almost cosmopolitan distribution. Historical estimates of diversity have reported approximately 350 species (Baker, 1945; Hubendick, 1955). This number was later updated by Meier-Brook (2002), who listed 300 species (200 of Planorbidae and 100 of "Ancylidae"). The most recent account (Strong et al., 2008) found ~250 species globally (100–200 Palearctic, 57 Nearctic, 59 Neotropical, 116 Afrotropical, 49 Oriental, 43 Australasian, and 8 on Pacific Oceanic Islands). These numbers, however, included Bulinidae and Burnupiidae (see chaps. 31, 32, this volume), which means a considerable reduction of the number of species (mostly in the Afrotropical region), leaving approximately 150 species globally.

Phylogenetic relationships among the Planorboidea are confused and controversially discussed. Traditionally, anatomical characters such as male copulatory organ structures or radulae were used to divide the Planorbidae into different subfamilies and tribes. In the traditional sense, the Planorbidae included bulinid and all ancylid (i.e., freshwater limpet) taxa (Hubendick, 1978), or Planorbinae and Bulininae excluding ancylid species (Burch, 1965; Brown, 1994), or including ancylid but no bulinine taxa (Starobogatov, 1967). This taxonomic instability reflects the knowledge on phylogenetic relationships of these taxa prior to the era of molecular phylogenetics. The most recent classification of freshwater gastropods (Bouchet et al., 2017), based on various phylogenetic analyses conducted during the past two decades, treats Bulinidae and Burnupiidae as separate families (see chaps. 31, 32, this volume). Thus, the Planorbidae consist of three subfamilies, namely Planorbinae Rafinesque, 1815, Ancylinae Rafinesque, 1815, and Miratestinae P. Sarasin & F. Sarasin, 1897.

Planorbidae are not only the most diverse and widespread freshwater pulmonate family; they also exploit the gastropod morphospace extensively with forms that are planispiral/discoidal (e.g., genera *Planorbis* Müller, 1773, *Planorbarius* Duméril, 1805, and *Biomphalaria* Preston, 1910), high-spired (e.g., *Amerianna* Strand, 1928, *Camptoceras* Benson, 1843, and

Distribution of
Planorbidae.

Glyptophysa Crosse, 1872), and patelliform/limpet-shaped (e.g., *Ancylus* Müller, 1773, *Gundlachia* Pfeiffer, 1850) with pseudodextral or sinistral orientation (Hubendick, 1978). They can be minute, such as *Choanomphalus kozhovi* and *C. pygmaeus* (Beckmann and Starobogatov, 1975; Röpstorf and Riedel, 2004) but also huge and heavy-shelled like *Miratesta celebensis* (e.g., Sarasin and Sarasin, 1898).

Peculiarities of planorbid gastropods include the possession of hemoglobin (see, e.g., Lieb et al., 2006, and references therein) and the tendency toward polyploidy in a few groups (e.g., Städler et al., 1995; Jarne et al., 2010). Planorbidae occur in all kinds of freshwater habitats, ranging from low energy temporary and permanent ponds, streams, rivers, and springs, to large lakes (Strong et al., 2008). In rare cases, they are even able to survive brackish-water conditions (see Clewing et al., 2015). This ecological flexibility is further reflected by their occurrence in extreme habitats such as water depths down to 400 m in Lake Baikal (*Choanomphalus bathybius bathybius*; Sitnikova et al., 2004) or high-mountain areas such as the Tibetan Plateau (von Oheimb et al., 2013; Clewing et al., 2015) or the northern and central Andes (Bößneck, 2012).

Planorbids can be found in all ancient lakes worldwide, and these special habitats have given rise to particular planorbid faunas (e.g., Albrecht and Wilke, 2008). Moreover, in some of these ancient lakes, so-called species flocks have evolved, as shown, e.g., for Lake Ohrid (*Gyraulus* spp., 5 species, Radoman, 1985; *Ancylus* spp., 4 species, Albrecht et al., 2006) and Lake Baikal (*Choanomphalus* spp., 22 species, Kantor et al., 2010). From a morphological perspective, the genus *Gyraulus* Charpentier, 1837, often develops extralimital shell shapes in ancient lakes, including pseudodextral shells and thalassoidy, which means unusual shell thickening and the development of hairs or keels (Meier-Brook, 1983; Gorthner, 1992). This latter phenomenon of thalassoidy, i.e., marine-like shell appearance in the most extreme form, is found in *Miratesta* Sarasin & Sarasin, 1897, of ancient Lake Poso on Sulawesi, Indonesia (Sarasin and Sarasin, 1898).

The cosmopolitan distribution of Planorbidae has been the result of a high dispersal capacity and ecological flexibility, including desiccation resistance that

Helisoma anceps (Menke, 1830). USA: Cass Lake, Pike Bay, Cass County, Minnesota. 9 mm. Poppe 825802.

is particularly important for the successful passive transport via (aerial) vectors. Such resistance and its possibly related high dispersal capacity may explain the colonization of isolated Caribbean islands along major bird migration routes (Baker, 1945). Moreover, life history characteristics such as hermaphroditism—as a single individual is often sufficient to establish a viable population—and large numbers of offspring are thought to account for colonization success (Meier-Brook, 2002). Nowadays, planorbid gastropod species occur basically everywhere on the globe except for Arctic and Antarctic regions with permanent ice covers, but also including very remote oceanic islands if permanent freshwater systems exist.

Though the family Planorbidae has reached a global distribution in its ~250 my of evolution (Zilch, 1959; Gray, 1988; Taylor, 1988; Tracey et al., 1993), there are marked differences in the three currently recognized subfamilies (Bouchet et al., 2017). In the Planorbinae, the tribe Planorbini Rafinesque, 1815, has an almost global distribution: the Planorbini and Segmentinini Baker, 1945, are of West Palearctic origin and might have colonized North America, Africa, Eurasia/Australia in Jurassic/Cretaceous times as suggested by Meier-Brook (1984). Segmentinini are absent in North America and comprise Palearctic,

Oriental, and Afrotropical species (Meier-Brook 1984); Drepanotrematini Zilch, 1959, occur in Central and South America. The Neoplanorbini Hannibal, 1912, represent a likely extinct taxon endemic to river systems in the southeast of the United States (e.g., Bogan, 2000; Cordeiro and Perez, 2012). The tribe Helisomatini Baker, 1928, includes Afrotropical and American taxa, while Coretini Gray, 1847, is primarily European and Camptoceratini Dall, 1870, occur in southern and eastern Asia (Hubendick, 1978).

Some planorbid snails, especially *Biomphalaria* spp. (see, e.g., DeJong et al., 2001, 2003; Morgan et al., 2002; Jørgensen et al., 2004), are of great significance for humans, as members of this group are intermediate hosts of blood flukes (schistosomes) that infected at least 250 million people in 2014, mainly in the tropics (WHO, 2016). Schistosomes cause a chronic disease in humans, schistosomiasis, which is considered the second most important tropical disease after malaria (Chitsulo et al., 2000). Also, 165 million cattle worldwide suffer from infections with schistosomes (De Bont and Vercruysse, 1998). The genus *Biomphalaria* Preston, 1910, is widespread in South America and Africa (Brown, 1994; Toledo and Fried, 2011), possibly as a result of the second breakup of Gondwana, as suggested by Meier-Brook (1984). However, recent studies revealed that African species of *Biomphalaria* are apparently as young as 2.5 my (DeJong et al., 2001) and might have colonized the continent from South America during the Plio-Pleistocene (Jørgensen et al., 2007).

Freshwater limpets of the subfamily Ancylinae Rafinesque, 1815, occur on all continents. The limpet shell form has evolved several times in planorboid snails, and at least two times in Planorbidae. Functional and evolutionary scenarios that have been proposed for the evolution of patelliform shells are discussed in Albrecht et al. (2004, 2007) and Vermeij (2016). The last comprehensive treatment of ancylid taxa lists seven genera as valid (Hubendick, 1964), but a taxonomic revision is still lacking. Basch (1963), who primarily dealt with North American ancylids, suggested two major groups within the Ancylidae, one comprising American *Ferrissia, Laevapex, Hebetancylus,* and the African genus *Burnupia* (which belongs to its own family Burnupiidae; see chap. 32, this volume), while the second contains European *Ancylus* species and the North American genus *Rhodacmea*. Though the sister group relationship of *Rhodacmea* and *Ancylus* has recently been confirmed (Ó Foighil et al., 2011), a global phylogeny of Ancylinae is still not available (see Walther et al., 2006; Albrecht et al., 2007). Two tribes are used in the most current classification (Bouchet et al., 2017), namely Ancylini Rafinesque, 1815, and Laevapicini Hannibal, 1912.

The subfamily Miratestinae Sarasin & Sarasin, 1897, comprises Australian high-spired planorbid species (Walker, 1988; Albrecht et al., 2007) and taxa that occur in the Indo-Malayan archipelago, including ancient lakes on Sulawesi (*Miratesta*). Australian high-spired planorbids have been most controversially discussed for a long time (Walker, 1984, 1988), including scenarios of a Gondwanan origin (Walker, 1988). Morphologically strikingly different is the limpet-shaped genus *Protancylus* Duméril, 1805, an ancient lake endemic from Sulawesi with an aberrant life-history strategy and morphological peculiarities (Albrecht and Glaubrecht, 2006) that shows clear Australian affinities. However, further studies are required that should include another enigmatic calyptraeid taxon, *Patelloplanorbis* Hubendick, 1957, from the New Guinean Wissel Lakes. Such taxa and the use of molecular data are crucial to clarify biogeographical and phylogenetic relationships among this subfamily.

When studying the historical biogeography of Planorbidae, no clear-cut pattern can be drawn of either vicariance or dispersal, as both processes are often involved and several extinction events can be assumed throughout the taxon's existence. Comparatively recent and repeated long-distance dispersal events obviously represent a major contribution to the present wide range of several genera and species. Some planorbids are known as notorious global invaders, for example, the limpet species *Ferrissia fragilis* (e.g., Walther et al., 2006; Albrecht et al., 2014), which, however, should be named *F. californica* (Rowell, 1863) for nomenclatural reasons (Christensen, 2016). Further examples of invasive species include the originally North American discoidal species *Gyraulus parvus* (Meier-Brook, 2002; Pointier et al., 2005) and the buliniform species *Amerianna carinata* that spread widely from its Australian origin (Madsen and Frandsen, 1989).

Whereas some species are globally invasive, others are declining or are at the brink of extinction. For the majority of the 266 planorbid species listed, the IUCN red list included 118 Data Deficient species by 2016, and another 105 species were of Least Concern. A total of 15 species were assessed Endangered, 6 Critically Endangered, 12 Vulnerable, 4 Near Threatened, and 6 Extinct. These species represent either species with very restricted ranges in largely modified river systems (*Rhodacmea, Neoplanorbis,* and *Amphigyra*), single lakes (e.g., *Ancylus ashangiensis*), or islands (e.g., *Ancylastrum*). Large faunas, however, remain unassessed, and it can be presumed that even more species of Planorbidae (and other freshwater gastropods) are endangered or even lost already (Lydeard et al., 2004; Regnier et al., 2015).

LITERATURE CITED

Albrecht, C., and M. Glaubrecht. 2006. Brood care among basommatophorans: A unique reproductive strategy in the freshwater limpet snail *Protancylus* (Heterobranchia: Protancylidae), endemic to ancient lakes on Sulawesi, Indonesia. Acta Zoologica 87: 49-58.

Albrecht, C., and T. Wilke. 2008. Ancient Lake Ohrid: Biodiversity and evolution. Hydrobiologia 615: 103-140.

Albrecht, C., T. Wilke, K. Kuhn, and B. Streit. 2004. Convergent evolution of shell shape in freshwater limpets: The African genus *Burnupia*. Zoological Journal of the Linnean Society 140: 577-586.

Albrecht, C., S. Trajanovski, K. Kuhn, B. Streit, and T. Wilke. 2006. Rapid evolution of an ancient lake species flock: Freshwater limpets (Gastropoda: Ancylidae) in the Balkan Lake Ohrid. Organisms, Diversity & Evolution 6: 294-307.

Albrecht, C., K. Kuhn, and B. Streit. 2007. A molecular phylogeny of Planorboidea (Gastropoda, Pulmonata): Insights from enhanced taxon sampling. Zoologica Scripta 36: 27-39.

Albrecht, C., K. Föller, T. Hauffe, C. Clewing, and T. Wilke. 2014. Invaders versus endemics: Alien gastropod species in ancient Lake Ohrid. Hydrobiologia 739: 163-174.

Baker, F.C. 1945. The Molluscan Family Planorbidae. University of Illinois Press, Urbana.

Basch, P.F. 1963. A review of the recent freshwater limpet snails of North America (Mollusca: Pulmonata). Bulletin of the Museum Comparative Zoology 129: 401-461.

Beckman, M.Y., and Ya.I. Starobogatov. 1975. Baikal deepwater molluscs and related forms: New data on Baikal fauna [in Russian]. Proceedings of the Limnological Institute SB AS USSR, Novosibirsk, Part I, 18: 92-111.

Bogan, A.E. 2000. *Neoplanorbis carinatus.* The IUCN Red List of Threatened Species 2000: e.T14553A4444062. http://dx.doi.org/10.2305/IUCN.UK.2000.RLTS.T14553A4444062.en. Accessed 14 October 2016.

Bößneck, U. 2012. Leben am Limit: Besiedlung von Süßwasser-Habitaten extremer Hochlagen Asiens, Amerikas und Afrikas durch Mollusken (Mollusca: Bivalvia & Gastropoda). Pp. 103-106 in Biodiversität und Naturausstattung im Himalaya IV—Biodiversity and Natural Heritage of the Himalaya IV (M. Hartmann and J. Weipert, eds.). Verein der Freunde & Förderer des Naturkundemuseums Erfurt e.V., Erfurt, Germany.

Bouchet, P., J.P. Rocroi, B. Hausdorf, A. Kaim, Y. Kano, A. Nützel, P. Parkhaev, M. Schrödl, and E.E. Strong. 2017. Revised classification, nomenclator and typification of gastropod and monoplacophoran families. Malacologia 61: 1-526.

Brown, D.S. 1994. Freshwater Snails of Africa and Their Medical Importance. 2nd ed. Taylor & Francis, London.

Burch, J.B. 1965. Chromosome numbers and systematics in euthyneuran snails. Proceedings of the First European Malacological Congress 1962: 215-241.

Chitsulo, L., D. Engels, A. Montresor, and L. Savioli. 2000. The global status of schistosomiasis and its control. Acta Tropica 77: 41-51.

Christensen, C.C. 2016. Change of status and name for a Hawaiian freshwater limpet: *Ancylus sharpi* Sykes, 1900, is the invasive North American *Ferrissia californica* (Rowell, 1863), formerly known as *Ferrissia fragilis* (Tryon, 1863) (Gastropoda: Planorbidae: Ancylinae). Bishop Museum Occasional Papers 118: 5-8.

Clewing, C., F. Riedel, T. Wilke, and C. Albrecht. 2015. Ecophenotypic plasticity leads to extraordinary gastropod shells found on the 'Roof of the World.' Ecology and Evolution 5: 2966-2979.

Cordeiro, J., and K. Perez. 2012. *Neoplanorbis tantillus.* The IUCN Red List of Threatened Species 2012: e.T14556A546008. http://dx.doi.org/10.2305/IUCN.UK.2012.RLTS.T14556A546008.en. Accessed 14 October 2016.

De Bont, J., and J. Vercruysse. 1998. Schistosomiasis in cattle. Advances in Parasitology 41: 285-364.

DeJong, R.J., J.A.T. Morgan, W.L. Paraense, J.-P. Pointier, M. Amarista, K.P.F.K. Ayeh, A. Babiker, et al. 2001. Evolutionary relationships and biogeography of *Biomphalaria* (Gastropoda: Planorbidae) with implications regarding its role as host of the human bloodfluke, *Schistosoma mansoni.* Molecular Biology and Evolution 18: 2225-2239.

DeJong, R.J., J.A.T. Morgan, W.D. Wilson, M.H. Al-Jaser, C.C. Appleton, G. Coulibaly, P.S. D'Andrea, et al. 2003. Phylogeography of *Biomphalaria glabrata* and *B. pfeifferi,* important intermediate hosts of *Schistosoma mansoni* in the New and Old World tropics. Molecular Ecology 12: 3041-3056.

Gorthner, A. 1992. Bau, Funktion und Evolution komplexer Gastropodenschalen in Langzeit-Seen: Mit einem Beitrag von *Gyraulus 'multiformis'* im Steinheimer Becken. Stuttgarter Beiträge zur Naturkunde 190: 1-173.

Gray, J. 1988. Evolution of the freshwater ecosystem: The fossil record. Palaeogeography, Palaeoclimatology, Palaeoecology 62: 1-214.

Hubendick, B. 1955. Phylogeny in the Planorbidae. Transactions of the Zoological Society of London 28: 453-542.

Hubendick, B. 1964. Studies on Ancylidae, the subgroups. Meddelanden från Göteborgs Musei Zoologiska Avdelning 137: 1-71.

Hubendick, B. 1978. Systematics and comparative morphology of the Basommatophora. Pp. 1-47 in Pulmonates, vol. 2A, Systematics, Evolution and Ecology (V. Fretter and J. Peake, eds.). Academic Press, London.

Jarne, P., J.P. Pointier, P. David, and J.M. Koene. 2010. Basommatophoran gastropods. Pp. 173-196 in The Evolution of Primary Sexual Characters in Animals (A. Córdoba-Aguilar and J.L. Leonard, eds.). Oxford University Press, New York.

Jørgensen, A., T.K. Kristensen, and J.R. Stothard. 2004. An investigation of the 'Ancyloplanorbidae' (Gastropoda, Pulmonata, Hygrophila): Preliminary evidence from DNA sequence data. Molecular Phylogenetics and Evolution 32: 778-787.

Jørgensen, A., T.K. Kristensen, and J.R. Stothard. 2007. Phylogeny and biogeography of African *Biomphalaria* (Gastropoda: Planorbidae), with emphasis on endemic species of the great East African lakes. Zoological Journal of the Linnean Society 151: 337-349.

Kantor, Y., M. Vinarski, A. Schileyko, and A. Sysoev. 2010. Catalogue of the continental mollusks of Russia and adjacent territories, v.2.3.1. Available at www.ruthenica.com /documents/Continental_Russian_molluscs_ver2-3-1.pdf.

Lieb, B., K. Dimitrova, H.-S. Kang, S. Braun, W. Gebauer, A. Martin, B. Hanelt, S.A. Saenz, C.M. Adema, and J. Markl. 2006. Red blood with blue-blood ancestry: Intriguing structure of a snail hemoglobin. Proceedings of the National Academy of Sciences USA 103: 12011-12016.

Lydeard, C., R.H. Cowie, W.F. Ponder, A.E. Bogan, P. Bouchet, S.A. Clark, K.S. Cummings, et al. 2004. The global decline of nonmarine mollusks. BioScience 54: 321-330.

Madsen, H., and F. Frandsen. 1989. The spread of freshwater snails including those of medical and veterinary importance. Acta Tropica 46: 139-146.

Meier-Brook, C. 1983. Taxonomic studies on *Gyraulus* (Gastropoda: Planorbidae). Malacologia 24: 1-113.

Meier-Brook, C. 1984. A preliminary biogeography of freshwater pulmonate gastropods. Pp. 23-37 in World-Wide Snails: Biogeographical Studies on Non-Marine Mollusca (A. Solem and A.C. van Bruggen, eds.). Brill, Leiden.

Meier-Brook, C. 2002. What makes an aquatic ecosystem susceptible to mollusc invasions? Pp. 405-417 in Collectanea malacologica: Festschrift für Gerhard Falkner (M. Falkner, K. Groh, and M.C.D. Speight, eds.). Conch-Books, Hackenheim, Germany.

Morgan, J.A.T., R.J. DeJong, Y. Jung, K. Khallayoune, S. Kock, G.M. Mkoji, and E.S. Loker. 2002. *Schistosoma* parasites of humans and domestic animals did not cospeciate with their planorbid snail hosts, a story of host switching and recent diversification. Molecular Phylogenetics and Evolution 25: 477-488.

Ó Foighil, D., J. Li, T. Lee, P. Johnson, R. Evans, and J.B. Burch. 2011. Conservation genetics of a critically endangered limpet genus and rediscovery of an extinct species. PLoS ONE 6: e20496.

Pointier, J.-P., P. David, and P. Jarne. 2005. Biological invasion: The case of planorbid snails. Journal of Helminthology 79: 249-256.

Radoman, P. 1985. Hydrobioidea, A Superfamily of Prosobranchia (Gastropoda). II. Origin, Zoogeography, Evolution in the Balkans and Asia Minor. Monographs, vol. 1, Institute of Zoology No. 1. Faculty of Science, Department of Biology, Belgrade.

Régnier, C., G. Achaz, A. Lambert, R.H. Cowie, P. Bouchet, and B. Fontaine. 2015. Mass extinction in poorly known taxa. Proceedings of the National Academy of Sciences USA 112 (25): 7761-7766.

Röpstorf, P., and F. Riedel. 2004. Deep-water gastropods endemic to Lake Baikal—an SEM study on protoconchs and radulae. Journal of Conchology 38: 253-282.

Sarasin, P., and F. Sarasin. 1898. Die Süßwassermollusken von Celebes. Kreidel, Wiesbaden.

Sitnikova, T.Y., Ya.I. Starobogatov, A.A. Shirokaya, I.V. Shibanova, N.V. Korobkova, and F. Adov. 2004. Gastropod molluscs (Gastropoda) [in Russian]. Pp. 937-1002 in Index of Animal Species Inhabiting Lake Baikal and Its Catchment Area (O.A. Timoshkin, ed.). Nauka, Novosibirsk, Russia.

Städler, T., S. Weisner, and B. Streit. 1995. Outcrossing rates and correlated matings in a predominately selfing freshwater snail. Proceedings of the Royal Society of London B 262: 119-125.

Starobogatov, Ya.I. 1967. On the Systematization of Freshwater Pulmonate Molluscs [in Russian]. Nauka, Leningrad.

Strong, E.E., O. Gargominy, W.F. Ponder, and P. Bouchet. 2008. Global diversity of gastropods (Gastropoda: Mollusca) in freshwater. Hydrobiologia 595: 149-166.

186

Christian Albrecht, Björn Stelbrink, and Catharina Clewing

Taylor, D.W. 1988. Aspects of freshwater mollusc ecological biogeography. Palaeogeography, Palaeoclimatology, Palaeoecology 62: 511-576.

Toledo, R., and B. Fried. 2011. *Biomphalaria* Snails and Larval Trematodes. Springer-Verlag, New York.

Tracey, S., J.A. Todd, and D.H. Erwin. 1993. Mollusca: Gastropoda. Pp. 131-167 in The Fossil Record (M.J. Benton, ed.). Chapman and Hall, London.

Vermeij, G.J. 2016. The limpet form in gastropods: Evolution, distribution, and implications for the comparative study of history. Biological Journal of the Linnean Society 120: 22-37.

Von Oheimb, P.V., C. Albrecht, F. Riedel, U. Bössneck, H. Zhang, and T. Wilke. 2013. Testing the role of the Himalaya Mountains as a dispersal barrier in freshwater gastropods (*Gyraulus* spp.). Biological Journal of the Linnean Society 109: 526-534.

Walker, J.C. 1984. Geographical relationships of the buliniform planorbids of Australia. Pp. 189-197 in World-Wide Snails: Biogeographical Studies on Non-Marine Mollusca (A. Solem and A.C. van Bruggen, eds.). Brill, Leiden, Netherlands.

Walker, J.C. 1988. Classification of Australian buliniform planorbids (Mollusca: Pulmonata). Records of the Australian Museum 40: 61-89.

Walther, A.C., T. Lee, J.B. Burch, and D. Ó Foighil. 2006. E Pluribus Unum: A phylogenetic and phylogeographic reassessment of *Laevapex* (Pulmonata: Ancylidae), a North American genus of freshwater limpets. Molecular Phylogenetics and Evolution 40: 501-516.

WHO, 2016. Fact sheet on schistosomiasis. Updated February 2016. Available at www.who.int/mediacentre/factsheets/fs115/en.

Zilch, A. 1959-1960. Gastropoda (Euthyneura). Pp. 91-102 in Handbuch der Paläozoologie, vol. 6 (O.H. Schindewolf, ed.). Gebrüder Bornträger, Berlin.

35 Cyrenidae Gray, 1840

RÜDIGER BIELER AND PAULA M. MIKKELSEN

Members of the family Cyrenidae (syn. Corbiculidae, Geloinidae, Polymesodinae, Serrilaminulinae) live in shallow freshwater, brackish-water, and nearshore marine (especially mangrove) habitats in temperate to tropical regions. With the likewise freshwater/brackish families Cyrenoididae and Glauconomidae, cyrenids form a monophyletic clade, Cyrenoidea, in the higher heterobranch (Imparidentia: Neoheterodontei) bivalves (Taylor et al., 2009, Bieler et al., 2014). A previously assumed close relationship with Sphaeriidae was not confirmed by recent analyses (e.g., Combosch et al., 2017).

Morphologically, the adult cyrenid shell ranges in size from approximately 15 to 150 mm and is thin-walled to solid, rounded trigonal to oval in shape, nearly equivalve, equilateral, and moderately inflated. Exteriorly, cyrenids are brown or olive green to brightly colored in shades of pink, yellow, and purple, covered by a thick, fibrous, often glossy periostracum. Sculpture is commarginal, ranging from simple growth lines to strong ridges. Interiorly, the shell is nonnacreous and in some species brightly colored. The pallial line is either entire or has a narrow sinus. The inner shell margins are smooth. The hinge plate is narrow and heterodont, with usually three (often bifid) cardinal teeth in each valve; anterior and posterior lateral teeth are strong, long, and smooth or serrated. The ligament is strong, parivincular, and opisthodetic. The shell muscles are isomyarian or weakly heteromyarian (anterior adductor muscle slightly smaller); anterior and posterior pedal retractor muscles are present. Posterior excurrent and incurrent siphons are short and separate. The mantle margins are not fused ventrally. The foot is strong and trigonal; a byssal groove and byssus are absent in the adult. The ctenidia are eulamellibranch (synaptorhabdic) and homorhabdic or heterorhabdic; the demibranchs are united to each other and to the siphonal septum, separating infra- and suprabranchial chambers. The anatomy of several species is well documented (e.g., *Corbicula fluminea* by Britton and Morton, 1982, *Polymesoda floridana* by Simone et al., 2015).

No comprehensive modern systematic treatment of

Distribution of Cyrenidae.

the Cyrenidae exists. More than 500 nominal species have been introduced (Counts, 2006; Glaubrecht et al., 2007; Huber, 2015), with some authors recognizing large numbers of locally restricted species (e.g., Glaubrecht et al., 2003; Korniushin, 2004; Lee et al., 2005), whereas others saw few wide-ranging taxa (e.g., Morton, 1986). Recent estimates are of approximately 100–120 extant cyrenid species worldwide (Mikkelsen and Bieler, 2007; Huber, 2015). All are infaunal in soft substrata, with one notable exception, *Corbicula anomioides,* a cementing form originally described as a monotypic genus (*Posostrea* Bogan & Bouchet, 1998) but subsequently shown to be genetically close to a noncementing species in the same Indonesian lake (Bogan and Bouchet, 1998; Rintelen and Glaubrecht, 2006). Morphospecies delineations in this group are problematic, with intraspecific variability in shell shape and color and ontogenetic changes in periostracal appearance. Molecular evidence exists for hybridization among morphotypes in *Corbicula* von Mühlfeld, 1811 (Pfenninger et al., 2002), and Huber (2015) assumed that some of the nominal species in cyrenids were based on such hybrid specimens. Worldwide genus-level treatments vary (Glaubrecht et al., 2007; Graf, 2013; Huber, 2015) and the extant species are currently grouped in six genera (MolluscaBase, 2018), largely based on morphological features such as shell shape and sculpture, the serration of lateral hinge teeth, the depth of the pallial sinus, and the development of the siphons. Three of these genera comprise smaller-shelled species (*Corbicula* von Mühlfeld, 1811, *Cyanocyclas* Blainville, 1818, and *Villorita* Gray, 1833), with the first two largely occurring in freshwater habitats; the other three genera (*Polymesoda* Gray, 1853, *Geloina* Gray, 1842, and *Batissa* Gray, 1853) include larger-shelled species that mostly live in estuarine or nearshore mangrove environments. Earliest recognized members of Cyrenidae include species of *Hemicorbicula, Paracorbicula,* and *Tetoria* from the latest Jurassic and earliest Cretaceous of northwestern Europe and Japan (Casey, 1955; Nishida et al., 2013; Matsukawa and Koarai, 2017).

Corbicula is by far the largest and most well-known genus, with a current morphology-based estimate of 60–70 living species (Huber, 2015). The majority live in riverine and lacustrine habitats, with *Corbicula*

Corbicula fluminea (Müller, 1774). USA: South Fork Sangamon River, Kincaid, Christian County, Illinois. 38 mm. INHS 7276.

s. s. assumed as having a natural range from Africa to Asia, with subsequent introductions into Europe and the Americas. Smaller morphologically defined groups currently treated as subgenera include *Corbiculina* Dall, 1903 (with three species in Australia), monotypic *Cyrenodonax* Dall, 1903 (China), and *Cyrenobatissa* Suzuki & Oyama, 1943 (Vietnam to China). *Cyanocyclas* (= *Neocorbicula* Fischer, 1887) is a genus of approximately six currently recognized species restricted to the freshwater habitats of the Neotropics. Monotypic *Villorita* occurs in brackish/estuarine waters of India, and *Polymesoda,* with about 18 species, inhabits estuarine and nearshore mangrove habitats of the Americas. *Geloina* and *Batissa,* with 13–15 species each in the tropical Pacific, have members with the largest shell sizes in the family, reaching up to 150 mm.

Members of *Corbicula* s. s. have attracted considerable attention because some forms have become hyperinvasive and spread to large parts of the world since the beginning of the twentieth century (Mouthon, 1981; Counts, 1986), with invasions of North America dated to 1924 (Counts, 1981), South America in the 1970s (Ituarte, 1994), and Europe since the early 1980s (Mouthon, 1981; Araujo et al., 1993). The invaders have become major biofoulers of industrial and domestic water supply systems (e.g., Isom, 1986), have affected river oxygen budget and nutrient cycling (Pigneur et al., 2014b), and have negatively impacted endemic unionoid bivalves (Cherry et al., 2005; Sousa et al., 2008). *Corbicula* is now among the most common freshwater mollusks in the Americas. The identity and number of invading species and their source regions have been objects of intense study and debate. Although there was early evidence of more than one morphotype introduced into North America (Hillis and Patton, 1982), many authors have adopted a single species name for it, *C. fluminea* (e.g., Counts, 1986; Morton, 1986; Foster et al., 2017). Four distinct morphotypes (A–D) are currently recognized in the Americas and have been characterized by molecular data (Siripattrawan et al., 2000; Lee et al., 2005; Tiemann et al., 2017). Likewise, at least two morphotypes—often assigned to the nominal species *C. fluminea* and *C. fluminalis*—have been characterized as invasive in European rivers (e.g., Araujo et al., 1993; Renard et al., 2000; Paunović et al., 2007).

Reproductive biology in Cyrenidae is diverse and complex (for a detailed summary, see Pigneur et al., 2012). Marine/brackish taxa such as *Polymesoda* and a couple of lacustrine and brackish-water *Corbicula* species retain the ancestral bivalve traits of dioecy, broadcast spawning, and dispersal of planktonic veligers. The majority of freshwater *Corbicula* clams are hermaphroditic and brood their larvae in their gills (e.g., Korniushin and Glaubrecht, 2003; Glaubrecht et al., 2006), with brooding allowing the retention of offspring in river currents. *Corbicula* is the only known bivalve genus exhibiting biflagellate, unreduced sperm in some species (e.g., Konishi et al., 1998), whereas others (presumably all nonclonal forms) produce "normal" monoflagellate, reduced sperm. Viviparity apparently evolved in parallel in both South American *Cyanocyclas* and Old World *Corbicula* (Glaubrecht et al., 2006).

No comprehensive phylogenetic study exists, and the status of and relationships among nominal cyrenid genera are in need of investigation. A few analyses using mitochondrial DNA focused on *Corbicula,* and mostly placed haplotypes of invasive forms among those of native Asian taxa (e.g., Siripattrawan et al., 2000; Glaubrecht et al., 2003, 2006; Lee et al., 2005; Pigneur et al., 2011). As discussed by Graf (2013), these studies discovered multiple invasive lineages and challenged an earlier paradigm that global *Corbicula* is represented by only two widespread species (Morton, 1986; Kijviriya et al., 1991). However, resolving the phylogeny of the Cyrenidae

will not be an easy task and will require an approach utilizing not only morphology and standard molecular markers, but also techniques that can explore more complex patterns of gene transfer among taxa. *Corbicula* includes both sexually reproducing and primarily asexual lineages. Androgenesis (asexual male reproduction) appears to have arisen repeatedly in the genus (Hedtke et al., 2008). Several *Corbicula* species have been discovered to be polyploid (e.g., Komaru et al., 1997; Qiu et al., 2001) and capable of self-fertilization as simultaneous hermaphrodites. In that process, the maternal nuclear DNA of the egg is discarded and replaced by that of the biflagellate unreduced sperm. This results in the transfer of the maternal mitochondrial DNA and paternal nuclear DNA to the next generation, making the offspring paternal nuclear clones (Pigneur et al., 2012). Hedtke et al. (2011) demonstrated that occasional outcrossing among *Corbicula* species can lead to the rare capture of maternal nuclear DNA from other species and thus new androgenetic lines. Interestingly, the sexually reproducing forms have remained restricted to the native Old World regions, whereas only a few androgenetic lineages have developed into the now widely distributed invasive pests (Pigneur et al., 2014a). As stated by Graf (2013: 145), "The complex evolutionary history of the Cyrenidae is surely one of the Great Unanswered Questions on the evolution of freshwater bivalves."

Individual species often show wide ranges of salinity tolerance, with the freshwater species *Corbicula fluminea* known to be able to osmoregulate below 13 ppt (Morton and Tong, 1985), and the estuarine *Cyrenobatissa subsulcata* to do so in a range of 0–24 ppt (Morton, 1992). A species of *Geloina* living at the base of Southeast Asian mangrove trees was described by Morton (1976) as ideally adapted to a life in a tropical habitat that is essentially transitional between the sea, freshwater, and the land; covered only by spring tides and at other times inundated by rainwater, it can withstand long periods of exposure, during which time it can use subterranean water contained in the burrow and employ aerial respiration via the mantle margin.

Interest in *Corbicula* and its control has spawned a large body of technical literature, international symposia, newsletters, and special volumes (e.g., Britton and Prezant, 1986; Mackie, 2007). Cyrenids are important targets of regional commercial and artisanal fisheries, for human consumption, bait, and uses of the shell material. Examples include species of *Batissa* and *Geloina* (as "*Polymesoda*") in the Philippines and the Solomon Islands (e.g., Mayor and Ancog, 2016; Dolorosa and Dangan-Galon, 2014; Carter, 2014), *Polymesoda* in Colombia (Rueda and Urban, 1998), and *Villorita* in India, where *V. cyprinoides* represents the most important clam harvested in the country (Suja and Mohamed, 2010). Cyrenid species are widely available in the aquarium trade (potentially leading to new introductions; *C. fluminea* is marketed in the American aquarium trade as the "Golden Freshwater Clam"). A species of *Corbicula* has been implicated as a vector of echinosomiasis in humans in the Philippines (Centers for Disease Control and Prevention, 2017).

The IUCN Red List (2017) includes 49 species of Cyrenidae, often with names currently considered synonyms or members of different genera. Sixteen of these are ranked as of Least Concern, 31 as Data Deficient, and 2, Indonesian *Corbicula possoensis* and Malagasy *C. madagascariensis,* as Endangered. Extensive efforts will be needed to address the status of native taxa and associated conservation needs, e.g., for the South American species of *Cyanocyclas* (Pereira et al., 2014).

LITERATURE CITED

Araujo, R., D. Moreno, and M.A. Ramos. 1993. The Asiatic clam *Corbicula fluminea* (Müller, 1774) (Bivalvia: Corbiculidae) in Europe. American Malacological Bulletin 10: 39-49.

Bieler, R., P.M. Mikkelsen, T.M. Collins, E.A. Glover, V.L. González, D.L. Graf, E.M. Harper, et al. 2014. Investigating the Bivalve Tree of Life—an exemplar-based approach combining molecular and novel morphological characters. Invertebrate Systematics 28: 32-115. DOI: 10.1071/IS13010.

Bogan, A., and P. Bouchet. 1998. Cementation in the freshwater bivalve family Corbiculidae (Mollusca: Bivalvia): A new genus and species from Lake Poso, Indonesia. Hydrobiologia 389: 131-139. DOI: 10.1023/A:1003562017200.

Britton, J.C., and B. Morton. 1982. A dissection guide, field and laboratory manual for the introduced bivalve *Corbicula fluminea*. Malacological Review, Supplement 3: i-vi, 1-82.

Britton, J.C., and R.S. Prezant, eds. 1986. Proceedings of the Second International *Corbicula* Symposium. American Malacological Bulletin, Special Edition No. 2: i-ii, 1-239.

Carter, M. 2014. Subsistence shell fishing in NW Santa Isabel, Solomon Islands: Ethnoarchaeology and the identification of two *Polymesoda* (Solander 1786) species. Ethnoarchaeology 6: 40-60. DOI: 10.1179/1944289013Z.00 000000013.

Casey, R. 1955. The pelecypod family Corbiculidae in the Mesozoic of Europe and the Near East. Journal of the Washington Academy of Sciences 45: 366-372.

Centers for Disease Control and Prevention. 2017. DPDx—Laboratory Identification of Parasitic Diseases of Public Health Concern: Echinostomiasis. https://www.cdc.gov /dpdx/echinostomiasis/index.html. Last page update 28 December 2017, accessed 23 May 2018.

Cherry, D.S., J.L. Scheller, N.L. Cooper, and J.R. Bidwell. 2005. Potential effects of Asian clam (*Corbicula fluminea*) die-offs on native freshwater mussels (Unionidae) I: Water-column ammonia levels and ammonia toxicity. Journal of the North American Benthological Society 24: 369-380. DOI: 10.1899/04-073.1.

Combosch, D.J., T.M. Collins, E.A. Glover, D.L. Graf, E.M. Harper, J.M. Healy, G.Y. Kawauchi, et al. 2017. A family-level Tree of Life for bivalves based on a Sanger-sequencing approach. Molecular Phylogenetics and Evolution 107: 191-208. DOI: 10.1016/j.ympev.2016.11.003.

Counts, C.L. III. 1981. *Corbicula fluminea* (Bivalvia: Sphaeriacea) in British Colombia. Nautilus 95: 12-13.

Counts, C.L., III. 1986. The zoogeography and history of the invasion of the United States by *Corbicula fluminea* (Bivalvia: Corbiculidae). American Malacological Bulletin, Special Edition No. 2: 7-39.

Counts, C.L. III. 2006. *Corbicula,* an annotated bibliography. www.carnegiemnh.org/mollusks/corbicula.pdf. 436 pages.

Dolorosa, R.G., and F. Dangan-Galon. 2014. Population dynamics of the mangrove clam *Polymesoda erosa* (Bivalvia: Corbiculidae) in Iwahig, Palawan, Philippines. International Journal of Fauna and Biological Studies 1: 11-15.

Foster, A.M., P. Fuller, A. Benson, S. Constant, D. Raikow, J. Larson, and A. Fusaro. 2017. *Corbicula fluminea*. USGS Nonindigenous Aquatic Species Database, Gainesville, Florida. https://nas.er.usgs.gov/queries/FactSheet.aspx ?speciesID=92, revision date: 8 May 2017.

Glaubrecht, M., T. von Rintelen, and A.V. Korniushin. 2003. Toward a systematic revision of brooding freshwater Corbiculidae in Southeast Asia (Bivalvia, Veneroida): On shell morphology, anatomy and molecular phylogenetics of endemic taxa from islands in Indonesia. Malacologia 45: 1-40.

Glaubrecht, M., Z. Fehér, and T. von Rintelen. 2006. Brooding in *Corbicula madagascariensis* (Bivalvia, Corbiculidae) and the repeated evolution of viviparity in corbiculids. Zoologica Scripta 35: 641-654. DOI: 10.1111/j.1463-6409.2 006.00252.x.

Glaubrecht, M., Z. Fehér, and F. Köhler. 2007. Inventorizing an invader: Annotated type catalogue of Corbiculidae Gray, 1847 (Bivalvia, Heterodonta, Veneroidea), including Old World limnic *Corbicula* in the Natural History Museum Berlin. Malacologia 49: 243-272. DOI: 10.4002/0076-2997-49.2.243.

Graf, D.L. 2013. Patterns of freshwater bivalve global diversity and the state of phylogenetic studies on the Unionoida, Sphaeriidae, and Cyrenidae. American Malacological Bulletin 31: 135-153. DOI: 10.4003/006.031.0106.

Hedtke, S.M., K. Stanger-Hall, R.J. Baker, and D.M. Hillis. 2008. All-male asexuality: Origin and maintenance of androgenesis in the Asian clam *Corbicula*. Evolution 62: 1119-1136. DOI: 10.1111/ j.1558-5646.2008.00344.x.

Hedtke, S.M., M. Glaubrecht, and D.M. Hillis. 2011. Rare gene capture in predominantly androgenetic species. Proceedings of the National Academy of Sciences 108: 9520-9524. DOI: 10.1073/pnas.1106742108.

Hillis, D.M., and J.C. Patton. 1982. Morphological and electrophoretic evidence for two species of *Corbicula* (Bivalvia: Corbiculidae) in North America. American Midland Naturalist 108 (1): 74- 80. DOI: 10.2307/2425294.

Huber, M. 2015. Compendium of Bivalves 2. ConchBooks, Harxheim, Germany.

Isom, B.G. 1986. Historical review of Asiatic clam (*Corbicula*) invasion and biofouling of waters and industries in the Americas. American Malacological Bulletin, Special Edition 2: 1-5.

Ituarte, C.F. 1994. *Corbicula* and *Neocorbicula* (Bivalvia, Corbiculidae) in the Paraná, Uruguay, and Rio de La Plata Basins. Nautilus 107: 129-135.

IUCN Red List of Threatened Species. 2017. Version 2017-1. www.iucnredlist.org. Accessed 16 May 2017.

Kijviriya, V., E.S. Upatham, V Viyanant, and D.S. Woodruff. 1991. Genetic studies of Asian clams, *Corbicula,* in Thailand: Allozymes of 21 nominal species are identical. American Malacological Bulletin 8: 97-106.

Komaru, A., K. Konishi, I. Nakayama, T. Kobayashi, H. Sakai, and K. Kawamura. 1997. Hermaphroditic freshwater clams in the genus *Corbicula* produce non-reductional spermatozoa with somatic DNA content. Biological Bulletin 193: 320-323.

Konishi, K., K. Kawamura, H. Furuita, and A. Komaru. 1998. Spermatogenesis of the freshwater clam *Corbicula* aff. *fluminea* Müller (Bivalvia: Corbiculidae). Journal Shellfish Research 17: 185-189.

Korniushin, A. 2004. A revision of some Asian and African freshwater clams assigned to *Corbicula fluminalis* (Müller, 1774) (Mollusca: Bivalvia: Corbiculidae), with a review of anatomical characters and reproductive features based on museum collections. Hydrobiologia 529: 251-270. DOI: 10.1007/s10750-004-9322-x.

Korniushin, A.V., and M. Glaubrecht. 2003. Novel reproductive modes in freshwater clams: Brooding and larval morphology in Southeast Asian taxa of *Corbicula* (Mollusca, Bivalvia, Corbiculidae). Acta Zoologica 84: 293-315.

Lee, T., S. Siripattrawan, C.F. Ituarte, and D. Ó Foighil. 2005. Invasion of the clonal clams: *Corbicula* lineages in the New World. American Malacological Bulletin 20: 113-122.

Mackie, G.L. 2007. Biology of freshwater corbiculid and sphaeriid clams of North America. Ohio Biological Survey Bulletin, new series 15 (3): ix, 1-436.

Mayor, A.D., and R. Ancog. 2016. Fishery status of freshwater calm (*Batissa violacea,* Corbiculidae) (Bivalvia) (Lamarck, 1818) in Cagayan River, northern Philippines. International Journal of Fisheries and Aquatic Studies 4: 500-506.

Matsukawa, M., and K. Koarai. 2017. Late Mesozoic bivalve faunas from the Tetori Group, Japan. Cretaceous Research 71: 145-165. DOI: 10.1016/j.cretres.2016.11.016.

Mikkelsen, P.M., and R. Bieler. 2007. Seashells of Southern Florida: Living Marine Mollusks of the Florida Keys and Adjacent Regions: Bivalves. Princeton University Press. Princeton, New Jersey.

MolluscaBase. 2018. Cyrenidae Gray, 1840. www .molluscabase.org/aphia.php?p=taxdetails&id=238370. Accessed 23 May 2018.

Morton, B. 1976. The biology and functional morphology of the Southeast Asian mangrove bivalve, *Polymesoda (Geloina) erosa* (Solander, 1786) (Bivalvia: Corbiculidae). Canadian Journal of Zoology 54: 482-500.

Morton, B. 1986. *Corbicula* in Asia—an updated synthesis. American Malacological Bulletin, Special Edition No. 2: 113-124.

Morton, B. 1992. The salinity tolerance of *Cyrenobatissa subsulcata* (Bivalvia: Corbiculoidea) from China. Malacological Review 25: 103-108.

Morton, B., and K.Y. Tong. 1985. The salinity tolerance of *Corbicula fluminea* (Bivalvia: Corbiculoidea) from Hong Kong. Malacological Review 18: 91-95.

Mouthon, J. 1981. Sur la présence en France et au Portugal de *Corbicula* (Bivalvia, Corbiculidae) originaire d'Asie. Basteria 45: 109-116.

Nishida, N., A. Shirai, K. Koarai, K. Nakada, and M. Matsukawa. 2013. Paleoecology and evolution of Jurassic-Cretaceous corbiculoids from Japan. Palaeogeography, Palaeoclimatology, Palaeoecology 369: 239-252. DOI: 10.1016/j.palaeo.2012.10.030.

Paunović, M., B. Csányi, S. Knežević, V. Simić, D. Nenadić, D. Jakovčev-Todorović, B. Stojanović, and P. Cakić. 2007. Distribution of Asian clams *Corbicula fluminea* (Müller, 1774) and *C. fluminalis* (Müller, 1774) in Serbia. Aquatic Invasions 2: 99-106. DOI: 10.3391/ai.2007.2.2.3.

Pereira, D., M.C.D. Mansur, L.D.S. Duarte, A.S. de Oliveira, D.M. Pimpão, C.T. Callil, C. Ituarte, et al. 2014. Bivalve distribution in hydrographic regions in South America: Historical overview and conservation. Hydrobiologia 735 (1): 15-44. DOI: 10.1007/s10750-013-1639-x.

Pfenninger, M., F. Reinhardt, and B. Streit. 2002. Evidence for cryptic hybridization between different evolutionary lineages of the invasive clam genus *Corbicula* (Veneroida, Bivalvia). Journal of Evolutionary Biology 15: 818-829. DOI: 10.1046/j.1420-9101.2002.00440.x.

Pigneur, L.-M., J. Marescaux, K. Roland, E. Etoundi, J.-P. Descy, and K. Van Doninck. 2011. Phylogeny and androgenesis in the invasive *Corbicula* clams (Bivalvia, Corbiculidae) in western Europe. BMC Evolutionary Biology 11: 147. DOI: 10.1186/1471-2148-11-147.

Pigneur, L-M., S.M. Hedtke, E. Etoundi, and K. Van Doninck. 2012. Androgenesis: A review through the study of the selfish shellfish *Corbicula* spp. Heredity 108: 581-591. DOI: 10.1038/hdy.2012.3.

Pigneur, L.-M., E. Etoundi, D.C. Aldridge, J. Marescaux, N. Yasuda, and K. Van Doninck. 2014a. Genetic uniformity and long-distance clonal dispersal in the invasive androgenetic *Corbicula* clams. Molecular Ecology 23: 5102-5116. DOI: 10.1111/mec.12912.

Pigneur, L.-M., E. Falisse, K. Roland, E. Everbecq, J.-F. Delière, J.S. Smitz, K. Van Doninck, and J.-P. Descy. 2014b. Impact of invasive Asian clams, *Corbicula* spp., on a large river ecosystem. Freshwater Biology 59: 573-583. DOI: 10.1111/fwb.12286.

Qiu, A., A. Shi, and A. Komaru. 2001. Yellow and brown shell color morphs of *Corbicula fluminea* (Bivalvia: Corbiculidae) from Sichuan Province, China, are triploids and tetraploids. Journal of Shellfish Research 20: 323-328.

Renard, E., V. Bachmann, M.L. Cariou, and J.C. Moreteau. 2000. Morphological and molecular differentiation of invasive freshwater species of the genus *Corbicula* (Bivalvia, Corbiculidea) suggest the presence of three taxa in French rivers. Molecular Ecology 9: 2009-2016. DOI: 10.1046/j.1365-294X.2000.01104.x.

Rintelen, T. von, and M. Glaubrecht. 2006. Rapid evolution of sessility in an endemic species flock of the freshwater bivalve *Corbicula* from ancient lakes on Sulawesi, Indonesia. Biology Letters 2: 73-77. DOI: 10.1098/rsbl.2005.0410.

Rueda, M., and H.-J. Urban. 1998. Population dynamics and fishery of the fresh-water clam *Polymesoda solida* (Corbiculidae) in Cienaga Poza Verde, Salamanca Island, Colombian Caribbean. Fisheries Research 39: 75-86.

Simone, L.R.S., P.M. Mikkelsen, and R. Bieler. 2015. Comparative anatomy of selected marine bivalves from the Florida Keys, with notes on Brazilian congeners (Mollusca: Bivalvia). Malacologia 58:1-127. DOI: 10.4002/040.058.0201.

Siripattrawan, S., J.K. Park, and D. Ó Foighil. 2000. Two lineages of the introduced Asian freshwater clam *Corbicula* occur in North America. Journal of Molluscan Studies 66: 423-429. DOI: 10.1093/mollus/66.3.423.

Sousa, R., C. Antunes, and L. Guilhermino. 2008. Ecology of the invasive Asian clam *Corbicula fluminea* (Müller, 1774) in aquatic ecosystems: An overview. International Journal of Limnology 44: 85-94. DOI: 10.1051/limn:2008017.

Suja, N., and K.S. Mohamed. 2010. The Black Clam, *Villorita cyprinoides*, fishery in the State of Kerala, India. Marine Fisheries Review 72: 48-61.

Taylor, J.D., E.A. Glover, and S.T. Williams. 2009. Phylogenetic position of the bivalve family Cyrenoididae— removal from (and further dismantling of) the superfamily Lucinoidea. Nautilus 123: 9-13.

Tiemann, J.S., A.E. Haponski, S.A. Douglass, T. Lee, K.S. Cummings, M.A. Davis, and D. Ó Foighil. 2017. First record of a putative novel invasive *Corbicula* lineage discovered in the Illinois River, Illinois, USA. BioInvasions Records 6: 159-166. DOI: 10.3391/bir.2017.6.2.12.

36 Dreissenidae Gray, 1840

NATHANIEL T. MARSHALL AND CAROL A. STEPIEN

The heterodont bivalve family Dreissenidae Gray, 1840 belongs to the Order Myida Stoliczka, 1870. Dreissenidae comprises three extant genera, which inhabit fresh and/or brackish waters, including *Dreissena* Van Beneden, 1835 (native to Eurasia), *Congeria* Partsch, 1835 (endemic to karst caverns along the Adriatic Sea), and *Mytilopsis* Conrad, 1857 (native to the Americas). *Dreissena* and *Mytilopsis* are notorious invasive species, which have spread throughout waterways through shipping transport and canals, and intercontinentally via ballast water (see van der Velde et al., 2010; Stepien et al., 2013). Life history and morphological characters of Dreissenidae include (1) free-swimming larvae (veligers) for dispersal, (2) a byssus for attachment, (3) byssal retractor muscles, and (4) reduced pedal retractor muscles that attach the byssal threads to substrate (the foot is not used for extensive locomotion, which was an ancestral trait) (Nuttall, 1990; Morton, 1993).

Ancestral Dreissenidae are believed to have originated in freshwater/brackish basins of the Tethys Ocean in Pangaea during the Triassic period, around the time when North and South America, Europe, and Africa began to break apart (Morton, 1993; Stepien et al., 2013). Phylogenetic analyses based on DNA sequences show that *Mytilopsis* and *Congeria* are sister taxa, comprising the sister group to *Dreissena* (Therriault et al., 2004; Stepien et al., 2013; Bilandžija et al., 2013).

Phylogenetic analyses using the mtDNA COI gene conformed to a molecular clock prediction, suggesting that dreissenid diversification patterns are relevant to evolutionary time (Stepien et al., 2013). Dreissenidae radiation and evolution occurred in the ancient Paratethys Sea, which once expanded across central Europe and western Asia, covering today's Black, Azov, Caspian, and Aral Seas (Starobogatov, 1994; Harzhauser and Mandic, 2010). Two primary taxon diversification events took place; the first, with isolation of the Paratethys Sea ~18 mya, and the second with its change from brackish to freshwater (~12.7 mya; Harzhauser and Mandic, 2010).

The genus *Mytilopsis* dates to the oldest diversification of the Dreissenidae, from the Oligiocene

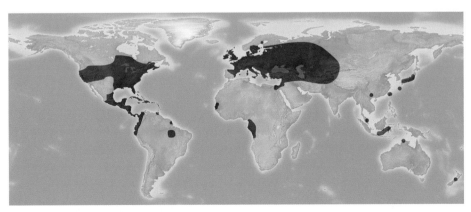

Distribution of Dreissenidae.

to mid-Miocene (Harzhauser and Mandic, 2010). Ancestral *Congeria* first appeared in the fossil record during the late-Miocene Epoch ~11.5 mya, when the central Paratethys Sea was replaced by the brackish-water Lake Pannon. The earliest fossil record of the genus *Dreissena* also dates to Lake Pannon, ~10.6 mya (Harzhauser and Mandic, 2010), and by the end of the Miocene Epoch (~6.2 to 5.8 mya), the *Pontodreissena* lineage had diverged from the common ancestral clade of *Dreissena/Carinodreissena* (Stepien et al., 2013). During the early-Pliocene Epoch, salinity fluctuations caused by glacial advances and saltwater incursions (Briggs, 1974), presumably led to the diversification of today's extant *Dreissena* species (Mordukhai-Boltovskoi, 1960).

Dreissena polymorpha (Pallas, 1771). USA: Des Plaines River, Gurnee, Lake County, Illinois. 40 mm. INHS 84526.

The number of extant species in the brackish-water genus *Mytilopsis* is debated (i.e., Fernandes et al., 2018); its fossils once inhabited the Old World, where it became extinct (Harzhauser and Mandic, 2010). *Mytilopsis* has been widespread by shipping; e.g., the dark false mussel *M. leucophaeata* (Conrad, 1831), which is native to the North American south-eastern Atlantic and Gulf coasts, is invasive in Europe (van der Velde et al., 1992; Zhulidov et al., 2018) and the black-striped mussel *M. sallei* (Récluz, 1849), which is native in the Gulf of Mexico to Colombia, has invaded Southeast Asia and the Mediterranean Sea (Tan and Morton, 2006; Galil and Bogi, 2009).

Congeria once contained a wide diversity of Old World coastal species that radiated during the late-Miocene Epoch; however, just three species are extant: *C. kusceri* Bole, 1962, *C. mulaomerovici* Morton and Bilandžija, 2013, and *C. jalzici* Morton and Bilandžija, 2013 (Bilandžija et al., 2013). These inhabit select karst caves in the southern European Dinaric Alps (Morton et al., 1998; Bilandžija et al., 2013). *Congeria* differs from other dreissenids in its shell form and microstructure, and in reproductive strategies (Morton, 1993; Morton et al., 1998; Morton and Pujas, 2013). *Congeria* has a small foot and short thick byssal threads and is more sedentary than other dreissenids (Morton et al., 1998). Like the other dreissenids, it forms large density clusters (Glavaš et al., 2017). Unlike other dreissenids, it is long-lived (30–40 years), with low recruitment and low adult mortality (Morton et al., 1998; Stepien et al., 2001).

Dreissena contains three subgenera, with *Pontodreis-*

sena Logvinenko & Starobogatov, 1966, forming the sister group to a clade containing *Dreissena* and *Carinodreissena* L'vova & Starobogatov, 1982 (Stepien et al., 2013). Each subgenus contains two extant species (Stepien et al., 2013). *Pontodreissena* includes the quagga mussel *D. rostrifomis* (Deshayes, 1838), which is native to the Ponto-Caspian region and is invasive throughout much of the rest of Europe and in North America, and *D. caputlacus* Schütt, 1993, which is endemic throughout Turkey. The subgenus *Dreissena* contains the highly successful invasive zebra mussel *D. polymorpha* (Pallas, 1771) and *D. anatolica* Locard, 1893, which is endemic to Turkish lakes. The two species in the subgenus *Carinodreissena* inhabit ancient Balkan freshwater lakes that possess high endemic biodiversity (Wilke et al., 2010), with *D. carinata* (Dunker, 1853) in the northern and central Balkan Peninsula, and *D. blanci* (Westerlund, 1890) in the south (Albrecht et al., 2007).

Dreissena polymorpha and *D. rostriformis* are two of the most notorious aquatic invasive species, due to their tremendous fouling and filtering capabilities and their transformation of soft to hard benthos, which exert ecological community changes (Mills et al., 1996; Zhu et al., 2006; Ward and Ricciardi, 2007). These dreissenids have fundamentally altered abiotic and biotic interactions, thereby exerting severely deleterious ecological (Vanderploeg et al., 2002; Strayer, 2009; Higgins and Vander Zanden, 2010) and economic (Pimentel et al., 2005) effects in invasive habitats. For the past 200 years, they spread throughout Europe via shipping and canals (van der

Velde et al., 2010; Zhulidov et al., 2010), were transported in the mid-1980s to the North American Great Lakes in ballast water (Hebert et al., 1989; Mills et al., 1996), and have been transported widely via trailered boats across much of North America (Bossenbroek et al., 2007).

Population genetic studies using DNA markers are useful for tracing the source(s) of an invasion. Such studies indicate that the North America zebra and quagga mussel invasions were founded from multiple European sources, with the zebra mussel tracing to the Baltic Sea and northern Europe, and the quagga mussel to northern Black Sea estuaries (Stepien et al., 2005; 2013; Brown and Stepien, 2010). Similar genetic diversity levels in North American invasive populations and their Eurasian sources indicate that they were founded by large numbers of propagules.

Due to the invasive ability of *Mytilopsis* and some *Dreissena,* most Dreissenidae are listed as taxa of Least Concern, except for *D. blanci* and *Congeria* spp., which are listed as Vulnerable, due to their limited ranges and endemism (IUCN, 2017). Continued surveying will aid prevention of future invasions of some taxa (see Peñarrubia et al., 2016; De Ventura et al., 2017), and provide more complete description of poorly understood ones (i.e., *Mytilopsis* and *Congeria*).

LITERATURE CITED

Albrecht, C., R. Schultheis, T. Kevrekidis, B. Streit, and T. Wilke. 2007. Invaders or endemics? Molecular phylogenetics, biogeography and systematics of *Dreissena* in the Balkans. Freshwater Biology 52: 1525-1536.

Bilandžija, H., B. Morton, M. Podnar, and H. Ćetković. 2013. Evolutionary history of relict *Congeria* (Bivalia: Dreissenidae): Unearthing the subterranean biodiversity of the Dinaric Karst. Frontiers in Zoology 10: 5.

Bossenbroek, J.M., L.E. Johnson, B. Peters, and D.M. Lodge. 2007. Forecasting the expansion of zebra mussels in the United States. Conservation Biology 21: 800-810.

Briggs, J.C. 1974. Marine Zoogeography. McGraw-Hill, New York.

Brown, J.E., and C.A. Stepien. 2010. Population genetic history of the dreissenid mussel invasion: Expansion patterns across North America. Biological Invasions 12: 3687-3710.

De Ventura, L., K. Kopp, K. Seppälä, and J. Jokela. 2017. Tracing the quagga mussel invasion along the Rhine River system using eDNA markers: Early detection and surveillance of invasive zebra and quagga mussels. Management of Biological Invasions 8: 101-112.

Fernandes, M.R., F. Salgueiro, I.C. Miyahira, and C.H.S. Caetano. 2018. mtDNA analysis of *Mytilopsis* (Bivalvia, Dreissenidae) invasion in Brazil reveals the existence of two species. Hydrobiologia 1: 97-110.

Galil, B.S., and C. Bogi. 2009. *Mytilopsis sallei* (Mollusca: Bivalvia: Dreissenidae) established on the Mediterranean coast of Israel. Marine Biodiversity Records 2: e73.

Glavaš, O.J., B. Jalžić, and H. Bilandžija. 2017. Population density, habitat dynamic and aerial survival of relict cave bivalves from genus *Congeria* in the Dinaric karst. International Journal of Speleology 46: 13-22.

Harzhauser, M., and O. Mandic. 2010. Neogene dreissenids in central Europe: Evolutionary shifts and diversity changes. Pp. 11-28 in The Zebra Mussel in Europe (G. van der Velde, S. Rajagopal, and A. Bij de Vaate, eds.). Backhuys, Leiden, Netherlands.

Hebert, P.D.N., B.W. Muncaster, and G.L. Mackie. 1989. Ecological and genetic studies on *Dreissena polymorpha* (Pallas): A new mollusc in the Great Lakes. Canadian Journal of Fisheries and Aquatic Sciences 46: 1587-1591.

Higgins, S.N., and M.J. Vander Zanden. 2010. What a difference a species makes: A meta-analysis of dreissenid mussel impacts on freshwater ecosystems. Ecological monographs 80: 179-196.

IUCN Red List of Threatened Species. Version 2016-3. www.iucnredlist.org. July 2017.

Mills, E.L., G. Rosenberg, A.P. Spidle, M. Ludyanskiy, Y. Pligin, and B. May. 1996. A review of the biology and ecology of the quagga mussel (*Dreissena bugensis*) a second species of freshwater dreissenid introduced to North America. American Zoologist 36: 271-286.

Mordukhai-Boltovskoi, F.D. 1960. Caspian Fauna in the Azov and Black Sea Basins. USSR AS Press, Moscow-Leningrad, Russia.

Morton, B. 1993. The anatomy of *Dreissena polymorpha* and the evolution and success of the heteromyarian form in the Dreissenoidea. Pp. 185-215 in Zebra Mussels: Biology, Impacts, and Control (T.F. Nalepa and D.W. Schloesser, eds.). CRC Press, Boca Raton, Florida.

Morton, B., F. Velkovrh, and B. Sket. 1998. Biology and anatomy of the "living fossil" *Congeria kusceri* (Bivalvia: Dreissenidae) from subterranean rivers and caves in the Dinaria karst of the former Yugoslavia. Journal of Zoology London 245: 147-174.

Morton, B., and S. Pujas. 2013. Life-history strategy, with ctenidial and pallial larval brooding, of the troglodytic 'living fossil' *Congeria kusceri* (Bivalvia: Dreissenidae) from the subterranean Dinaric Alpine karst of Croatia. Biological Journal of the Linnean Society 108(2): 294-314.

Nuttall, C.P. 1990. Review of the Caenozoic heterodont bivalve superfamily Dreissenacea. Palaeontology 33: 707-737.

Peñarrubia, L., C. Alcaraz, A. bij de Vaate, N. Sanz, C. Pla, O. Vidal, and J. Viñas. 2016. Validated methodology for quantifying infestation levels of dreissenid mussels in environmental DNA (eDNA) samples. Scientific Reports 6: 39067.

Pimentel, D., R. Zuniga, and D. Morrison. 2005. Update on the environmental and economic costs associated with alien-invasive species in the United States. Ecological Economics 52: 273-288.

Starobogatov, Y.I. 1994. Taxonomy and palaeontology. Pp. 47-55 in Freshwater Zebra Mussel *Dreissena polymorpha* (J.I. Starobogatov, ed.). Nauka, Moscow.

Stepien, C.A., B. Morton, K.A. Dabrowska, R.A. Guarnera, T. Radja, and B. Radja. 2001. Genetic diversity and evolutionary relationships of the troglodytic 'living fossil' *Congeria kusceri* (Bivalvia: Dreissenidae). Molecular Ecology 10: 1873-1879.

Stepien, C.A., J.E. Brown, M.E. Neilson, and M.A. Tumeo. 2005. Genetic diversity of invasive species in the Great Lakes versus their Eurasian source populations: Insights for risk analysis. Risk Analysis 25: 1043-1060.

Stepien, C.A., I.A. Grigorovich, M.A. Gray, T.J. Sullivan, S. Yerga-Woolwine, and G. Kalayci. 2013. Evolutionary, biogeographic, and population genetic relationships of dreissenid mussels, with revision of component taxa. Pp. 403-444 in Quagga Mussels and Zebra Mussels: Biology, Impacts, and Control (T.F. Nalepa and D.N Schloesser, eds.). CRC Press, Boca Raton, Florida.

Strayer, D.L. 2009. Twenty years of zebra mussels: Lessons from the mollusk that made headlines. Frontiers in Ecology and the Environment 7: 135-141.

Tan, K.S., and B. Morton. 2006. The invasive Caribbean bivalve *Mytilopsis sallei* (Dreissenidae) introduced into Malaysia and Johor Bahru Singapore. Raffles Bulletin of Zoology 54: 429-434.

Therriault, T.W., M.F. Docker, M.I. Orlova, D.D. Heath, and H.J. MacIsaac. 2004. Molecular resolution of the family Dreissenidae (Mollusca: Bivalvia) with emphasis on Ponto-Caspian species, including the first report of *Mytilopsis leucophaeata* in the Black Sea basin. Molecular Phylogenetics and Evolution 30: 479-489.

Vanderploeg, H.A., T.F. Nalepa, D.J. Jude, E.L. Mills, K.T. Holeck, J.R. Liebig, I.A. Grigorovich, and H. Ojaveer. 2002. Dispersal and emerging ecological impacts of Ponto-Caspian species in the Laurentian Great Lakes. Canadian Journal of Fisheries and Aquatic Sciences 59 (7): 1209-1228.

Van der Velde, G., K. Hermus, M. van der Gaag, and H. A. Jenner. 1992. Cadmium, zinc and copper in the body, byssus and shell of the mussels, *Mytilopsis leucophaeata* and *Dreissena polymorpha* in the brackish Noordzeekanaal of The Netherlands. Pp. 213-226 in The Zebra Mussel *Dreissena polymorpha* (D. Neumann and H.A. Jenner, eds.). Gustav Fischer Verlag, New York.

Van der Velde, G., S. Rajagopal, and A. Bij de Vaate. 2010. From zebra mussels to quagga mussels: An introduction to the Dreissenidae. Pp. 1-10 in The Zebra Mussel in Europe (G. van der Velde, S. Rajagopal, and A. Bij de Vaate, eds.). Backhuys, Leiden, Netherlands.

Ward, J.M., and A. Ricciardi. 2007. Impacts of *Dreissena* invasions on benthic macroinvertebrate communities: A meta-analysis. Diversity and Distributions 13: 155-165.

Wilke, T., R. Schultheiß, C. Albrecht, N. Bornmann, S. Trajanovski, and T. Kevrekidis. 2010. Native *Dreissena* freshwater mussels in the Balkans: In and out of ancient lakes. Biogeoscience 7: 3051-3065.

Zhu, B., D.G. Fitzgerald, C.M. Mayer, L.G. Rudstam, and E.L. Mills. 2006. Alteration of ecosystem function by zebra mussels in Oneida Lake: Impacts on submerged macrophytes. Ecosystems 9: 1017-1028.

Zhulidov, A.V., A.V. Kozhara, G.H. Scherbina, T.F. Nalepa, A. Protasov, S.A. Afanasiev, E.G. Pryanichnikova, D.A. Zhulidov, T.Yu. Gurtovaya, and D.F. Pavlov. 2010. Invasion history, distribution, and relative abundances of *Dreissena bugensis* in the old world: A synthesis of data. Biological Invasions 12: 1923-1940.

Zhulidov, A.V., A.V. Kozhara, G. van der Velde, R.S. Leuven, M.O. Son, T.Y. Gurtovaya, D.A. Zhulidov, T.F. Nalepa, V.J.R. Santiago-Fandino, and Y.S. Chuikov. 2018. Status of the invasive brackish water bivalve *Mytilopsis leucophaeata* (Conrad, 1831) (Dreissenidae) in the Ponto-Caspian region. Biological Invasions Records 7: 111-120.

37 Sphaeriidae Deshayes, 1855 (1820)

TAEHWAN LEE

The Sphaeriidae, commonly known as fingernail, pea, pill, or nut clams, has a cosmopolitan distribution in a variety of freshwater habitats, including rivers, streams, ponds, lakes, and even ephemeral pools (Burch, 1975; Kuiper, 1983). Although sphaeriid clams are the smallest freshwater bivalves, rarely exceeding 25 mm in shell length, they are often numerically dominant and play a key role in energy and nutrient cycling (Cummings and Graf, 2009). Recent phylogenetic studies (Dreher-Mansur and Meier-Brook, 2000; Korniushin and Glaubrecht, 2002; Lee and Ó Foighil, 2003) distinguish two primary subgroups, Euperinae and Sphaeriinae. The subfamily Euperinae is comprised of the genera *Byssanodonta* d'Orbigny, 1846, and *Eupera* Bourguignat, 1854, and has a restricted geographic distribution. *Byssanodonta* is represented by a single species, *B. paranensis,* endemic to the Paraná River, Argentina (Dreher-Mansur and Ituarte, 1999). *Eupera* occurs in Central and South America and Africa, although one species has been introduced to the southern United States (Heard, 1965b), and its range has recently expanded into the upper Missis-

sippi River basin (Sneen et al., 2009). The Sphaeriinae traditionally has been represented by three cosmopolitan genera, *Sphaerium* Scopoli, 1777, *Musculium* Link, 1807, and *Pisidium* Pfeiffer, 1821. However, a series of molecular studies (Lee and Ó Foighil, 2003; Schultheiß et al., 2008; Clewing et al., 2013) have consistently recovered *Musculium* as a subgenus nested within *Sphaerium,* and former subgeneric assemblages of *Pisidium* (*Afropisidium* Kuiper, 1962, *Odhneripisidium* Kuiper, 1962, *Cyclocalyx* Dall, 1903, and *Pisidium* s. str.) as generic level clades. Although relative branching order is unstable, a feature common to all the gene trees is that the *Afropisidium* is sister to the remainder of the Spaeriinae, either alone or in conjunction with the *Odhneripisidium.*

Convincing marine outgroups for the Sphaeriidae are presently unknown. Taxonomists (Keen and Casey, 1969; Boss, 1982) often grouped the Sphaeriidae with Cyrenidae (= Corbiculidae; Bieler et al., 2010), which has not only freshwater but brackish/marine representatives, into the superfamily Corbiculoidea based on shared shell and life history

Distribution of Sphaeriidae.

characteristics. However, the affinity between these two families has been rejected by both morphological (Dreher-Mansur and Meier-Brook, 2000) and molecular (Park and Ó Foighil, 2000; Taylor et al., 2007) studies. Similar brooding characters observed in sphaeriids and freshwater cyrenids are thought to represent convergent adaptations to freshwater habitats (Park and Ó Foighil, 2000). According to recent phylogenetic analyses (Taylor et al., 2007; Bieler et al., 2014), the Sphaeriidae forms a basal lineage within the unranked Neoheterodontei together with various marine bivalve groups, such as Myida, Mactroidea, Cyamiidae, Gaimardiidae, Ungulinidae, and Veneroida.

Sphaeriid taxonomy has been hampered by the paucity and plasticity of shell characters that contributed to a dramatic overestimation of species diversity (Lee, 2004). Kuiper (1983), for example, tallied more than 1000 published nominal species for *Pisidium s. lat.* but estimated that only about 80 species could actually be diagnosed. Revision of the North American sphaeriid fauna by Herrington (1962) greatly consolidated the early nomenclature of more than 330 names, and Burch (1975) listed only 37 species for North America. Sphaeriinae systematics has also been confounded by the extensive geographic ranges displayed by many taxa. For instance, *Cyclocalyx casertanum* is found on every continent except Antarctica (Burch, 1975; Kuiper, 1983) and almost half of North American sphaeriids co-occur in Europe and Asia (Burch, 1975). Passive dispersal by insects, birds, and fishes (Darwin, 1882; Rees, 1965; Brown, 2007) coupled with the ability to self-replicate (Meier-Brook, 1970) may account for their widespread distribution. Another problem has been the use of a strikingly divergent system of classification by the Russian taxonomic school (Korniushin, 1998). In general, Russian workers have attributed significant weight to all levels of morphological variation and have erected elaborate taxonomic rankings; Western taxonomists have tended to recognize larger and more inclusive taxa. For example, an estimated 200–500 sphaeriid species were extrapolated for the former Soviet Union (Scarlato, 1981), whereas fewer than 60 species are recognized in North America and Europe collectively (Bowden and Heppell, 1968; Burch, 1975). Currently, a total of 227 sphaeriid species are

Sphaerium simile (Say, 1817). USA: Menomonee River, Menomonee Falls, Waukesha County, Wisconsin. 19 mm. INHS 76579.

recognized in every continent except for Antarctica (Graf, 2013), although lines of evidence indicate that this may be an underestimation. Extended sampling efforts in mountainous areas and ancient lakes have recovered multiple genetically and morphologically distinct cryptic lineages (Guralnick, 2005; Schultheiß et al., 2008; Clewing et al., 2013).

Sphaeriid clams have complex reproductive and developmental features, some of which may be associated with evolutionary colonization of freshwater habitats. All sphaeriids studied to date are simultaneous hermaphrodites, with each individual producing eggs and sperm concurrently (Heard, 1965a; Ituarte, 1997). It is generally believed that sphaeriids typically reproduce by cross-fertilization but can facultatively self-fertilize (Heard, 1965a; Araujo and Ramos, 1999). Sphaeriid clams brood their direct-developing young within the gills until they are released as juveniles (Heard, 1965a, 1977; Mackie et al., 1974; Ituarte, 1997). The planktonic veliger larval stage, which is phylogenetically widespread in marine bivalves, has been completely lost (Dreher-Mansur and Meier-Brook, 2000), as in the freshwater cyrenids and unionids (Cummings and Graf, 2009). Sphaeriid taxa, however, show different degrees of complexity in the details of how brooding is achieved. The simplest form, synchronous brooding, is found in the subfamily Euperinae, where all embryos in a brood are spawned at the same time and develop as a single cohort (Heard, 1965b; Dreher-Mansur and Ituarte, 1999; Dreher-Mansur and Meier-Brook, 2000). Species of *Pisidium s. lat.* are also synchronous brooders. Unlike in the Euperinae genera *Byssanodonta* and *Eupera,* the embryos of *Pisidium s. lat.* develop within a distinct brood-sac, which is linked to the parental tissues (Heard, 1965a, 1977; Mackie et al., 1974). In contrast, *Sphaerium* (including subgenus *Musculium*) species are sequential brooders, where multiple subsets of embryos resulting from different spawning events are simultaneously present within separate brood-sacs (Heard, 1977; Mackie, 1978). Molecular phylogenetic studies (Cooley and Ó Foighil, 2000; Lee and Ó Foighil, 2003) suggest the evolutionary elaboration in parental care complexity from the relatively simple synchronous brooding, to the origin of marsupial sacs and parental embryonic nourishment, and eventually to the sequential brooding.

Cytogenetic studies (Lee, 1999; Kořínková and Morávkoá, 2010) of various species belong to the subfamily Sphaeriinae have added a genomic dimension to inferences of reproduction, cladogenesis, and evolution in the family. These authors revealed a remarkable degree of polyploidization, ranging from 30 to 247 mitotic chromosomes. Of ~25 species observed, only 4 species—3 that are closely related in Europe (*Sphaerium corneum:* 2n=30 or 36, *S. nucleus:* 2n=30, and *S. solidum:* 2n=30) and 1 in North America (*S. rhomboieum:* 2n=44)—had diploid chromosome complements (Keyl, 1956; Petkevičiūtė et al., 2007; Kořínková and Král, 2011; Stunžėnas et al., 2011). All the remaining taxa displayed well over 100 chromosomes, many more than the normal range known for the Bivalvia. Not only is genome amplification prevalent, but significant variation in ploidy levels (2n-13n) occurs in the subfamily. Although polyploidy is rare in animals, particularly in bivalve mollusks, it has been recorded in multiple asexual lineages of freshwater *Corbicula* (Komaru et al., 1998) and of marine *Lasaea* (Ó Foighil and Thiriot-Quiévreux, 1991). Pronounced polyploidy is frequently associated with asexuality (Mogie, 1986), and the similarities, such as polyploidy, hermaphroditism, and prolonged parental care, observed in the Sphaeriinae and in asexual *Corbicula* and *Lasaea,* are striking. Molecular evidence (Lee and Ó Foighil, 2002) suggests that ancestral patterns of polyploidization may underlay much of the present diversity of this freshwater clam radiation, although the evolutionary origins of genome duplication are not yet clear.

LITERATURE CITED

Araujo, R., and M.A. Ramos. 1999. Histological description of the gonad, reproductive cycle, and fertilization of *Pisidium amnicum* (Müller, 1774) (Bivalvia: Sphaeriidae). Veliger 42: 124-131.

Bieler, R., J.G. Carter, and E.V. Coan, 2010. Classification of bivalve families. In Nomenclator of Bivalve Families (P. Bouchet and J.-P. Rocroi, eds.). Malacologia 52: 113-133.

Bieler, R., P.M. Mikkelsen, T.M. Collins, E.A. Glover, V.L. González, D.L. Graf, E.M. Harper, et al. 2014. Investigating the Bivalve Tree of Life—an exemplar-based approach combining molecular and novel morphological characters. Invertebrate Systematics 28: 32-115. DOI: 10.1071/IS13010.

Boss, K.J. 1982. Mollusca. Pp. 945-1166 in Synopsis and Classification of Living Organisms, vol. 1 (Parker, S.P., ed.). McGraw Hill, New York.

Bowden, J., and D. Heppell, 1968. Revised list of British Mollusca. 2. Unionacea—Cardiacea. Journal of Conchology 26: 237-272.

Brown, R.J. 2007. Freshwater mollusks survive fish gut passage. Arctic 60: 124-128.

Burch, J.B. 1975. Freshwater Sphaeriacean Clams (Mollusca: Pelecypoda) of North America. Rev. ed. Malacological Publications, Hamburg, Michigan.

Clewing, C., U. Bössneck, P.V. von Oheimb, and C. Albrecht, 2013. Molecular phylogeny and biogeography of a high mountain bivalve fauna: The Sphaeriidae of the Tibetan Plateau. Malacologia 56: 231-252.

Cooley, L.R., and D. Ó Foighil, 2000. Phylogenetic analysis of the Sphaeriidae (Mollusca: Bivalvia) based on partial mitochondrial 16S rDNA gene sequences. Invertebrate Biology 119: 299-308.

Cummings, K.S., and D.L. Graf. 2009. Mollusca: Bivalvia. Pp. 309-384 in Ecology and Classification of North American Freshwater Invertebrates (J.H. Thorp and A.P. Covich, eds.). 3rd ed. Academic Press, Elsevier, New York.

Darwin, C. 1882. On the dispersal of freshwater bivalves. Nature 649: 529-530.

Dreher-Mansur, C.D., and C. Ituarte, 1999. Morphology of *Eupera elliptica* Ituarte & Dreher-Mansur 1993, with comments on the status of the genera within the Euperinae (Bivalvia: Sphaeriidae). Malacological Review, Supplement 8: 59-68.

Dreher-Mansur, C.D., and C. Meier-Brook, 2000. Morphology of *Eupera* Bourguignat 1854, and *Byssanodonta*

Orbigny 1846 with contributions to the phylogenetic systematics of Sphaeriidae and Corbiculidae (Bivalvia: Veneroida). Archiv für Molluskenkunde 128: 1–59.

Graf, D.L. 2013. Patterns of freshwater bivalve global diversity and the state of phylogenetic studies on the Unionoida, Sphaeriidae, and Cyrenidae. American Malacological Bulletin 31: 135–153. DOI: 10.4003/006.031.0106.

Guralnick, R.P. 2005. Combined molecular and morphological approaches to documenting regional biodiversity and ecological patterns in problematic taxa: A case study in the bivalve group Cyclocalyx (Sphaeriidae, Bivalvia) from western North America. Zoological Scripta 34: 469–482.

Heard, W.H. 1965a. Comparative life histories of North American pill clams (Sphaeriidae: Pisidium). Malacologia 2: 381–411.

Heard, W.H. 1965b. Recent Eupera (Pelecypoda: Sphaeriidae) in the United States. American Midland Naturalist 74: 309–317.

Heard, W.H. 1977. Reproduction of fingernail clams (Sphaeriidae: Sphaerium and Musculium). Malacologia 16: 421–455.

Herrington, H.B. 1962. A revision of the Sphaeriidae of North America (Mollusca: Pelecypoda). Miscellaneous Publications of the Museum of Zoology, University of Michigan, 118: 1–74.

Ituarte, C.F. 1997. The role of the follicular epithelium in the oosorption process in Eupera platensis Doello Jurado, 1921 (Bivalvia: Sphaeriidae): A light microscope approach. Veliger 40: 47–54.

Keen, M., and H. Casey, 1969. Superfamily Corbiculacea. Pp. N664–N669 in Treatise on Invertebrate Paleontology, part N, vol. 2, Mollusca 6, Bivalvia (R.C. Moore, ed.). Geological Society of America and University of Kansas Press, Lawrence.

Keyl, H.G. 1956. Boobachtungen über die ♂-meiose der Muschel Sphaerium corneum. Chromosoma 8: 12–17.

Komaru, A., T. Kawagishi, and K. Konishi, 1998. Cytological evidence of spontaneous androgenesis in the freshwater clam Corbicula leana Prime. Development Genes and Evolution 208: 46–50.

Kořínková, T., and J. Král. 2011. Structure and meiotic behaviour of B chromosomes in Sphaerium corneum/S. Nucleus complex. Genetica 139: 155–165.

Kořínková, T., and A. Morávkoá. 2010. Does polyploidy occur in central European species of the family Sphaeriidae (Mollusca: Bivalvia)? Central European Journal of Biology 5: 777–784.

Korniushin, A.V. 1998. Review of the studies on freshwater bivalve mollusc systematics carried out by the Russian taxonomic school. Malacological Review, Supplement 7: 65–82.

Korniushin, A.V., and M. Glaubrecht, 2002. Phylogenetic analysis based on the morphology of viviparous freshwater clams of the family Sphaeriidae (Mollusca, Bivalvia, Veneroida). Zoologica Scripta 31: 415–459.

Kuiper, J.G.J. 1983. The Sphaeriidae of Australia. Basteria 47: 3–52.

Lee, T. 1999. Polyploidy and meiosis in the freshwater clam Sphaerium striatinum (Lamarck) and chromosome numbers in the Sphaeriidae (Bivalvia, Veneroida). Cytologia 64: 247–252.

Lee, T. 2004. Morphology and phylogenetic relationships of genera of North American Sphaeriidae (Bivalvia, Veneroida). American Malacological Bulletin 19: 1–13.

Lee, T., and D. Ó Foighil. 2002. 6-Phosphogluconate dehydrogenase (PGD) allele phylogeny is incongruent with a recent origin of polyploidization in some North American Sphaeriidae (Mollusca, Bivalvia). Molecular Phylogenetics and Evolution 25: 112–124.

Lee, T., and D. Ó Foighil, 2003. Phylogenetic structure of the Sphaeriinae, a global clade of freshwater bivalve molluscs, inferred from nuclear (ITS-1) and mitochondrial (16S) ribosomal gene sequences. Zoological Journal of the Linnean Society 137: 245–260.

Mackie, G.L. 1978. Are sphaeriid clams ovoviviparous or viviparous? Nautilus 92: 145–147.

Mackie, G.L., S.U. Qadri, and A.H. Clarke, 1974. Development of brood sacs in Musculium securis Bivalvia: Sphaeriidae. Nautilus 88: 109–111.

Meier-Brook, C. 1970. Untersuchungen zur biologie einiger Pisidium-Arten (Mollusca; Eulamellibranchiata; Sphaeriidae). Archiv für Hydrobiologie, Supplement 38: 73–150.

Mogie, M. 1986. On the relationship between asexual reproduction and polyploidy. Journal of Theoretical Biology 122: 493–498.

Ó Foighil, D., and C. Thiriot-Quiévreux, 1991. Ploidy and pronuclear interaction in northeastern Pacific Lasaea clones. Biological Bulletin 181: 222–231.

Park, J-K., and D. Ó Foighil, 2000. Sphaeriid and corbiculid clams represent separated heterodont bivalve radiations into freshwater environments. Molecular Phylogenetics and Evolution 14: 75–88.

Petkevičiūtė, R., G. Stanevičiūtė, V. Stunžėnas, T. Lee, and D. Ó Foighil. 2007. Pronounced karyological divergence of the North American congeners Sphaerium rhomboideum and S. occidentale (Bivalvia, Veneroida, Sphaeriidae). Journal of Molluscan Studies 73: 315–321.

Rees, W.J. 1965. The aerial dispersal of Mollusca. Proceedings of the Malacological Society of London 36: 269–282.

Scarlato, O.A. 1981. Research of the Soviet malacologists in the recent years. Venus 40: 160–176.

Schultheiß, R., C. Albrecht, U. Bößneck, and T. Wilke, 2008.

The neglected side of speciation in ancient lakes: Phylogeography of an inconspicuous mollusc taxon in lakes Ohrid and Prespa. Hydrobiologia 615: 141–156.

Sneen, M.E., K.S. Cummings, T. Minarik Jr., and J. Wasik, 2009. The discovery of the nonindigenous, mottled fingernail clam, *Eupera cubensis* (Prime, 1865) (Bivalvia: Sphaeriidae) in the Chicago Sanitary and Ship Canal (Illinois River drainage), Cook County, Illinois. Journal of Great Lakes Research 35: 627–629.

Stunžėnas, V., R. Petkevičiūtė, and G. Stanevičiūtė. 2011. Phylogeny of *Sphaerium solidum* (Bivalvia) based on karyotype and sequences of 16S and ITS1 rDNA. Central European Journal of Biology 6: 105–117.

Taylor, J.D., S.T. Williams, E.A. Glover, and P. Dyal, 2007. A molecular phylogeny of heterodont bivalves (Mollusca: Bivalvia: Heterodonta): New analyses of 18S and 28S rRNA genes. Zoologica Scripta 36: 587–606.

38 Unionidae Rafinesque, 1820, and the General Unionida

DANIEL L. GRAF AND KEVIN S. CUMMINGS

The bivalve family Unionidae Rafinesque, 1820, is the most species-rich of the six extant families of the order Unionida. The family is composed of 713 Recent species in 136 genera, classified among six subfamilies. The Unionidae is also the most geographically widespread family of the order, occurring in all major continental biogeographical regions, with diversity hotspots in North America and southeastern Asia. Invasive species, habitat degradation, and overexploitation have led to worldwide freshwater mussel population reductions, although the global magnitude of these declines is poorly documented in most regions.

The radiation of the order Unionida (aka freshwater mussels) into inland fresh waters was facilitated by two evolutionary novelties: parental care and larval parasitism (Graf, 2013). Parental care evolved separately in each of the three major freshwater bivalve clades: Unionida, Sphaeriidae, and Cyrenidae (the latter two Venerida), as evidenced by each lineage sharing a more recent common ancestor with marine bivalves than with each other (Bieler et al., 2014; Combosch et al., 2017). Freshwater bivalves in general brood their larvae within marsupia derived from the demibranchs of their ctenidia, opting away from the planktonic larvae of marine bivalves and allowing for protected embryonic development in hypotonic flowing waters. Freshwater mussels in particular complete their larval metamorphosis while encysted in the gills or fins of a host vertebrate—usually a fish, rarely an amphibian (Cummings and Graf, 2009). The anatomical traits associated with these reproductive specializations are valuable for recognizing family-group level lineages of freshwater mussels (Graf and Cummings, 2006).

Two factors make the biogeography of the Unionida distinct from other freshwater mollusks and invertebrates generally: longevity and poor dispersal ability. Many freshwater mussels are long-lived, with years to the age of first reproduction and life spans measured in decades (Haag, 2009). The parasitic period, while larvae are associated with their hosts, is the primary dispersal phase for these otherwise sedentary mollusks. As a result, freshwater mussels

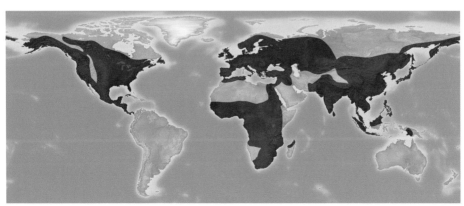

Distribution of Unionidae.

Lampsilis fasciola Rafinesque, 1820. USA: North Fork Vermilion River, Alvin, Vermilion County, Illinois. 70 mm. INHS 5486.

have no vagility across terrestrial or marine barriers, and species (and higher taxa) tend to be distributed among adjacent, confluent basins. Geographical disjunctions within clades are indicative of past confluence (Graf, 2002). The typically high spatial and temporal turnover rates of other aquatic invertebrates have led many biogeographers to rely on vertebrates to establish higher order patterns of freshwater diversity (e.g., ecoregions) (Abell et al., 2008). However, freshwater mussels are not like most invertebrates. Patterns of freshwater mussel richness and endemism are correlated with general aquatic community diversity and indicative of long-term, large-scale environmental processes (Graf and Cummings, 2011). The contemporary distributions of freshwater mussel taxa are the products of both current environmental conditions and geographical history.

The Recent Unionida is represented by six families: Unionidae, Margaritiferidae Haas, 1940, Hyriidae Swainson, 1840, Iridinidae Swainson, 1840, Mycetopodidae Gray, 1840, and Etheriidae Deshayes, 1830. The Unionidae is by far the most species-rich family of the order, with 713 of 918 total species (78%). That tally is surpassed only by the Hydrobiidae among freshwater mollusks (chap. 18, this volume). Graf and Cummings (2007) reviewed the global species richness of freshwater mussels, and that work has been updated on the MUSSEL Project Web Site (www .mussel-project.net/) and the Integrated Taxonomic Information System (www.itis.gov/). Both of these

websites provide complete and largely coordinated synonymies of all species and genera. All three of these sources provide references to the relevant literature. The Unionida is well supported as monophyletic and sister to the marine order Trigoniida (= *Neotrigonia* Cossmann, 1912) (Bieler et al., 2014; Combosch et al., 2017). The orders Unionida and Trigoniida comprise the bivalve subclass Palaeoheterodonta. The other families are discussed in the chapters that follow.

The Unionidae is monophyletic and classified with the Margaritiferidae in the superfamily Unionoidea Rafinesque, 1820 (Bieler et al., 2010). Unionids are morphologically diagnosable by the presence of a supra-anal aperture and the absence of mantle fusion between the incurrent and excurrent apertures (Graf and Cummings, 2006). Both margaritiferids and unionids exhibit unhooked-type glochidium larvae (Pfeiffer and Graf, 2015) and larval brooding in marsupia simultaneously occupying both inner and outer pairs of demibranchs (Graf and Ó Foighil, 2000), although these characteristics have been modified within the Unionidae (see below). Other tree topologies, with regard to the monophyly of the Unionidae and its relationship to the Margaritiferidae, have been frequently recovered but poorly supported (reviewed by Graf, 2013).

The fossil record of the Unionidae is difficult to interpret because (1) the genus name *"Unio"* is still in wide usage for worldwide paleo-species going back to the Triassic (Haas, 1969b; Watters, 2001), though the Recent genus of that name shows no evidence of being either that old or widespread (Froufe et al., 2016a; Araujo et al., 2018), and (2) fossil taxa are typically recognized by the same shell characteristics that phylogenetic studies of Recent mussels have shown to be highly variable and convergent (Van Damme et al., 2015; Zieritz et al., 2015). As a result, it has been difficult to distinguish stem species of the Unionida from stem species of the Unionidae from crown unionids. Skawina and Dzik (2011) discovered that well preserved Late Triassic freshwater mussels possessed filibranch ctenidia, a condition unknown in the Unionida crown group but present among the Trigoniida. Thus, it is unclear if the modern family clades (including Unionidae) extend that far back into the Mesozoic. Graf et al. (2015) applied the oldest

Table 38.1. Generic and species richness and geographic distribution of the Unionidae

Taxon	Genera	Species	Distribution
Unioninae	30	148	Europe, northern Africa, and the Middle East; eastern and southern Africa; central and eastern Asia, including the Philippines, Japan, and Kamchatka; western North America and Central America; eastern North America
Ambleminae	55	334	eastern North America and Central America
Gonideinae	18	77	Europe, northern Africa, and the Middle East; southern and eastern Asia; western North America
Rectidentinae	11	39	southeastern Asia, including the Sunda and Philippine Islands; southern Asia; New Guinea
Modellnaiinae	1	1	southeastern Asia
Parreysiinae	14	103	southern and southeastern Asia; Africa
incertae sedis	7	11	Madagascar; southeastern Asia, including Borneo
Totals	136	713	

Sources: Graf and Cummings, 2007; MUSSEL Project website, www.mussel-project.net/; Integrated Taxonomic Information System, www.itis.gov/).

Additional references for geographical regions are given in the text.

freshwater mussels of the Morrison Formation of North America to establish a minimum clade age for the Unionidae. Their analysis estimated a Middle Jurassic divergence between the Unionidae and Margaritiferidae (see also Bieler et al., 2014; Bolotov et al., 2017a). The biogeography of the Recent Unionidae is generally consistent with an origin on the northern continents following the disintegration of Pangaea during the Jurassic (see below).

Twentieth-century attempts to divide the Unionidae into subfamilies were contradictory and phylogenetically meaningless, leaving Graf and Cummings (2006, 2007) to call shenanigans on all available classifications of Old World freshwater mussels (e.g., Modell, 1964; Haas, 1969a; Starobogatov, 1970). Bieler et al. (2010) proposed a novel classification for the subfamilies of the Unionidae, and that system has been mostly consistent with recent phylogenetic work (with some revisions). Six subfamilies are recognized herein: Unioninae Rafinesque, 1820, Ambleminae Rafinesque, 1820, Gonideinae Ortmann, 1916, Rectidentinae Modell, 1942, Modellnaiinae Brandt, 1974, and Parreysiinae Henderson, 1935. The monophyly of and relationships among the subfamilies remain incompletely understood, although a basal split between the Parreysiinae and clade composed of the other subfamilies has been consistently supported

(Whelan et al., 2011; Pfeiffer and Graf, 2015; Bolotov et al., 2017a). Table 38.1 summarizes the genus and species richness of the subfamilies and their geographical distributions. Recognition of these subfamilies as clades has only recently figured prominently in assessments of global and regional freshwater mussel diversity (Lopes-Lima et al., 2017a).

The Nearctic region has the highest richness of unionid species in the world: 298 native Recent species (42% of the family) in 56 genera (Williams et al., 1993, 2017; Cummings and Graf, 2009; Haag, 2012) and an introduced species from southeast Asia, *Sinanodonta woodiana* (Lea, 1834) (Sousa et al., 2014). Biogeographically, the unionids of the region are divided between two endemic assemblages: a Pacific, western Nearctic assemblage with phylogenetic affinities to eastern Asia, and another assemblage east of the Rocky Mountains, extending from the Arctic south to Mexico and east to the Atlantic coastal drainages.

The western assemblage is composed of only three genera: *Gonidea* Conrad, 1857, *Anodonta* Lamarck, 1799, and *Sinanodonta* Modell, 1945. The monotypic *Gonidea* is the type genus of the Gonideinae, which is otherwise known from the Palearctic and Indotropics (Pfeiffer and Graf, 2015; Lopes-Lima et al., 2017a). Five species classified as *Anodonta* (Unioninae, tribe Anodontini Rafinesque, 1820) are found from

Alaska south to Mexico. However, these species are widely disjunct from Eurasian *Anodonta* (Chong et al., 2008; Lopes-Lima et al., 2017a). Williams et al. (2017) recently reclassified a sixth species of *Anodonta* to *Sinanodonta*, *S. beringiana* (Middendorff, 1851). The generic nomen *Berginiana* Starobogatov in Zatravkin, 1983, is available for those anodontine species.

The overwhelming majority of Nearctic freshwater mussel species occur east of the Rocky Mountains (292 spp., 98%). The native, eastern unionid assemblage is composed of (1) an endemic radiation of 11 genera (45 spp.) of the Unioninae (tribe Anodontini), and (2) the subfamily Ambleminae, which is restricted to eastern North America (42 genera, 246 spp.) and adjacent Neotropical Central America (20, 91) (Frierson, 1927). The monophyly of the Ambleminae is well established, and the subfamily has been divided among four tribes: Amblemini Rafinesque, 1820, Quadrulini Ihering, 1901, Pleurobemini Hannibal, 1912, and Lampsilini Ihering, 1901 (Campbell et al., 2005; Campbell and Lydeard, 2012). The eastern North American anodontine radiation has also been found to be monophyletic (Lopes-Lima et al., 2017a). The (sub)tribe Alasmidontini Rafinesque, 1820, has been applied to these taxa (Clarke, 1981, 1985).

The Indotropical region (extending from the Indus River to the Yangtze, including the Sunda and Philippine islands) is the second hotspot of unionid richness, with 238 species (33%) (Brandt, 1974; Graf and Cummings, 2007; He and Zhuang, 2013; Zieritz et al., 2018). The taxa are classified among five subfamilies: Unioninae (15 genera, 54 spp.), Gonideinae (12, 64), Rectidentinae (10, 37), Modellnaiinae (1, 1), and Parreysiinae (7, 69), and we regard six more unionid genera in the region as *incertae sedis* (10 spp.). Until recently, there had been relatively few phylogenetic studies with sufficient character and taxon sampling of the Indotropical Unionidae to test relationships among subfamilies. Much more phylogenetic information is now available, and the relationships among the genera are in a state of flux (Pfeiffer and Graf, 2013, 2015; Lopes-Lima et al., 2017a; Bolotov et al., 2017a, b). *Haasodonta* McMichael, 1956 (2 spp.), is endemic to New Guinea (Walker et al., 2014). It is currently classified in the Rectidentinae and is the only unionid genus in the Australasian region. The MUSSEL Project website and the ITIS website (URLs above) provide lists of Indotropical genera and species classified by subfamily.

Palearctic Eurasia is relatively species-poor, with only 57 unionid species, and, like the Nearctic (above), the region is occupied by two distinct assemblages in the east and west. The eastern Palearctic (including the Amur River, Russian Maritime Provinces, Korea, Japan, Kamchatka, and the Kuril Islands) is largely populated by a northern extension of the Indotropical Unioninae (7 genera, 20 spp.) and Gonideinae (6, 8), with some widespread genera and species distributed in both regions (Graf, 2007; Kondo, 2008). Both subfamilies are represented in the western Palearctic (Europe, northern Africa, and the Middle East to central Asia), although with a distinct complement of genera: Unioninae (4 genera, 23 spp.) and Gonideinae (4, 7) (Froufe et al., 2016a, 2016b; Lopes-Lima et al., 2017b; Araujo et al., 2018). This tally also includes the introduced *S. woodiana* from eastern Asia (Sousa et al., 2014). Our assessment of Palearctic genus and species richness follows Graf (2007) with regard to synonymy of the microsplitting and resultant hyperinflation of diversity perpetrated by some former Soviet malacologists (Korniushin, 1998).

The Afrotropical freshwater mussel assemblage occupies the Nile River, sub-Saharan Africa, and possibly Madagascar. The Unionidae in Africa is represented by two subfamilies: Parreysiinae and Unioninae. Most species are classified in the former, as the monophyletic, Africa-endemic tribe Coelaturini Modell, 1942 (7 genera, 34 spp.). In addition to these, there are three species of the otherwise western Palearctic genus *Unio* (Graf and Cummings, 2011; Whelan et al., 2011; Graf et al., 2014). The *incertae sedis* genus *Germainaia* Graf & Cummings, 2009 (1 sp.), is reported from Madagascar, although the taxon (previously classified among the Coelaturini) and the provenance of the types are questionable (Graf and Cummings, 2009).

The Recent lineages of the Unionidae are predominantly found in North and Central America and Eurasia, i.e., northern continents derived from the breakup of Laurasia. Parreysiinae, *Haasodonta* (Rectidentinae), and certain African species of *Unio* (Unioninae), represent unionid taxa found on Gondwanan continents. The most extensive radiations of these southern taxa are the Parreysiinae in Africa

Table 38.2. Unionidae species richness and Red List conservation status

Region	Species richness	Threatened (EX+CR+EN+VU)	Stable (NT+LC)	Unknown (DD+NE)
Nearctic	299 (42%)	110 (37%)	109 (36%)	80 (27%)
Neotropical	94 (13%)	2 (2%)	2 (2%)	90 (96%)
Afrotropical	38 (5%)	11 (29%)	21 (55%)	6 (16%)
Palearctic	58 (8%)	9 (16%)	13 (22%)	36 (62%)
Indotropical	241 (34%)	19 (8%)	77 (32%)	145 (60%)
Australasian	2 (<1%)	0	0	2 (100%)
Total	713			

Sources: See text for geographic and taxonomic sources.

Note: Abbreviations for IUCN Red List (www.iucnredlist.org) categories: EX = Extinct, CR = Critically Endangered, EN = Endangered, VU = Vulnerable, NT = Near Threatened, LC = Least Concern, DD = Data Deficient, NE = Not Evaluated.

(Tribe Coelaturini) (Graf and Cummings, 2011) and India (tribes Parreysiini, Lamellidentini, and Oxynaiini) (Subba Rao, 1989; Nesemann et al., 2007). The biogeography of the family is hypothesized to reflect an evolutionary origin of the crown-Unionidae in the north during the Jurassic, followed by subsequent dispersal onto Gondwanan continents as those landmasses came into contact during the Cenozoic. This hypothesis is supported by Cenozoic fossils in Africa (Van Damme and Pickford, 2010) and India (Nesemann et al., 2005), clade age estimates based on molecular clocks (Graf et al., 2014, 2015), and the phylogeny/classification of Recent taxa (Lopes-Lima et al., 2017a; Bolotov et al., 2017a).

Freshwater mussels are widely regarded as among the most imperiled taxa in the world (Lydeard et al., 2004; Strayer et al., 2004). Of the 425 species of the Unionida that have been assessed for the Red List produced by the International Union for the Conservation of Nature (www.iucnredlist.org, accessed January 2018), 175 (41%) are regarded as threatened or extinct. However, that tally is an underestimate, as 493 of the 918 species (54%) are regarded as Data Deficient or have not been evaluated. Table 38.2 summarizes the Red List status of the family Unionidae among the six continental biogeographical regions. Except for those of North America and Africa, the majority of unionid species in most regions have not been evaluated or lack sufficient data to make an assessment.

Conservation status reviews are available for most regions of the world: North America (Williams et al.,

1993; Haag & Williams, 2014), South America (Pereira et al., 2014; Tognelli et al., 2016), Africa (Seddon et al., 2011), northern Eurasia (Cuttelod et al., 2011; Lopes-Lima et al., 2017b), tropical and eastern Asia (Aravind et al., 2011; Zieritz et al., 2018), and Australasia (Ponder, 1997; Walker et al., 2014). This is not a comprehensive list of references, but they provide a suitable entrée into the literature. The consensus view is that there has been a global decline in freshwater mussel diversity, resulting from the ongoing extinction "catastrophe" in fresh waters (Strayer and Dudgeon, 2010), which has itself resulted from anthropogenic habitat destruction and fragmentation, introduction of exotic species, overexploitation, and a litany of other insults to natural waterways (Strayer et al., 2004).

LITERATURE CITED

Abell, R.A., M.L. Thieme, C. Revenga, M. Bryer, M. Kottelat, N. Bogutskaya, B. Coad, et al. 2008. Freshwater ecoregions of the world: A new map of biogeographic units for freshwater biodiversity conservation. Bioscience 58: 403-414.

Araujo, R., D. Buckley, K.-O. Nagel, R. García-Jiménez, and A. Machordom. 2018. Species boundaries, geographic distribution and evolutionary history of the Western Palaearctic freshwater mussels *Unio* (Bivalvia: Unionidae). Zoological Journal of the Linnean Society 182: 275-299.

Aravind, N.A., N.A. Madhyastha, G.M. Rajendra, and A. Dey. 2011. The status and distribution of freshwater molluscs of the Western Ghats. Pp. 49-60 in The Status and Distribution of Freshwater Biodiversity in the Western Ghats, India (S. Molur, K.G. Smith, B.A. Daniel, and W.R.T. Darwall, compilers). IUCN, Cambridge, UK. 108 pp. + 3 appendices.

Bieler, R., J.G. Carter, and E.V. Coan. 2010. Classification of bivalve families. In Nomenclator of Bivalve Familes (P. Bouchet and J.-P. Rocroi, eds.). Malacologia 52: 113-133.

Bieler, R., P. Mikkelsen, T.M. Collins, E.A. Glover, V.L. González, D.L. Graf, E.M. Harper, et al. 2014. Investigating the Bivalve Tree of Life—an exemplar-based approach combining molecular and novel morphological characters. Invertebrate Systematics 28: 32-115. DOI: 10.1071/IS13010.

Bolotov, I.N., A.V. Kondakov, I.V. Vikhrev, O.V. Aksenova, Y.V. Bespalaya, M.Y. Gofarov, Y.S. Kolosova, et al. 2017a. Ancient river inference explains exceptional Oriental freshwater mussel radiations. Scientific Reports 7: 2135. DOI: 10.1038/s41598-017-02312-z.

Bolotov, I.N., I.V. Vikhrev, A.V. Kondakov, E.S. Konopleva, M.Y. Gofarov, O.V. Aksenova, and S. Tumpeesuwan. 2017b. New taxa of freshwater mussels (Unionidae) from a species-rich but overlooked evolutionary hotspot in Southeast Asia. Scientific Reports 7: 11573. DOI: 10.1038/s41598-017-11957-9.

Brandt, R.A.M. 1974. The non-marine aquatic Mollusca of Thailand. Archiv für Molluskenkunde 105: i-iv, 1-423.

Campbell, D.C., and C. Lydeard. 2012. The genera of Pleurobemini (Bivalvia: Unionidae: Ambleminae). American Malacological Bulletin 30: 19-38.

Campbell, D.C., J.M. Serb, J.E. Buhay, K.J. Roe, R.L. Minton, and C. Lydeard. 2005. Phylogeny of North American amblemines (Bivaliva, Unionoida): Prodigious polyphyly proves pervasive across genera. Invertebrate Biology 124: 131-164.

Chong, J.P., J.C. Brim Box, J.K. Howard, D. Wolf, T.L. Myers, and K.E. Mock. 2008. Three deeply divided lineages of the freshwater mussel genus *Anodonta* in western North America. Conservation Genetics 9: 1303-1309.

Clarke, A.H. 1981. The tribe Alasmidontini (Unionidae: Anodontinae). Part I. *Pegias, Alasmidonta,* and *Arcidens.* Smithsonian Contributions to Zoology 326: 1-101.

Clarke, A.H. 1985. The tribe Alasmidontini (Unionidae: Anodontinae). Part II: *Lasmigona* and *Simpsonaias.* Smithsonian Contributions to Zoology 399: 1-75.

Combosch, D.J., T.M. Collins, E.A. Glover, D.L. Graf, E.M. Harper, J.M. Healy, G.Y. Kawauchi, et al. 2017. A family-level Tree of Life for bivalves based on a Sanger-sequencing approach. Molecular Phylogenetics and Evolution 107: 191-208.

Cummings, K.S., and D.L. Graf. 2009. Mollusca: Bivalvia. Pp. 309-384 in Ecology and Classification of North American Freshwater Invertebrates (J.H. Thorp and A.P. Covich, eds.). 3rd ed. Academic Press, Elsevier, New York.

Cuttelod, A., M. Seddon, and E. Neubert. 2011. European Red List of Non-Marine Molluscs. Publications Office of the European Union, Luxembourg.

Frierson, L.S. 1927. A Classified and Annotated Check List of the North American Naiades. Baylor University Press, Waco, Texas.

Froufe, E., D.V. Gonçalves, A. Teixeira, R. Sousa, S. Varandas, M. Ghamizi, A. Zieritz, and M. Lopes-Lima. 2016a. Who lives where? Molcular and morphometric analyses clarify which *Unio* species (Unionida, Mollusca) inhabit the southwestern Palearctic. Organisms, Diversity & Evolution 16: 597-611.

Froufe, E., V. Prié, J. Faria, M. Ghamizi, D.V. Gonçalves, M.E. Gürlek, J. Karaouzas, et al. 2016b. Phylogeny, phylogeography, and evolution in the Mediterranean region: News from a freshwater mussel (*Potomida*, Unionida). Molecular Phylogenetics and Evolution 100: 322-332.

Graf, D.L. 2002. Historical biogeography and late glacial origin of the freshwater pearly mussel (Bivalvia: Unionidae) faunas of Lake Erie, North America. Occasional Papers on Mollusks 6: 175-211.

Graf, D.L. 2007. Palearctic freshwater mussel (Mollusca: Bivalvia: Unionoida) diversity and the comparatory method as a species concept. Proceedings of the Academy of Natural Sciences of Philadelphia 156: 71-88.

Graf, D.L. 2013. Patterns of freshwater bivalve global diversity and the state of phylogenetic studies on the Unionoida, Sphaeriidae, and Cyrenidae. American Malacological Bulletin 31: 135-153. DOI: 10.4003/006.031.0106.

Graf, D.L., and K.S. Cummings. 2006. Palaeoheterodont diversity (Mollusca: Trigonioida + Unionoida): What we know and what we wish we knew about freshwater mussel evolution. Zoological Journal of the Linnean Society 148: 343-394.

Graf, D.L., and K.S. Cummings. 2007. Review of the systematics and global diversity of freshwater mussel species (Bivalvia: Unionoida). Journal of Molluscan Studies 73: 291-314.

Graf, D.L., and K.S. Cummings. 2009. Actual and alleged freshwater mussels (Mollusca: Bivalvia: Unionoida) from Madagascar and the Mascarenes, with description of a new genus, *Germainaia*. Proceedings of the Academy of Natural Sciences of Philadelphia 158: 221-238.

Graf, D.L., and K.S. Cummings. 2011. Freshwater mussel (Mollusca: Bivalvia: Unionoida) richness and endemism in the ecoregions of Africa and Madagascar, based on comprehensive museum sampling. Hydrobiologia 678: 17-36.

Graf, D.L., and D.Ó Foighil. 2000. The evolution of brooding characters among the freshwater pearly mussels (Bivalvia: Unionoidea) of North America. Journal of Molluscan Studies 66: 157-170.

Graf, D.L., A.J. Geneva, J.M. Pfeiffer III, and A.D. Chilala. 2014. Phylogenetic analysis of *Prisodontopsis* Tomlin, 1928 and *Mweruella* Haas, 1936 (Bivalvia: Unionidae) from Lake Mweru (Congo basin) supports a Quaternary radiation in the Zambian Congo. Journal of Molluscan Studies 80: 303-314.

Graf, D.L., H. Jones, A.J. Geneva, J.M. Pfeiffer III, and M.A. Klunzinger. 2015. Molecular phylogenetic analysis supports a Gondwanan origin of the Hyriidae (Mollusca: Bivalvia: Unionida) and the paraphyly of Australasian taxa. Molecular Phylogenetics and Evolution 85: 1-9.

Haag, W.R. 2009. Extreme longevity in freshwater mussels revisited: Sources of bias in age estimates derived from mark-recapture experiments. Freshwater Biology 54: 1474-1486.

Haag, W.R. 2012. North American Freshwater Mussels: Natural History, Ecology, and Conservation. Cambridge University Press, Cambridge, UK.

Haag, W.R., and J.D. Williams. 2014. Biodiversity on the brink: An assessment of conservation strategies for North American freshwater mussels. Hydrobiologia 735: 45-60.

Haas, F. 1969a. Superfamilia Unionacea: Das Tierreich, Lief. 88. Walter de Gruyter, Berlin.

Haas, F. 1969b. Superfamily Unionacea Fleming, 1828. Pp. N411-N467 in Treatise on Invertebrate Paleontology, Part N, Mollusca 6: Bivalvia, vol. 1 (R.C. Moore, ed.). Geological Society of America and University of Kansas, Lawrence, Kansas.

He, J., and Z. Zhuang. 2013. The Freshwater Bivalves of China. ConchBooks, Harxheim, Germany.

Kondo, T. 2008. Monograph of Unionoida in Japan (Mollusca: Bivalvia). Special Publication of the Malacological Society of Japan, number 3.

Korniushin, A.V. 1998. Review of the studies on freshwater bivalve mollusc systematics carried out by the Russian taxonomic school. Malacological Review, Supplement 7: 65-82.

Lopes-Lima, M., E. Froufe, V.T. Do, M. Ghamizi, K.E. Mock, Ü. Kebapçı, O. Klishko, et al. 2017a. Phylogeny of the most species-rich freshwater bivalve family (Bivalvia: Unionida: Unionidae): Defining modern subfamilies and tribes. Molecular Phylogenetics and Evolution 106: 174-191.

Lopes-Lima, M., R. Sousa, J. Geist, D.C. Aldridge, R. Araujo, J. Bergengren, Y. Bespalaya, et al. 2017b. Conservation status of freshwater mussels in Europe: State of the art and future challenges. Biological Reviews 92: 572-607.

Lydeard, C., R.H. Cowie, W.F. Ponder, A.E. Bogan, P. Bouchet, S.A. Clark, K.S. Cummings, et al. 2004. The global decline of nonmarine mollusks. BioScience 54: 321-330.

Modell, H. 1964. Das natürliche System der Najaden. Archiv für Molluskenkunde 93: 71-129.

Nesemann, H., S. Sharma, G. Sharma, and R.K. Sinha. 2005. Illustrated checklist of large freshwater bivalves of the Ganga River System (Mollusca: Bivalvia: Solecurtidae, Unionidae, Ambleminae). Nachrichtenblatt der Ersten Vorarlberger Malakologischen Gesellschaft 13: 1-51.

Nesemann, H., S. Sharma, G. Sharma, S.N. Khanal, B. Pradhan, D.N. Shah, and R.D. Tachamo. 2007. Aquatic Invertebrates of the Ganga River System, vol. 1, Mollusca, Annelida, Crustacea (in part). H. Nesemann, Chandi Press, Kathmandu, Nepal.

Pereira, D., M.C.D. Mansur, L.D.S. Duarte, A.S. de Oliveira, D.M. Pimpão, C.T. Callil, C. Ituarte, et al. 2014. Bivalve distribution in hydrographic regions in South America: Historical overview and conservation. Hydrobiologia 735 (1): 15-44. DOI: 10.1007/s10750-013-1639-x.

Pfeiffer, J.M., and D.L. Graf. 2013. Re-analysis confirms the polyphyly of *Lamprotula* Simpson, 1900 (Bivalvia: Unionidae). Journal of Molluscan Studies 79: 249-256.

Pfeiffer, J.M., III, and D.L. Graf. 2015. Evolution of bilaterally asymmetrical larvae in freshwater mussels (Bivalvia: Unionoida: Unionidae). Zoological Journal of the Linnean Society 175: 307-318.

Ponder, W.F. 1997. Conservation status, threats and habitat requirements of Australian terrestrial and freshwater Mollusca. Memoirs of the Museum of Victoria 56: 421-430.

Seddon, M., C. Appleton, D. Van Damme, and D.L. Graf. 2011. Freshwater molluscs of Africa: Diversity, distribution, and conservation. Pp. 92-119 in The Diversity of Life in African Freshwaters: Under Water, Under Threat: An Analysis of the Status and Distribution of Freshwater Species throughout Mainland Africa (W.R.T. Darwall, K.G. Smith, D.J. Allen, R.A. Holland, I.J. Harrison, and E.G.E. Brooks, eds.). IUCN, Cambridge, UK.

Skawina, A., and J. Dzik. 2011. Umbonal musculature and relationships of the Late Triassic filibranch unionoid bivalves. Zoological Journal of the Linnean Society 163: 863-883.

Sousa, R., A. Novais, R. Costa, and D.L. Strayer. 2014. Invasive bivalves in fresh waters: Impacts from individuals to ecosystems and possible control strategies. Hydrobiologia 735: 233-251.

Starobogatov, Ya.I. 1970. Fauna Mollyuskov i Zoogeograficheskoe Raionirovanie Kontinental'nykh Vodoemov Zemnogo Shara. Nauka, Leningrad.

Strayer, D.L., and D. Dudgeon. 2010. Freshwater biodiversity conservation: Recent progress and future challenges. Journal of the North American Benthological Society 29: 344-358.

Strayer, D.L., J.A. Downing, W.R. Haag, T.L. King, J.B. Layzer, T.J. Newton, and S.J. Nichols. 2004. Changing perspectives on pearly mussels, North America's most imperiled animals. BioScience 54: 429-439.

Subba Rao, N.V. 1989. Handbook Freshwater Molluscs of India. Zoological Survey of India, Calcutta.

Tognelli, M.F., C.A. Lasso, C.A. Bota-Sierra, L.F. Jiménez-Segura, and N.A. Cox, eds. 2016. Estado de conservación y distribución de la biodiversidad de agua dulce en los Andes Tropicales. IUCN, Gland, Switzerland.

Van Damme, D., and M. Pickford. 2010. The Late Cenozoic bivalves of the Albertine Basin (Uganda-Congo). Geo-Pal Uganda 2: 1-121.

Van Damme, D., A.E. Bogan, and M. Dierick. 2015. A revision of the Mesozoic naiads (Unionoida) of Africa and the biogeographic implications. Earth-Science Reviews 147: 141-200.

Walker, K.F., H.A. Jones, and M.W. Klunzinger. 2014. Bivalves in a bottleneck: Taxonomy, phylogeography and conservation of freshwater mussels (Bivalvia: Unionoida) in Australasia. Hydrobiologia 735: 61-79.

Watters, G.T. 2001. The evolution of the Unionacea in North America, and its implications for the worldwide fauna. Pp. 281-307 in Ecology and Evolution of the Freshwater Mussels Unionoida (G. Bauer and K. Wächtler, eds.). Ecological Studies, vol. 145. Springer-Verlag, Berlin.

Whelan, N.V., A.J. Geneva, and D.L. Graf. 2011. Molecular phylogenetic analysis of tropical freshwater mussels (Mollusca: Bivalvia: Unionoida) resolves the position of *Coelatura* and supports a monophyletic Unionidae. Molecular Phylogenetics and Evolution 61: 504-514.

Williams, J.D., M.L. Warren Jr., K.S. Cummings, J.L. Harris, and R.J. Neves. 1993. Conservation status of the freshwater mussels of the United States and Canada. Fisheries 18: 6-22.

Williams, J.D., A.E. Bogan, R.S. Butler, K.S. Cummings, J.T. Garner, J.L. Harris, N.A. Johnson, and G.T. Watters. 2017. A revised list of the freshwater mussels (Mollusca: Bivalvia: Unionida) of the United States and Canada. Freshwater Mollusk Biology and Conservation 20: 33-58.

Zieritz, A., A.F. Sartori, A.E. Bogan, and D.C. Aldridge. 2015. Reconstructing the evolution of umbonal sculptures in the Unionida. Journal of Zoological Systematics and Evolutionary Research 53: 76-86.

Zieritz, A., A.E. Bogan, E. Froufe, O. Klishko, T. Kondo, U. Kovitvadhi, S. Kovitvadhi, et al. 2018. Diversity, biogeography and conservation of freshwater mussels (Bivalvia: Unionida) in East and Southeast Asia. Hydrobiologia 810: 29-44. DOI: org/10.1007/s10750-017-3104-8.

39 Margaritiferidae Henderson, 1929

DANIEL L. GRAF AND KEVIN S. CUMMINGS

The freshwater mussel family Margaritiferidae Henderson, 1929 (order Unionida), is composed of 12 Recent species of the genus *Margaritifera* Schumacher, 1816. Despite being the least species-rich freshwater mussel family, the Margaritiferidae is distributed widely (but discontinuously) across the northern continents.

The Margaritiferidae had traditionally been regarded as "primitive" among the freshwater mussels for lacking characters shared among the other five families, e.g., fusion of the distal ends of ctenidia to the adjacent mantle, arrangement of interlamellar connections into vertical septa, posterior mantle fusion (Ortmann, 1912; Smith, 1980). In the absence of these characters, margaritiferids superficially resemble *Neotrigonia* Cossmann, 1912 (Trigoniida), the marine sister group to freshwater mussels (Bieler et al., 2014; Combosch et al., 2017). These (and other) hypothesized "plesiomorphies" have confounded phylogenetic analyses of morphological characters (Graf, 2000; Hoeh et al., 2001), but molecular phylogenetic studies have typically recovered the Margari-

tiferidae in a clade with the Unionidae Rafinesque, 1820 (but see Graf, 2013). Margaritiferidae is currently classified with the Unionidae in the superfamily Unionoidea (Bieler et al., 2010), and the distinctive margaritiferid anatomy is regarded as derived and degenerate rather than indicative of the ancestral condition of the Unionida (Graf and Cummings, 2006).

Until recently, the taxonomy of the family was in a state of flux. Various genera and subgenera had been in use (Haas, 1969a; Smith, 2001), with conflicting phylogenetic support (Huff et al., 2004; Walker et al., 2006; Araujo et al., 2009; Inoue et al., 2013; Bolotov et al., 2015; Takeuchi et al., 2015; Araujo et al., 2017). Bolotov et al. (2016) recently tested the relationships of all but two species in the family. Their results supported classifying all species in a single genus, *Margaritifera,* with three subgenera: *M.* (*Margaritifera*), *M.* (*Margaritanopsis*) Haas, 1910, and *M.* (*Pseudunio*) Haas, 1910. The monotypic North American genus *Cumberlandia* Ortmann, 1912, was sister to *M. laosensis* (Lea, 1863), the type species of *Margaritanopsis*. Though it is one of the best-studied species in the family, the

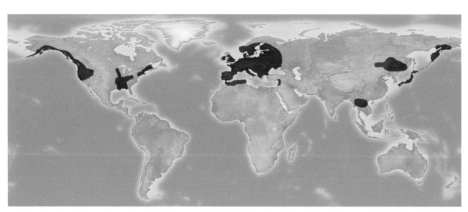

Distribution of Margaritiferidae.

Margaritifera monodonta (Say, 1829). USA: Mississippi River, Campbell Island, Rock Island County, Illinois. 140 mm. INHS 29297.

nomenclature of *Cumberlandia monodonta* (Say, 1829) should be updated to *Margaritifera monodonta* (Graf and Cummings, 2007).

The fossil margaritiferids have been reported to extend into the Triassic, with the oldest, most diverse fossil assemblages discovered in China (Haas, 1969b; Fang et al., 2009). Van Damme et al. (2015) reported that the Recent genus *Margaritifera* is a "living fossil," extending back to the Jurassic based on conchological characters. However, Graf et al. (2015) dated the Unionidae-Margaritiferidae split as Middle Jurassic, and Bolotov et al. (2016) dated the crown *Margaritifera*/Margaritiferidae as Cretaceous. As with freshwater mussels generally, the phylogenetic affinities of fossil taxa are difficult to interpret with the same rigor as Recent species, although there are muscle scar synapomorphies that diagnose the crown mar-

garitiferid clade (Smith, 1983; Graf and Cummings, 2006).

The family is widely distributed across the northern continents, with multiple regions of species endemism in (1) Europe and northern Africa (4 spp.), (2) Pacific East Asia and western North America (4 spp.), and (3) eastern North America (4 spp.) (Ziuganov et al., 1994; Smith, 2001; Bolotov et al., 2016). *M. margaritifera* (Linneaus, 1758) has an amphi-Atlantic distribution (occurring in both Europe and eastern North America), and *M. laosensis* (Lea, 1863) is the sole tropical margaritiferid in southeastern Asia (Bolotov et al., 2014). A close ecological association between margaritiferids and salmonid fishes (as hosts) has been hypothesized (Ziuganov et al., 1994). However, other host taxa are parasitized as well (e.g., *M. auricularia* (Spengler, 1793) with sturgeon, *Acipenser* Linneaus, 1758), and the hosts of several species are unknown (Bolotov et al., 2016).

References to conservation assessments of the Unionida are reviewed in chapter 38 (Graf and Cummings, this volume). The current conservation status assessments of margaritiferid species according to the IUCN Red List (www.iucnredlist.org, accessed January 2018) are provided in Table 39.1. Red List assessments were available for all species, but two were evaluated as Data Deficient. The majority of species are of threatened conservation status.

LITERATURE CITED

Araujo, R., C. Toledo, D. Van Damme, M. Ghamizi, and A. Machordom. 2009. *Margaritifera marocana* (Pallary, 1918): A valid species inhabiting Moroccan rivers. Journal of Molluscan Studies 75: 95-101.

Araujo, R., S. Schneider, K.J. Roe, D. Erpenbeck, and A. Machordom. 2017. The origin and phylogeny of Mar-

Table 39.1. Margaritiferidae species richness and Red List conservation status

Region	Species richness	Threatened (EX+CR+EN+VU)	Stable (NT+LC)	Unknown (DD+NE)
Nearctic	5 (42%)	4 (80%)	1 (20%)	0
Palearctic	7 (58%)	5 (71%)	0	2 (29%)
Indotropical	1 (8%)	1	0	0
Total	12			

Note: Abbreviations for IUCN Red List (www.iucnredlist.org) categories: EX = Extinct, CR = Critically Endangered, EN = Endangered, VU = Vulnerable, NT = Near Threatened, LC = Least Concern, DD = Data Deficient, NE = Not Evaluated.

garitiferidae (Bivalvia, Unionoida): A synthesis of molecular and fossil data. Zoologica Scripta 46: 289-307.

Bieler, R., J.G. Carter, and E.V. Coan. 2010. Classification of bivalve families. In Nomenclator of Bivalve Familes (P. Bouchet and J.-P. Rocroi, eds.). Malacologia 52: 113-133.

Bieler, R., P. Mikkelsen, T.M. Collins, E.A. Glover, V.L. González, D.L. Graf, E.M. Harper, et al. 2014. Investigating the Bivalve Tree of Life—an exemplar-based approach combining molecular and novel morphological characters. Invertebrate Systematics 28: 32-115. DOI: 10.1071/IS13010.

Bolotov, I., I. Vikhrev, Y. Bespalaya, V. Artamonova, M. Gofarov, J. Kolosova, A. Kondakov, et al. 2014. Ecology and conservation of the endangered Indochinese freshwater pearl mussel, *Margaritifera laosensis* (Lea, 1863) in the Nam Pe and Nam Long rivers, Northern Laos. Tropical Conservation Science 7: 706-719.

Bolotov, I.N., Y.V. Bespalaya, I.V. Vikhrev, O.V. Aksenova, P.E. Aspholm, M.Y. Gofarov, O.K. Klishko, et al. 2015. Taxonomy and distribution of freshwater pearl mussels (Unionoida: Margaritiferidae) of the Russian Far East. PLoS ONE 10: e0122408.

Bolotov, I., I. Vikhrev, Y.V. Bespalaya, M.Y. Gofarov, A.V. Kondakov, E.S. Konopleva, N.N. Bolotov, and A.A. Lyubas. 2016. Multi-locus fossil-calibrated phylogeny, biogeography and a subgeneric revision of the Margaritiferidae (Mollusca: Bivalvia: Unionoida). Molecular Phylogenetics and Evolution 103: 104-121.

Combosch, D.J., T.M. Collins, E.A. Glover, D.L. Graf, E.M. Harper, J.M. Healy, G.Y. Kawauchi, et al. 2017. A family-level Tree of Life for bivalves based on a Sanger-sequencing approach. Molecular Phylogenetics and Evolution 107: 191-208.

Fang, Z.-J., J.-H. Chen, C.-Z. Chen, J.-G. Sha, X. Lan, and S.-X. Wen. 2009. Supraspecific taxa of the Bivalvia first named, described, and published in China (1927-2007). University of Kansas Paleontological Contributions, new series, 17: 1-157.

Graf, D.L. 2000. The Etherioidea revisited: A phylogenetic analysis of hyriid relationships (Mollusca: Bivalvia: Paleoheterodonta: Unionoida). Occasional Papers of the Museum of Zoology, University of Michigan, 729: 1-21.

Graf, D.L. 2013. Patterns of freshwater bivalve global diversity and the state of phylogenetic studies on the Unionoida, Sphaeriidae, and Cyrenidae. American Malacological Bulletin 31: 135-153. DOI: 10.4003/006.031.0106.

Graf, D.L., and K.S. Cummings. 2006. Palaeoheterodont diversity (Mollusca: Trigonioida + Unionoida): What we know and what we wish we knew about freshwater mussel evolution. Zoological Journal of the Linnean Society 148: 343-394.

Graf, D.L., and K.S. Cummings. 2007. Review of the systematics and global diversity of freshwater mussel species (Bivalvia: Unionoida). Journal of Molluscan Studies 73: 291-314.

Graf, D.L., H. Jones, A.J. Geneva, J.M. Pfeiffer III, and M.A. Klunzinger. 2015. Molecular phylogenetic analysis supports a Gondwanan origin of the Hyriidae (Mollusca: Bivalvia: Unionida) and the paraphyly of Australasian taxa. Molecular Phylogenetics and Evolution 85: 1-9.

Haas, F. 1969a. Superfamilia Unionacea: Das Tierreich, Lief. 88. Walter de Gruyter, Berlin.

Haas, F. 1969b. Superfamily Unionacea Fleming, 1828. Pp. N411-N467 in Treatise on Invertebrate Paleontology, Part N, Mollusca 6: Bivalvia, vol. 1 (R.C. Moore, ed.). Geological Society of America and University of Kansas, Lawrence, Kansas.

Hoeh, W.R., A.E. Bogan, and W.H. Heard. 2001. A phylogenetic perspective on the evolution of morphological and reproductive characteristics in the Unionoida. Pp. 257-280 in Ecology and Evolution of the Freshwater Mussels Unionoida (G. Bauer and K. Wächtler, eds.). Ecological Studies, vol. 145. Springer-Verlag, Berlin.

Huff, S.W., D. Campbell, D.L. Gustafson, C. Lydeard, C.R. Altaba, and G. Giribet. 2004. Investigations into the phylogenetic relationships of freshwater pearl mussels (Bivalvia: Margaritiferidae) based on molecular data: Implications for their taxonomy and biogeography. Journal of Molluscan Studies 70: 379-388.

Inoue, K., E.M. Monroe, C.L. Elderkin, and D.J. Berg. 2013. Phylogeographic and population genetic analyses reveal Pleistocene isolation followed by high gene flow in a wide ranging, but endangered, freshwater mussel. Heredity 112: 282-290.

Ortmann, A.E. 1912. Notes upon the families and genera of the Najades. Annals of the Carnegie Museum 8: 222-365.

Smith, D.G. 1980. Anatomical studies on *Margaritifera margaritifera* and *Cumberlandia monodonta* (Mollusca: Pelecypoda: Margaritiferidae). Zoological Journal of the Linnean Society 69: 257-270.

Smith, D.G. 1983. On the so-called mantle muscle scars on the shells of the Margaritiferidae (Mollusca, Pelecypoda), with observations on the mantle-shell attachment in the Unionoida and Trigonioida. Zoologica Scripta 12: 67-71.

Smith, D.G. 2001. Systematics and distribution of the Recent Margaritiferidae. Pp. 34-49 in Ecology and Evolution of the Freshwater Mussels Unionoida (G. Bauer and K. Wächtler, eds.). Ecological Studies, vol. 145. Springer-Verlag, Berlin.

Takeuchi, M., A. Okada, and W. Kakino. 2015. Phylogenetic relationships of two freshwater pearl mussels, *Margaritifera laevis* (Haas, 1910) and *Margaritifera togakushiensis*

Kondo & Kobayashi, 2005 (Bivalvia: Margaritiferidae), in the Japanese archipelago. Molluscan Research 35: 218–226.

Van Damme, D., A.E. Bogan, and M. Dierick. 2015. A revision of the Mesozoic naiads (Unionoida) of Africa and the biogeographic implications. Earth-Science Reviews 147: 141–200.

Walker, J.M., J.P. Curole, D.E. Wade, E.G. Chapman, A.E.

Bogan, G.T. Watters, and W.R. Hoeh. 2006. Taxonomic distribution and phylogenetic utility of gender-associated mitochondrial genomes in the Unionoida (Bivalvia). Malacologia 48: 265–282.

Ziuganov, V., A. Zotin, L. Nezlin, and V. Tretiakov. 1994. The Freshwater Pearl Mussels and Their Relationships with Salmonid Fishes. VNIRO, Moscow.

40 Hyriidae Swainson, 1840

DANIEL L. GRAF AND KEVIN S. CUMMINGS

The Hyriidae Swainson, 1840, is a family of fresh-water mussels (order Unionida), consisting of 96 Recent species in 15 genera, classified in two subfamilies: Hyriinae s.s. and Velesunioninae Iredale, 1934. The range of the Hyriidae occupies three disjunct regions of endemism: (1) South America; (2) Australia, New Guinea, and adjacent islands; and (3) New Zealand. The phylogeny and biogeography of these mussels have figured prominently in discussions of general freshwater mussel evolution.

The Hyriidae is monophyletic and the sole Recent family of the superfamily Hyrioidea (Bieler et al., 2010; Graf et al., 2015). The family has been reclassified over the past century as different authorities have emphasized alternative characters to diagnose taxonomic affinities. Ortmann (1921) arranged the Hyriidae with the families Iridinidae Swainson, 1840, and Mycetopodidae Gray, 1840, based on adult soft anatomy. Key among these characters were (1) the presence of mantle fusion between the incurrent and excurrent apertures and (2) larval brooding restricted to the inner demibranchs of the ctenidia. In contrast, both the Unionidae and Margaritiferidae lack mantle fusion between the apertures, and larval brooding typically occupies all four demibranchs or only the outer pair.

Parodiz and Bonetto (1963) subsequently arranged the families of the Unionida according to larval rather than adult morphology. Freshwater mussels exhibit two distinct types of parasitic larvae: glochidia and lasidia. Glochidia are bivalved, with thin calcified shells and an adductor mussel, whereas lasidia lack a shell. Both types are brooded by the females within the demibranchs and both are parasitic upon freshwater fishes (Wächtler et al., 2001; Graf and Cummings, 2006). Parodiz and Bonetto (1963) grouped the glochidium-bearing Unionidae, Margaritiferidae, and Hyriidae together as the superfamily Unionoidea ("Unionacea"), and the three families with lasidia—Iridinidae, Mycetopodidae, and Etheriidae Deshayes, 1832—were united in the Etherioidea Deshayes, 1832 ("Mutelacea"). This two-superfamily system was

Distribution of Hyriidae.

Triplodon corrugatus (Lamarck, 1819). Brazil: Rio Xingu (Amazon Basin), Porto de Moz, Pará. 85 mm. INHS 46946.

followed in subsequent classifications leading into the cladistic era (e.g., Haas, 1969; Starobogatov, 1970; Boss, 1982; Vaught, 1989).

Little support has been found in phylogenetic analyses for the Parodiz and Bonetto (1963) classification, but different character sets have recovered the Hyriidae in various positions. Analyses of morphological characters have generally supported a clade composed of the Hyriidae and the lasidium-bearing families, while nucleotide characters typically placed the Hyriidae as sister to a clade of the other five families (Graf, 2013). The compromise has been to classify the Hyriidae in its own superfamily (Bieler et al., 2010; Graf et al., 2015). The most parsimonious interpretation would be that both sets of traditional adult and larval characteristics used to classify the Hyriidae are plesiomorphic synapomorphies of the Unionida, although the combination of (1) mantle fusion between the posterior apertures and (2) glochidia are diagnostic of the family (Graf and Cummings, 2006). Pfeiffer and Graf (2015) analyzed larval character evolution, but there has been no formal analysis of adult characters traced on the most up-to-date phylogenies.

The two hyriid subfamilies are generally well recognized based on molecular and morphological synapomorphies (Graf and Cummings, 2006; Graf et al., 2015), including the degree of umbo sculpturing. The Hyriinae is diagnosed by coarse "radial" or zigzag

sculpture, whereas the Velesunioninae have smooth or only weakly sculptured umbos (Zieritz et al., 2013). The Velesunioninae (16 spp.) is endemic to Australia (including Tasmania) and New Guinea, and the subfamily consists of the genera *Velesunio* Iredale, 1934, *Alathyria* Iredale, 1934, *Lortiella* Iredale, 1934, *Microdontia* Tapparone Cafefri, 1883, and *Westralunio* Iredale, 1934. The Hyriinae is divided among four tribes: Hyriini s.s., Castaliini Morretes, 1949, Rhipidodontini Starobogatov, 1970, and Hyridellini McMichael, 1956. The Hyridellini (*Hyridella* Swainson, 1840, *Cucumerunio* Iredale, 1934, and *Virgus* Simpson, 1900; 9 spp.) occurs in eastern Australia, New Guinea, and the Solomon Islands. In addition, the genus *Echyridella* McMichael & Hiscock, 1958 (3 spp.), is endemic to New Zealand, but no phylogenetic analysis has robustly placed it among other Australasian genera (Marshall et al., 2014). The other three tribes of the Hyriinae comprise a clade (without a formal name) and are endemic to South America: Hyriini (*Prisodon* Schumacher, 1817, *Triplodon* Spix & Wagner, 1827; 4 spp.), Castaliini (*Castalia* Lamarck, 1819, *Callonaia* Simpson, 1900, *Castaliella* Simpson, 1900; 17 spp.), and Rhipidodontini (*Diplodon* Spix & Wagner, 1827, 47 spp.). While the overall species richness is higher in the Neotropics (68 spp.) (Simone, 2006; Pereira et al., 2014) than Australasia (28 spp.) (McMichael and Hiscock, 1958; Walker et al., 2001; Walker et al., 2014), the latter has greater phylogenetic diversity.

The paraphyly of the Australasian hyriids relative to the three Neotropical tribes (Velesunioninae [Hyridellini, *Echyridella* (Hyriini, Castaliini, Rhipidodontini)]) reflects their ancient origins on Gondwana during the Mesozoic. Graf et al. (2015) dated the origins of the two subfamilies to the Middle Jurassic, substantially predating Tertiary isolation of South American from Australia by marine barriers (Sanmartín and Ronquist, 2004; Upchurch, 2008). Thus, these taxa were already differentiated in river systems at opposite ends of the supercontinent before terrestrial isolation was completed. The crown-Hyriidae was dated as Early Jurassic in origin. This chronology conflicts with Triassic fossils assigned to the Hyriidae in Australasia (Hocknull, 2000) and North America (Good, 1998). Given the difficulty of assigning Recent freshwater mussel

Table 40.1. Hyriidae species richness and Red List conservation status

Region	Species richness	Threatened (EX+CR+EN+VU)	Stable (NT+LC)	Unknown (DD+NE)
Neotropical	68 (71%)	5 (7%)	6 (9%)	57 (84%)
Afrotropical	28 (29%)	2 (7%)	5 (18%)	21 (75%)
Total	96			

Note: Abbreviations for IUCN Red List (www.iucnredlist.org) categories: EX = Extinct, CR = Critically Endangered, EN = Endangered, VU = Vulnerable, NT = Near Threatened, LC = Least Concern, DD = Data Deficient, NE = Not Evaluated.

species to higher taxa based solely on shell shape and sculpture (Zieritz et al., 2015), we find the classification of these fossils dubious. Triassic unionoids may represent stem-groups of the modern lineages (Skawina and Dzik, 2011).

References to conservation assessments of the Unionida are reviewed in chapter 38 (Graf and Cummings, this volume). The current conservation status assessments of the species of the Hyriidae according to the IUCN Red List (www.iucnredlist.org, accessed January 2018) are provided in Table 40.1. The majority of hyriid species lack any conservation status assessment on the Red List.

LITERATURE CITED

Bieler, R., J.G. Carter, and E.V. Coan. 2010. Classification of bivalve families. In Nomenclator of Bivalve Familes (P. Bouchet and J.-P. Rocroi, eds.). Malacologia 52: 113-133.

Boss, K.J. 1982. Mollusca. Pp. 945-1166 in Synopsis and Classification of Living Organisms, vol. 1 (Parker, S.P., ed.). McGraw Hill, New York.

Good, S.C. 1998. Freshwater bivalve fauna of the Late Triassic (Carnian-Norian) Chinle, Dockum, and Dolores formations of the southwest United States. Pp. 223-249 in Bivalves: An Eon of Evoluion—Paleobiological Studies Honoring Norman D. Newell (P.A. Johnston and J.W. Haggart, eds.). University of Calgary Press, Alberta, Canada.

Graf, D.L. 2013. Patterns of freshwater bivalve global diversity and the state of phylogenetic studies on the Unionoida, Sphaeriidae, and Cyrenidae. American Malacological Bulletin 31: 135-153. DOI: 10.4003/006.031.0106.

Graf, D.L., and K.S. Cummings. 2006. Palaeoheterodont diversity (Mollusca: Trigonioida + Unionoida): What we know and what we wish we knew about freshwater mussel evolution. Zoological Journal of the Linnean Society 148: 343-394.

Graf, D.L., H. Jones, A.J. Geneva, J.M. Pfeiffer III, and M.A. Klunzinger. 2015. Molecular phylogenetic analysis supports a Gondwanan origin of the Hyriidae (Mollusca: Bivalvia: Unionida) and the paraphyly of Australasian taxa. Molecular Phylogenetics and Evolution 85: 1-9.

Haas, F. 1969. Superfamilia Unionacea: Das Tierreich, Lief. 88. Walter de Gruyter, Berlin.

Hocknull, S.A. 2000. Mesozoic freshwater and estuarine bivalves from Australia. Memoirs of the Queensland Museum 45: 405-426.

Marshall, B.A., M.C. Fenwick, and P.A. Ritchie. 2014. New Zealand recent Hyriidae (Mollusca: Bivalvia: Unionida). Molluscan Research 34: 181-200.

McMichael, D.F., and I.D. Hiscock. 1958. A monograph of the freshwater mussels (Mollusca: Pelecypoda) of the Australian region. Australian Journal of Marine and Freshwater Research 9: 372-508.

Ortmann, A.E. 1921. South American Naiades; a contribution to the knowledge of the freshwater mussels of South America. Memoirs of the Carnegie Museum 8: 451-670.

Parodiz, J.J., and A.A. Bonetto. 1963. Taxonomy and zoogeographic relationships of the South American Naiades (Pelecypoda: Unionacea and Mutelacea). Malacologia 1: 179-214.

Pereira, D., M.C.D. Mansur, L.D.S. Duarte, A.S. de Oliveira, D.M. Pimpão, C.T. Callil, C. Ituarte, et al., 2014. Bivalve distribution in hydrographic regions in South America: Historical overview and conservation. Hydrobiologia 735 (1): 15-44. DOI: 10.1007/s10750-013-1639-x.

Pfeiffer, J.M., III, and D.L. Graf. 2015. Evolution of bilaterally asymmetrical larvae in freshwater mussels (Bivalvia: Unionoida: Unionidae). Zoological Journal of the Linnean Society 175: 307-318.

Sanmartín, I., and F. Ronquist. 2004. Southern hemisphere biogeography inferred by event-based models: Plants versus animal patterns. Systematic Biology 53: 216-243.

Simone, L.R.L. 2006. Land and Freshwater Molluscs of Brazil: An Illustrated Inventory of the Brazilian Malacofauna, Including Neighboring Regions of South America, Respect to the Terrestrial and Freshwater Ecosystems. EGB, Sao Paulo.

Skawina, A., and J. Dzik. 2011. Umbonal musculature and relationships of the Late Triassic filibranch unionoid bivalves. Zoological Journal of the Linnean Society 163: 863-883.

Starobogatov, Ya.I. 1970. Fauna Mollyuskov i Zoogeograficheskoe Raionirovanie Kontinental'nykh Vodoemov Zemnogo Shara [Fauna of molluscs and zoogeographic division of continental water bodies of the globe]. Nauka, Leningrad.

Upchurch, P. 2008. Gondwana break-up: Legacies of a lost world? Trends in Ecology and Evolution 23: 229-236.

Vaught, K.C. 1989. A Classification of the Living Mollusca. American Malacologists, Melbourne, Florida.

Wächtler, K., M.C.D. Mansur, and T. Richter. 2001. Larval types and early postlarval biology in Naiads (Unionoida). Pp. 93-125 in Ecology and Evolution of the Freshwater Mussels Unionoida (G. Bauer and K. Wächtler, eds.). Springer-Verlag, Berlin.

Walker, K.F., M. Byrne, C.W. Hickey, and D.S. Roper. 2001. Freshwater mussels (Hyriidae, Unionidae) of Australasia. Pp. 3-31 in Ecology and Evolution of the Freshwater Mussels Unionoida (G. Bauer and K. Wächtler, eds.). Springer-Verlag, Berlin.

Walker, K.F., H.A. Jones, and M.W. Klunzinger. 2014. Bivalves in a bottleneck: Taxonomy, phylogeography and conservation of freshwater mussels (Bivalvia: Unionoida) in Australasia. Hydrobiologia 735: 61-79.

Zieritz, A., A.F. Sartori, and M.W. Klunzinger. 2013. Morphological evidence shows that not all Velesunioninae have smooth umbos. Journal of Molluscan Studies 79: 277-282.

Zieritz, A., A.F. Sartori, A.E. Bogan, and D.C. Aldridge. 2015. Reconstructing the evolution of umbonal sculptures in the Unionida. Journal of Zoological Systematics and Evolutionary Research 53: 76-86.

41 Etheriidae Deshayes, 1832

DANIEL L. GRAF AND KEVIN S. CUMMINGS

The Etheriidae Deshayes, 1832 (+ Mulleriidae Deshayes, 1832) is a family of four freshwater mussel species (Order Unionida) distributed widely on South America, Africa, and India. The monophyly of the family is contentious. The species classified in the Etheriidae exhibit varying degrees of "oysterization," tending toward cementing habits and monomyarian adductor muscles.

Four geographically disjunct, monotypic genera comprise the Etheriidae: *Etheria* Lamarck, 1807, *Acostaea* d'Orbigny, 1851, *Pseudomulleria* R. Anthony, 1907, and *Bartlettia* H. Adams, 1867 (Anthony, 1907; Pain and Woodward, 1961; Graf and Cummings, 2007). *Etheria elliptica* Lamarck, 1807, is distributed throughout much of Africa as well as in Madagascar (Pilsbry and Bequaert, 1927; Graf and Cummings, 2011), and recent life history and phylogeographic research has indicated multiple cryptic lineages (Bauer, 2013; Elderkin et al., 2016). *Etheria elliptica* cements to its substrate, with a superficially oyster-like shell, although both adductor muscles are retained (Yonge, 1962; Heard and Vail, 1976). *Acostaea rivolii* (Deshayes,

1827) is endemic to the Rio Magdalena in Colombia. This species also cements to its substrate and has taken the oyster habit further by losing its anterior adductor (Yonge, 1978). In the Western Ghats of India, *Pseudomulleria dalyi* (E. A. Smith, 1898) has a shell and soft anatomy strikingly similar to *Acostaea* (Woodward, 1898). *Bartlettia stefanensis* (Moricand, 1856), from the Upper Amazon and Paraná basins, is not strictly a cementer. Instead, the animal secretes its shell to wedge into a tight burrow (Mansur and da Silva, 1990).

During the twentieth century, it was occasionally hypothesized that some species of the Etheriidae shared closer evolutionary affinities to taxa in other families (e.g., Prashad, 1932; Parodiz and Bonetto, 1963; Morrison, 1973; Bonetto, 1997). In general, these authors argued in favor of convergent evolution of the oyster habit from local freshwater mussel lineages, as opposed to broad geographical disjunctions that, at the time, lacked a credible explanation. However, plate tectonics and continental drift—explicitly, the breakup of Gondwana during the Mesozoic

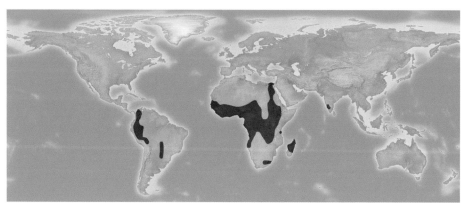

Distribution of Etheriidae.

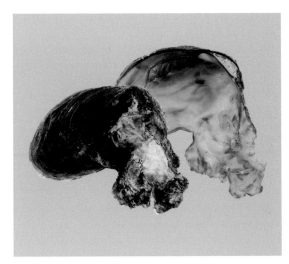

Bartlettia stefanensis (Moricand, 1856). Peru: Río Alto Yurúa (Amazon Basin), at the Brazilian border, Ucayali Region. 75 mm. INHS 35130.

(Sanmartín and Ronquist, 2004; Upchurch, 2008)—provide a sufficient rationale for vicariance as the mechanism of cladogenesis, despite the near lack of a fossil record for the family (Haas, 1969).

Phylogenetic analyses of molecular and morphological characters have recovered conflicting results for the core trio of cementing genera: *Etheria, Acostaea,* and *Pseudomulleria.* Cladistic analysis of morphology supported their monophyly (Graf, 2000; Graf and Cummings, 2006). In molecular analyses, *Etheria* and/or *Acostaea* have been typically recovered in a clade with the Mycetopodidae, and *Pseudomulleria* was placed among the Unionidae (Bogan and Hoeh, 2000; Walker et al., 2006; Hoeh et al., 2009). Resolution of *Pseudomulleria* among the Unionidae is based on a single sequence of cytochrome oxidase

subunit I, a gene fragment with limited utility for recovering deep phylogenetic divergences. Moreover, what is known of the anatomy of *Pseudomulleria* is incompatible with the Unionidae (Graf and Cummings, 2010). The posterior mantle of *P. dalyi* is fused between the incurrent and excurrent apertures, unlike any unionid (Graf and Cummings, 2006). This condition matches that of other etheriid and mycetopodid species. However, the basis of our anatomical knowledge of *Pseudomulleria* is Woodward's (1898) observations of two specimens.

The evidence for the monophyly of the cementing genera of the Etheriidae is based upon the shared presence of the cementing habit, plicate ctenidia, the reduction of the foot, and characteristics of the shell microstructure (Graf and Cummings, 2006). Moreover, *Acostaea* and *Pseudomulleria* share highly similar morphology (Yonge, 1978). *Bartlettia stefanensis* has never been included in a phylogenetic analysis, and its placement among the Etheriidae follows Mansur and da Silva (1990). Molecular phylogenetic analyses based on multiple mitochondrial and nuclear loci have recovered *Etheria* in a clade with the families Iridinidae and Mycetopodidae (Pfeiffer and Graf, 2015; Combosch et al., 2017), the superfamily Etherioidea Deshayes, 1832.

References to conservation assessments of the Unionida are reviewed in chapter 38 (Graf and Cummings, this volume). The current conservation status assessments of etheriid species according to the IUCN Red List (www.iucnredlist.org, accessed January 2018) are provided in Table 41.1. Red List assessments are available for all four species of the Etheriidae. Both *Acostaea rivolii* and *Pseudomulleria dalyi* were evaluated in threatened status categories.

Table 41.1. Etheriidae Species Richness and Red List Conservation Status

Region	Species richness	Threatened (EX+CR+EN+VU)	Stable (NT+LC)	Unknown (DD+NE)
Neotropical	2	1	1	0
Afrotropical	1	0	1	0
Indotropical	1	1	0	0
Total	4			

Note: Abbreviations for IUCN Red List (www.iucnredlist.org) categories: EX = Extinct, CR = Critically Endangered, EN = Endangered, VU = Vulnerable, NT = Near Threatened, LC = Least Concern, DD = Data Deficient, NE = Not Evaluated.

LITERATURE CITED

Anthony, R. 1907. Etude monographique des Aetheriidae (anatomie, morphogénie, systématique). Annales de la Societe Royal Zoologique et Malacologique de Belgique 41: 322-428.

Bauer, G. 2013. Reproductive biology of naiads in the upper Blue Nile. Malacologia 56: 321-328.

Bogan, A.E., and W.R. Hoeh. 2000. On becoming cemented: Evolutionary relationships among the genera in the freshwater bivalve family Etheriidae (Bivalia: Unionoida). Pp. 159-168 in The Evolutionary Biology of the Bivalvia (E.M. Harper, J.D. Taylor, and J.A. Crame, eds.). Geological Society, London, Special Publications 177. Geological Society of London, London.

Bonetto, A.A. 1997. Las 'Ostras de Agua Dulce' (Muteloidea: Mutelidae): Su taxonomia y distribucion geographica en el conjunto de las naiades del Mundo. Biociências 5: 113-142.

Combosch, D.J., T.M. Collins, E.A. Glover, D.L. Graf, E.M. Harper, J.M. Healy, G.Y. Kawauchi, et al. 2017. A family-level Tree of Life for bivalves based on a Sanger-sequencing approach. Molecular Phylogenetics and Evolution 107: 191-208.

Elderkin, C.L., C. Clewing, O. Wembo Ndeo, and C. Albrecht. 2016. Molecular phylogeny and DNA barcoding confirm cryptic species in the African freshwater oyster Etheria elliptica Lamarck, 1807 (Bivalvia: Etheriidae). Biological Journal of the Linnean Society 118: 369-381.

Graf, D.L. 2000. The Etherioidea revisited: A phylogenetic analysis of hyriid relationships (Mollusca: Bivalvia: Paleoheterodonta: Unionoida). Occasional Papers of the Museum of Zoology, University of Michigan, 729: 1-21.

Graf, D.L., and K.S. Cummings. 2006. Palaeoheterodont diversity (Mollusca: Trigonioida + Unionoida): What we know and what we wish we knew about freshwater mussel evolution. Zoological Journal of the Linnean Society 148: 343-394.

Graf, D.L., and K.S. Cummings. 2007. Review of the systematics and global diversity of freshwater mussel species (Bivalvia: Unionoida). Journal of Molluscan Studies 73: 291-314.

Graf, D.L., and K.S. Cummings. 2010. Comments on the value of COI for family-level freshwater mussel systematics: A reply to Hoeh, Bogan, Heard and Chapman. Malacologia 52: 191-197.

Graf, D.L., and K.S. Cummings. 2011. Freshwater mussel (Mollusca: Bivalvia: Unionoida) richness and endemism in the ecoregions of Africa and Madagascar, based on comprehensive museum sampling. Hydrobiologia 678: 17-36.

Haas, F. 1969. Superfamily Unionacea Fleming, 1828. Pp. N411-N467 in Treatise on Invertebrate Paleontology, Part N, Mollusca 6: Bivalvia, vol. 1 (R.C. Moore, ed.). Geological Society of America and University of Kansas, Lawrence, Kansas.

Heard, W.H., and V.A. Vail. 1976. Anatomical systematics of Etheria elliptica (Pelecyopoda: Mycetopodidae). Malacological Review 9: 15-24.

Hoeh, W.R., A.E. Bogan, W.H. Heard, and E.G. Chapman. 2009. Palaeoheterodont phylogeny, character evolution and phylogenetic classification: A reflection on methods of analysis. Malacologia 51: 307-317.

Mansur, M.C.D., and M.G.O. da Silva. 1990. Morfologia e microanatomia comparada de Bartlettia stefanensis (Morricand, 1856) e Anodontites tenebricosus (Lea, 1834) (Bivalvia, Unionoida, Muteloidea). Amazoniana 11: 147-166.

Morrison, J.P.E. 1973. The families of the pearly freshwater mussels. Bulletin of the American Malacological Union 1973: 45-46.

Pain, T., and F.R. Woodward. 1961. A revision of the freshwater mussels of the family Etheriidae. Journal of Conchology 25: 2-8.

Parodiz, J.J., and A.A. Bonetto. 1963. Taxonomy and zoogeographic relationships of the South American Naiades (Pelecypoda: Unionacea and Mutelacea). Malacologia 1: 179-214.

Pfeiffer, J.M., III, and D.L. Graf. 2015. Evolution of bilaterally asymmetrical larvae in freshwater mussels (Bivalvia: Unionoida: Unionidae). Zoological Journal of the Linnean Society 175: 307-318.

Pilsbry, H.A., and J. Bequaert. 1927. The aquatic mollusks of the Belgian Congo: With a geographical and ecological account of Congo malacology. Bulletin of the American Museum of Natural History 53: 69-602.

Prashad, B. 1932. VIII. Some noteworthy examples of parallel evolution in the molluscan faunas of south-eastern Asia and South America. Proceedings of the Royal Society of Edinburgh 51: 42-53.

Sanmartín, I., and F. Ronquist. 2004. Southern hemisphere biogeography inferred by event-based models: Plants versus animal patterns. Systematic Biology 53: 216-243.

Upchurch, P. 2008. Gondwana break-up: Legacies of a lost world? Trends in Ecology and Evolution 23: 229-236.

Walker, J.M., J.P. Curole, D.E. Wade, E.G. Chapman, A.E. Bogan, G.T. Watters, and W.R. Hoeh. 2006. Taxonomic distribution and phylogenetic utility of gender-associated mitochondrial genomes in the Unionoida (Bivalvia). Malacologia 48: 265-282.

Woodward, M.F. 1898. On the anatomy of *Mulleria dalyi*, Smith. Proceedings of the Malacological Society of London 3: 87-91.

Yonge, C.M. 1962. On *Etheria elliptica* Lam. and the course of evolution, including assumption of monomyarianism, in the Family Etheriidae (Bivalvia: Unionacea).

Philosophical Transactions of the Royal Society B 244: 423-458.

Yonge, C.M. 1978. On the monomyarian, *Acostaea rivoli* and evolution in the family Etheriidae (Bivaliva: Unionacea). Journal of Zoology, London 184: 429-448.

42 Mycetopodidae Gray, 1840

DANIEL L. GRAF AND KEVIN S. CUMMINGS

The Mycetopodidae Gray, 1840, is a family of freshwater mussels (Order Unionida) found throughout the Neotropics, from Mexico south to Argentina. The 53 species classified in the Mycetopodidae are divided among four subfamilies: Mycetopodinae s.s., Anodontitinae Modell, 1942, Leilinae Morretes, 1949, and Monocondylaeinae Modell, 1942.

There is no evidence against the monophyly of the Mycetopodidae, but taxon sampling in phylogenetic studies to date has been too sparse for a robust test. Morphological (Graf, 2000) and molecular (Bogan and Hoeh, 2000; Walker et al., 2006; Whelan et al., 2011; Combosch et al., 2017) analyses have included representatives of at most two subfamilies at a time. Graf and Cummings (2006) identified only a single synapomorphy to distinguish the family from the other families of the Etherioidea Deshayes, 1832: pedal elevator scars that are "inconspicuous." Otherwise, mycetopodids have lasidium-type parasitic larvae and mantle fusion between their incurrent and excurrent apertures (homologous with the conditions seen in Etheriidae Deshayes, 1832, and Iridinidae Swainson,

1840). Mycetopodid shells are largely edentulous except in the genera of the Monocondylaeinae.

No well-sampled phylogeny exists for any mycetopodid subfamily or genus, and the classification of those taxa remains largely unchanged from Parodiz and Bonetto (1963). Haas (1969). Simone (2006), Graf and Cummings (2007), and Pereira et al. (2014) provided syntheses of Neotropical species richness of the Mycetopodidae. The most species-rich subfamily is the Anodontitinae, composed of the genera *Anodontites* Bruguière, 1792 (26 spp.) and *Lamproscapha* Swainson, 1840 (2 spp.). The species of the subfamily are mostly restricted to South America, but five species of *Anodontites* occur in Central America, three of which extend as far north as Mexico (Frierson, 1927). The Monocondylaeinae has the highest generic richness of the family: *Monocondylaea* d'Orbigny, 1835 (8 spp.), *Fossula* Lea, 1870 (1), *Haasica* Strand, 1932 (1), *Iheringella* Pilsbry, 1893 (2), *Diplodontites* Marshall, 1922 (3), and *Tamsiella* Haas, 1931 (3). The first four genera are restricted to the Paraná-La Plata Basin and, with one exception, southern tributaries of the

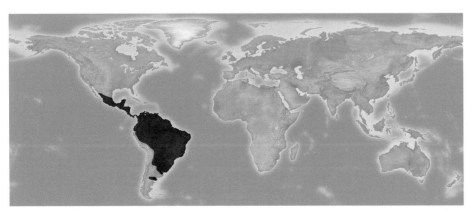

Distribution of Mycetopodidae.

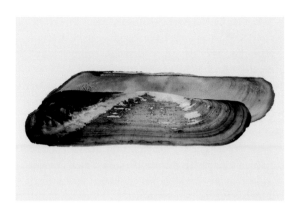

Mycetopoda soleniformis d'Orbigny 1835. Peru: Río Alto Yurúa (Amazon Basin), between Breu and Victoria, Ucayali Region. 160 mm. INHS 35118.

Amazon, whereas *Tamsiella* and *Diplodontites* are distributed from the Amazon northward. *Mycetopoda* d'Orbigny, 1835 (4 spp.), and the monotypic *Mycetopodella* Marshall, 1927, comprise the Mycetopodinae. *Mycetopodella falcata* (Higgins, 1868) is endemic to the Upper Amazon, but *Mycetopoda* extends from Central America south to Argentina (Paraná-La Plata). *Leila* Gray, 1840 (2 spp.), is the sole genus of the Leilinae, widespread in the Amazon and Paraná-La Plata basins. Previous speculation of a phylogenetic affinity between *Leila* and the Iridinidae (Bonetto, 1963; Heard and Vail, 1976) was based on convergence of larger larval size and deeply impressed pedal elevator scars, as well as superficially similar siphons (Veitenheimer, 1973).

The Mycetopodidae presumably arose on South America following the separation from Africa and the Iridinidae during the Mesozoic (Sanmartín and Ronquist, 2004; Upchurch, 2008). The oldest Neotropical fossil that we are aware of is *Anodontites freitasi* Mezzalira, 1974, from the Upper Cretaceous of Brazil (Simone and Mezzalira, 1993). A single Triassic

fossil from North America, *Mycetopoda diluculi* Pilsbry in Wanner, 1921 (Modell, 1957), is widely disjunct in space and time. The generic identification is based on the elongate outline of the shell, with no indication of the shell interior and the structure of the hinge. Shell forms of similar elongation have evolved multiple times in the Unionida and other bivalve orders (Anderson, 2014), and we regard that alleged mycetopodid fossil record as spurious.

References to conservation assessments of the Unionida are reviewed in chapter 38 (Graf and Cummings, this volume). The current conservation status assessments of the species of the Mycetopodidae according to the IUCN Red List (www.iucnredlist.org, accessed January 2018) are provided in Table 42.1. Too few mycetopodid species have been evaluated to draw any conclusions about their general conservation status.

LITERATURE CITED

Anderson, L.C. 2014. Ultra-elongate freshwater pearly mussels (Unionida): Roles for function and constraint in multiple morphologic convergences with marine taxa. Pp. 21-47 in Experimental Approaches to Understanding Fossil Organisms: Lessons from the Living (D.I. Hembree, B.F. Platt, and J.J. Smith, eds.). Topics in Geobiology, vol. 41. Springer, Netherlands.

Bogan, A.E., and W.R. Hoeh. 2000. On becoming cemented: Evolutionary relationships among the genera in the freshwater bivalve family Etheriidae (Bivalvia: Unionoida). Pp. 159-168 in The Evolutionary Biology of the Bivalvia (E.M. Harper, J.D. Taylor, and J.A. Crame, eds.). Geological Society, London, Special Publications 177. Geological Society of London, London.

Bonetto, A.A. 1963. Contribución al conocimiento de *Leila blainvilleana* (Lea) (Mollusca: Pelecypoda). Physis 24: 11-16.

Combosch, D.J., T.M. Collins, E.A. Glover, D.L. Graf, E.M. Harper, J.M. Healy, G.Y. Kawauchi, et al. 2017. A

Table 42.1. Mycetopodidae Species Richness and Red List Conservation Status

Region	Species richness	Threatened (EX+CR+EN+VU)	Stable (NT+LC)	Unknown (DD+NE)
Neotropical	53	1 (2%)	5 (9%)	47 (89%)

Note: Abbreviations for IUCN Red List (www.iucnredlist.org) categories: EX = Extinct, CR = Critically Endangered, EN = Endangered, VU = Vulnerable, NT = Near Threatened, LC = Least Concern, DD = Data Deficient, NE = Not Evaluated.

family-level Tree of Life for bivalves based on a Sanger-sequencing approach. Molecular Phylogenetics and Evolution 107: 191-208.

Frierson, L.S. 1927. A Classified and Annotated Check List of the North American Naiades. Baylor University Press, Waco, Texas.

Graf, D.L. 2000. The Etherioidea revisited: A phylogenetic analysis of hyriid relationships (Mollusca: Bivalvia: Paleoheterodonta: Unionoida). Occasional Papers of the Museum of Zoology, University of Michigan, 729: 1-21.

Graf, D.L., and K.S. Cummings. 2006. Palaeoheterodont diversity (Mollusca: Trigonioida + Unionoida): What we know and what we wish we knew about freshwater mussel evolution. Zoological Journal of the Linnean Society 148: 343-394.

Graf, D.L., and K.S. Cummings. 2007. Review of the systematics and global diversity of freshwater mussel species (Bivalvia: Unionoida). Journal of Molluscan Studies 73: 291-314.

Haas, F. 1969. Superfamilia Unionacea: Das Tierreich, Lief. 88. Walter de Gruyter, Berlin.

Heard, W.H., and V.A. Vail. 1976. Anatomical systematics of *Etheria elliptica* (Pelecyopoda: Mycetopodidae). Malacological Review 9: 15-24.

Modell, H. 1957. Die fossilen Najaden Nordamerikas. Archiv für Molluskenkunde 86: 183-200.

Parodiz, J.J., and A.A. Bonetto. 1963. Taxonomy and zoogeographic relationships of the South American Naiades (Pelecypoda: Unionacea and Mutelacea). Malacologia 1: 179-214.

Pereira, D., M.C.D. Mansur, L.D.S. Duarte, A.S. de Oliveira, D.M. Pimpão, C.T. Callil, C. Ituarte, et al. 2014. Bivalve distribution in hydrographic regions in South America: Historical overview and conservation. Hydrobiologia 735 (1): 15-44. DOI: 10.1007/s10750-013-1639-x.

Sanmartín, I., and F. Ronquist. 2004. Southern hemisphere biogeography inferred by event-based models: Plants versus animal patterns. Systematic Biology 53: 216-243.

Simone, L.R.L. 2006. Land and Freshwater Molluscs of Brazil: An Illustrated Inventory of the Brazilian Malacofauna, Including Neighboring Regions of South America, Respect to the Terrestrial and Freshwater Ecosystems. EGB, Sao Paulo.

Simone, L.R.L., and S. Mezzalira. 1993. Vestígios de partes moles em um Bivalve fóssil (Unionoida, Mycetopodidae) do Grupo Bauru (Cretáceo Superior), São Paulo, Brasil. Anais-Academia Brasileira de Ciencias 65: 155-159.

Upchurch, P. 2008. Gondwana break-up: Legacies of a lost world? Trends in Ecology and Evolution 23: 229-236.

Veitenheimer, I.L. 1973. Contribuição ao estudo do gênero *Leila* Gray, 1840 (Mycetopodidae-Bivalvia). Iheringia Série Zoologia 42: 64-89.

Walker, J.M., J.P. Curole, D.E. Wade, E.G. Chapman, A.E. Bogan, G.T. Watters, and W.R. Hoeh. 2006. Taxonomic distribution and phylogenetic utility of gender-associated mitochondrial genomes in the Unionoida (Bivalvia). Malacologia 48: 265-282.

Whelan, N.V., A.J. Geneva, and D.L. Graf. 2011. Molecular phylogenetic analysis of tropical freshwater mussels (Mollusca: Bivalvia: Unionoida) resolves the position of *Coelatura* and supports a monophyletic Unionidae. Molecular Phylogenetics and Evolution 61: 504-514.

43 Iridinidae Swainson, 1840

DANIEL L. GRAF AND KEVIN S. CUMMINGS

The freshwater mussel (Order Unionida) family Iridinidae Swainson, 1840 (+ Mutelidae Gray, 1847), is endemic to sub-Saharan Africa, including the Nile. Forty-one species comprise the family, classified among six genera in two subfamilies: Iridininae s.s. and Aspathariinae Modell, 1942. The family may not be monophyletic.

Phylogenetic studies with any iridinids have been infrequent, and only a few have included at least one representative from both subfamilies (Graf and Cummings, 2006b; Walker et al., 2006; Hoeh et al., 2009; Whelan et al., 2011; Combosch et al., 2017). Graf and Cummings (2006b) recovered the Iridinidae as monophyletic based largely on three morphological characters: well-impressed pedal elevator scars, a tubular excurrent "siphon," and a modified lasidium-type parasitic larval stage known as a haustorium (Fryer, 1961; Wächtler et al., 2001). These traditional characters, however, were based on scant observations, and more character-rich molecular data sets have placed the two subfamilies separately in various positions relative to the other families of the Etheri-oidea Deshayes, 1832 (i.e., Mycetopodidae Gray, 1840, and Etheriidae s.s.) (Walker et al., 2006; Whelan et al., 2011; Combosch et al., 2017).

The monophyly of each of the two subfamilies has not been comprehensively tested (Graf and Cummings, 2006b), but their distributions in Africa are well understood (Mandahl-Barth, 1988; Daget, 1998; Graf and Cummings, 2011). The Iridininae includes the genera *Mutela* Scopoli, 1777 (14 spp.), *Pleiodon* Conrad, 1834 (2 spp.), and *Chelidonopsis* Ancey, 1887 (1 sp.). All three genera exhibit well developed excurrent and incurrent siphons, with ventral mantle fusion extending anteriorly as far as the foot. *Mutela* is the most widespread genus of the Iridininae, extending from the Zambezi, Okavango, and Cunene rivers, north to western Africa and the Nile (excluding eastern Africa). The other two genera have more restricted, disjunct distributions in western Africa and Lake Tanganyika (*Pleiodon*) and in the Congo Basin (*Chelidonopsis*). *Aspatharia* Bourguignat, 1885 (12 spp.), *Chambardia* Bourguignat in Servain, 1890 (11 spp.), and *Moncetia* Bourguignat,

Distribution of Iridinidae.

1885 (1 sp.), comprise the Aspathariinae. These genera typically lack or have only minimal fusion ventral to the incurrent aperture/siphon (Graf and Cummings, 2006b). *Aspatharia* has a distribution similar to *Mutela*, from the Zambezi northward, and *Moncetia anceyi* Bourguignat, 1885, is endemic to Lake Tanganyika. *Chambardia* is found throughout sub-Saharan Africa, including southern and eastern Africa. *Chambardia wahlbergi* (Krauss, 1848) is a widespread species-complex, with recognized subspecies across its range (Mandahl-Barth, 1988; Daget, 1998). As future revisions update species circumscriptions, we expect that this complex will be divided into multiple species (Graf and Cummings, 2006a).

The Iridinidae is part of a clade of freshwater mussels diagnosed by the presence of lasidium-type larvae (= Etherioidea = Etheriidae + Mycetopodidae + Iridinidae). The lasidium-bearing mussels are largely restricted to South America and Africa, although *Pseudomulleria* Anthony, 1907, is endemic to India. These disjunctions on the southern continents are consistent with an origin that predates the separation of South America from Africa during the Cretaceous (Sanmartín and Ronquist, 2004; Upchurch, 2008). The oldest iridinid fossil in Africa is *Iridina aswanensis* Cox, 1955, described from the "Nubian Sandstone" of Egypt (Upper Cretaceous) (Van Damme et al., 2015). Fossils attributed to the Iridinidae have been reported from North and South America (Haas, 1969; Morris and Williamson, 1988; Simone and Mezzalira, 1997) from limited material. It is more likely that those specimens from outside Africa are due to convergent evolution of the pseudotaxonodont hinge dentition seen in African *Pleiodon* spp. rather than the result of an interesting biogeographic pattern.

References to conservation assessments of the Unionida are reviewed in chapter 38 (Graf and Cum-

Mutela hargeri E. A. Smith, 1908. Zambia: Lake Bangweulu, Luapula Province. 70 mm. INHS 30877.

mings, this volume). The current conservation status assessments of the species of the Iridinidae according to the IUCN Red List (www.iucnredlist.org, accessed January 2018) are provided in Table 43.1. A systematic effort has been made to assess most Afrotropical freshwater mollusks (Seddon et al., 2011). Of the 40 iridinid species, only 5 lack evaluations, though 8 species remain Data Deficient. Half of species are ranked in stable categories.

LITERATURE CITED

Combosch, D.J., T.M. Collins, E.A. Glover, D.L. Graf, E.M. Harper, J.M. Healy, G.Y. Kawauchi, et al. 2017. A family-level Tree of Life for bivalves based on a Sanger-sequencing approach. Molecular Phylogenetics and Evolution 107: 191–208.

Daget, J. 1998. Catalogue raisonné des mollusques bivalves d'eau douce africains. Backhuys Publishers, Orstom, Leiden.

Fryer, G. 1961. The developmental history of *Mutela bour-*

Table 43.1. Iridinidae species richness and Red List conservation status

Region	Species richness	Threatened (EX+CR+EN+VU)	Stable (NT+LC)	Unknown (DD+NE)
Afrotropical	40	7 (18%)	20 (50%)	13 (33%)

Note: Abbreviations for IUCN Red List (www.iucnredlist.org) categories: EX = Extinct, CR = Critically Endangered, EN = Endangered, VU = Vulnerable, NT = Near Threatened, LC = Least Concern, DD = Data Deficient, NE = Not Evaluated.

guignati (Ancey) Bourguignat (Mollusca: Bivalvia). Philosophical Transactions of the Royal Society B 244: 259-298.

Graf, D.L., and K.S. Cummings. 2006a. Freshwater mussels (Mollusca: Bivalvia: Unionoida) of Angola, with description of a new species, *Mutela wistarmorrisi*. Proceedings of the Academy of Natural Sciences of Philadelphia 155: 163-194.

Graf, D.L., and K.S. Cummings. 2006b. Palaeoheterodont diversity (Mollusca: Trigonioida + Unionoida): What we know and what we wish we knew about freshwater mussel evolution. Zoological Journal of the Linnean Society 148: 343-394.

Graf, D.L., and K.S. Cummings. 2011. Freshwater mussel (Mollusca: Bivalvia: Unionoida) richness and endemism in the ecoregions of Africa and Madagascar, based on comprehensive museum sampling. Hydrobiologia 678: 17-36.

Haas, F. 1969. Superfamily Unionacea Fleming, 1828. Pp. N411-N467 in Treatise on Invertebrate Paleontology, Part N, Mollusca 6: Bivalvia, vol. 1 (R.C. Moore, ed.). Geological Society of America and University of Kansas, Lawrence, Kansas.

Hoeh, W.R., A.E. Bogan, W.H. Heard, and E.G. Chapman. 2009. Palaeoheterodont phylogeny, character evolution and phylogenetic classification: A reflection on methods of analysis. Malacologia 51: 307-317.

Mandahl-Barth, G. 1988. Studies on African Freshwater Bivalves. Danish Bilharziasis Laboratory, Charlottenlund.

Morris, P.J., and P.G. Williamson. 1988. *Pleiodon* (Conrad) (Bivalvia: Mutelidae: Pleiodontinae) from the Late Cretaceous of Montana: A first North American record for the Mutelidae. Journal of Paleontology 62: 758-765.

Sanmartín, I., and F. Ronquist. 2004. Southern hemisphere biogeography inferred by event-based models: Plants versus animal patterns. Systematic Biology 53: 216-243.

Seddon, M., C. Appleton, D. Van Damme, and D. Graf. 2011. Freshwater molluscs of Africa: Diversity, distribution, and conservation. Pp. 92-119 in The Diversity of Life in African Freshwaters: Under Water, Under Threat: An Analysis of the Status and Distribution of Freshwater Species throughout Mainland Africa (W.R.T. Darwall, K.G. Smith, D.J. Allen, R.A. Holland, I.J. Harrison, and E.G.E. Brooks, eds.). IUCN, Cambridge, UK.

Simone, L.R.L., and S. Mezzalira. 1997. A posição sistemática de alguns bivalves Unionoidea do Grupo Bauru (Cretáceo Superior) do Brasil. Revista Brasileira de Geociências 2: 63-65.

Upchurch, P. 2008. Gondwana break-up: Legacies of a lost world? Trends in Ecology and Evolution 23: 229-236.

Van Damme, D., A.E. Bogan, and M. Dierick. 2015. A revision of the Mesozoic naiads (Unionoida) of Africa and the biogeographic implications. Earth-Science Reviews 147: 141-200.

Wächtler, K., M.C.D. Mansur, and T. Richter. 2001. Larval types and early postlarval biology in Naiads (Unionoida). Pp. 93-125 in Ecology and Evolution of the Freshwater Mussels Unionoida (G. Bauer and K. Wächtler, eds.). Springer-Verlag, Berlin.

Walker, J.M., J.P. Curole, D.E. Wade, E.G. Chapman, A.E. Bogan, G.T. Watters, and W.R. Hoeh. 2006. Taxonomic distribution and phylogenetic utility of gender-associated mitochondrial genomes in the Unionoida (Bivalvia). Malacologia 48: 265-282.

Whelan, N.V., A.J. Geneva, and D.L. Graf. 2011. Molecular phylogenetic analysis of tropical freshwater mussels (Mollusca: Bivalvia: Unionoida) resolves the position of *Coelatura* and supports a monophyletic Unionidae. Molecular Phylogenetics and Evolution 61: 504-514.

Glossary of Systematic Terms

CLADE. A group that has a common ancestor and clusters together on a phylogenetic tree.

MONOPHYLETIC. An ancestor and all of its descendants. They are thought to be natural groups (i.e., groups that truly exist in nature). For example, the families examined in this book are currently thought to be monophyletic, but they can still be subject to further testing.

MONOTYPIC. A single type. For example, a monotypic genus has just one species in the genus.

PARAPHYLETIC. An unnatural "group." Systematists avoid recognizing paraphyletic groups because they do not depict evolutionary history. For example, the gastropod group Prosobranchia is no longer recognized because systematic studies revealed it was paraphyletic and consequently unnatural.

PHYLOGENY. The evolutionary history of a group or lineage. Can be depicted in an evolutionary or phylogenetic tree.

PLESIOMORPHIC. A basal or "primitive" state of a character.

SISTER GROUP. A species or higher level taxonomic group that is the most closely related group to the one being examined.

SYNAPOMORPHY. Shared, derived character. Systematists look for morphological or molecular synapomorphies to denote monophyletic groups.

TAXON (pl. TAXA). A generic term for a species or a higher taxonomic group.

Contributors

CHRISTIAN ALBRECHT
Justus Liebig University Giessen
Department of Animal Ecology and Systematics
Heinrich-Buff-Ring 26-32, 35392
Giessen, Germany

RÜDIGER BIELER
Field Museum of Natural History
Curator and Head of Invertebrates
1400 S. Lake Shore Dr.
Chicago, Illinois 60605-2496, USA

BERT VAN BOCXLAER
French National Centre for Scientific Research (CNRS)
UMR 8198—Evolution, Ecology, Paleontology
Lille University
F-59000 Lille, France

DAVID C. CAMPBELL
Gardner-Webb University
Department of Natural Sciences
Boiling Springs, North Carolina 28017, USA

STEPHANIE A. CLARK
Director, Invertebrate Identification
481a Great Western Highway
Faulconbridge NSW 2776 Australia

CATHARINA CLEWING
Justus Liebig University Giessen
Department of Animal Ecology and Systematics
Heinrich-Buff-Ring 26-32, 35392
Giessen, Germany

ROBERT H. COWIE
University of Hawaii
Pacific Biosciences Research Center
3050 Maile Way, Gilmore 408
Honolulu, Hawaii 96822, USA

KEVIN S. CUMMINGS
Illinois Natural History Survey
Prairie Research Institute
University of Illinois at Urbana-Champaign
607 E. Peabody Dr.
Champaign, Illinois 61820, USA

DIANA DELICADO
Justus Liebig University Giessen
Department of Animal Ecology and Systematics
Heinrich-Buff-Ring 26-32 (IFZ) 35392
Giessen, Germany

HIROSHI FUKUDA
Okayama University
Conservation of Aquatic Biodiversity
Faculty of Agriculture
Tsushima-naka 1-1-1
Kita-ku, Okayama 700-8530, Japan

HIROAKI FUKUMORI
The University of Tokyo
Department of Marine Ecosystems Dynamics
Atmosphere and Ocean Research Institute
Kashiwa, Chiba 277-8564, Japan

MATTHIAS GLAUBRECHT
Universität Hamburg
Centrum für Naturkunde (CeNak)
Zoologisches Museum
Martin-Luther-King-Platz 3
D-20146 Hamburg, Germany

Daniel L. Graf
University of Wisconsin–Stevens Point
Biology Department
800 Reserve Street
Stevens Point, Wisconsin 54481, USA

Diego E. Gutiérrez Gregoric
Universidad Nacional de La Plata
Museo de La Plata
División Zoología Invertebrados
Paseo del Bosque s/n°–La Plata
Buenos Aires, B1900WFA, Argentina

Kenneth A. Hayes
Pacific Center for Molecular Biodiversity
Bernice Pauahi Bishop Museum
1525 Bernice Street
Honolulu, HI 96817, USA

Yasunori Kano
The University of Tokyo
Department of Marine Ecosystems Dynamics
Atmosphere and Ocean Research Institute
Kashiwa, Chiba 277-8564, Japan

Taehwan Lee
Department of Ecology and Evolutionary Biology
Museum of Zoology
University of Michigan
1109 Geddes Avenue
Ann Arbor, Michigan 48109-1079, USA

Charles Lydeard
Department of Biology and Chemistry
Morehead State University
103 Lappin Hall
Morehead, Kentucky 40351, USA

Nathaniel T. Marshall
Department of Environmental Sciences
University of Toledo
2801 Bancroft Street
Toledo, Ohio 43606, USA

Paula M. Mikkelsen
Integrative Research Center
Field Museum of Natural History
1400 South Lake Shore Drive
Chicago, Illinois 60605-2496, USA

Marco T. Neiber
Universität Hamburg
Centrum für Naturkunde (CeNak)
Zoologisches Museum
Martin-Luther-King-Platz 3
D-20146 Hamburg, Germany

Timea P. Neusser
Ludwig-Maximilian-University of Munich
Department Biology II
Großhaderner Str. 2
82152 Planegg-Martinsried, Germany

Winston Ponder
Australian Museum Research Institute
1 William Street
Sydney, NSW 2010, Australia

Michael Schrödl
Ludwig-Maximilians-University of Munich
Department Biology II
Großhaderner Str. 2
82152 Planegg-Martinsried, Germany

Staatliche Naturwissenschaftliche Sammlungen Bayerns
(SNSB)
Bavarian State Collection of Zoology
Münchhausenstr. 21
81247 Munich, Germany

Alena A. Shirokaya
Limnological Institute
Siberian Branch Russian Academy of Sciences
3 Ulan-Batorskaya St.
PO Box 4199, 664033
Irkutsk, Russia

Björn Stelbrink
Justus Liebig University Giessen
Department of Animal Ecology and Systematics
Heinrich-Buff-Ring 26-32, 35392
Giessen, Germany

Carol A. Stepien
NOAA Pacific Marine Environmental Lab
Ocean Environment Research Division
7600 Sand Point Way, NE
Seattle, Washington 98115, USA

ELLEN E. STRONG
Department of Invertebrate Zoology
National Museum of Natural History
Smithsonian Institution
PO Box 37012
Washington, DC 20013-7012, USA

MAXIM V. VINARSKI
Saint-Petersburg State University
7/9 Universitetskaya Emb.
Saint Petersburg, 199034, Russia

AMY R. WETHINGTON
Chowan University
1 University Place
Murfreesboro, North Carolina 27855, USA

THOMAS WILKE
Justus Liebig University Giessen
Department of Animal Ecology and Systematics
Heinrich-Buff-Ring 26-32 (IFZ) 35392
Giessen, Germany

Index